Symbiosis in Nature

Edited by Everlon Cid Rigobelo

Published in London, United Kingdom

Symbiosis in Nature
http://dx.doi.org/10.5772/intechopen.105293
Edited by Everlon Cid Rigobelo

Contributors
Solange da Fonseca Rutz, Marcelo Santos Carielo, Pallavi Mansotra, Heike Bücking, Dhruv Aditya Srivastava, Debatosh Das, Salina Tripathi, Prema Mutyala, Malik A. Aziz, Shayesta Islam, Zaffar M. Dar, Amjad Masood, Ravinder K. Goyal, Jemenah Z. Habtewold, Aqsa Tariq, Ambreen Ahmed, Gayatri Sharma, Maria Daniela Artigas Ramirez, Jean Louise Cocson Damo, Takayoshi Koike, Yannan Wang, Laiye Qu, Yutaka Tamai, Toshihiro Watanabe, Lamine Baba-Moussa, Ricardos M. Aguégué, Corentin Akpode, Tomabu Adjobimey, Haziz Sina, Sylvestre A. Assogba, Abdel D. Koda, Nadège A. Agbodjato, Adolphe Adjanohoun, Olubukola O. Babalola, Nestor R Ahoyo Adjovi, Olaoluwa Oluwafunto Ajayi, M. Dianda, Viktoriia Oliferchuk, Nataliia Kendzora, Igor Shukel, Mariia Samarska, Oksana Olejniuk-Puchniak

First published in London, United Kingdom, 2023 by IntechOpen
IntechOpen is the global imprint of INTECHOPEN LIMITED, registered in England and Wales, registration number: 11086078, 5 Princes Gate Court, London, SW7 2QJ, United Kingdom

British Library Cataloguing-in-Publication Data
A catalogue record for this book is available from the British Library

Additional hard and PDF copies can be obtained from orders@intechopen.com

Symbiosis in Nature
Edited by Everlon Cid Rigobelo
p. cm.
Print ISBN 978-1-83768-637-7
Online ISBN 978-1-83768-638-4
eBook (PDF) ISBN 978-1-83768-639-1

For EU product safety concerns:
IN TECH d.o.o., Prolaz Marije Krucifikse Kozulić 3, 51000 Rijeka, Croatia, info@intechopen.com or visit our website at intechopen.com.

Meet the editor

Dr. Everlon Cid Rigobelo graduated from the Agronomy School, São Paulo State University Jaboticabal Campus, Brazil, in 2000. He received an MSc in Animal Science Microbiology and a Ph.D. in Microbiology from the same university, where he also worked as a researcher. Dr. Rigobelo has experience in genetics and epidemiology and is active in microbial biotechnology, molecular genetics, and bacterial genomics. He works with plant growth-promoting rhizobacteria and previously worked with probiotic strains to reduce the spread of *Escherichia coli* in animal production. He has edited several books.

Contents

Preface

Plants have the unique ability to interact and coexist with numerous microorganisms that inhabit their roots, surfaces, and internal tissues. These interactions can occur at different levels, and their outcomes can be beneficial or harmful to the plant. When the interaction is beneficial, the microorganisms promote plant growth by facilitating nutrient availability, such as nitrogen and phosphorus, stimulating root and aerial growth, and suppressing pathogenic microorganisms. However, when microorganisms compete with the plant for nutrients or cause disease, the interaction can be detrimental. The key challenge for plants during these interactions is to differentiate between beneficial and harmful microorganisms by detecting specific signals. This recognition process is crucial for the plant's survival and is necessary to ensure that the plant is not exploited by harmful microorganisms. In summary, plants engage in complex interactions with microorganisms that can have both positive and negative effects on plant growth and survival. The ability of the plant to distinguish between beneficial and harmful microorganisms is critical for maintaining a healthy and thriving ecosystem.

One of the most significant beneficial interactions between plants and microorganisms is known as symbiosis. Symbiosis is a complex phenomenon that involves a mutualistic relationship between the plant and microorganisms, leading to changes in their physiology and anatomy. This evolutionary process has developed overtime, resulting in a beneficial association where both organisms offer each other what they need. One of the most well-known symbiotic processes is biological nitrogen fixation (BNF), where certain microorganisms provide the plant with nitrogen, which is essential for its growth and survival. In return, the plant provides the microorganisms with a source of energy and carbon. Overall, symbiosis is a sophisticated and mutually beneficial relationship that has evolved over time between plants and microorganisms. The BNF process is just one example of how this relationship can provide significant benefits to both parties involved.

In the Earth's atmosphere, nitrogen is the most abundant gas, although it is exceptionally unreactive. BNF, the process by which gaseous N2 is converted into ammonia (NH3) via the enzyme nitrogenase is crucial for the availability of nitrogen in a terrestrial ecosystem. Some bacteria have the remarkable capacity to fix atmospheric nitrogen to ammonia under ambient conditions, a reaction only mimicked on an industrial scale by a chemical process. This microbiological process converts atmospheric nitrogen into a plant-usable form, thus decreasing the need to use chemical fertilizers in crop production. In many systems, nitrogen fixers have driven the accumulation of fixed nitrogen on long-time scales, bringing nitrogen supply dose to equilibrium with other potentially limiting resources. BNF includes symbiotic (nodule formation) and free-living bacteria nitrogen fixation, defined as nitrogen fixation occurring without formal plant–microbe symbioses. The ability of microorganisms to use nitrogen gas as the sole nitrogen source and engage in symbioses with host plants confers many ecological advantages, but it also incurs physiological penalties. BNF is highly regulated at the transcriptional level by sophisticated regulatory networks that respond to multiple environmental cues.

Another crucial example of symbiosis between plants and microorganisms is the relationship between arbuscular mycorrhizal fungi and plants. This interaction occurs between the roots of plants and a fungus, typically of the genus *Glomus*. During this interaction, the fungus provides the plant with essential nutrients such as water, phosphorus, and nitrogen, while the plant supplies the fungus with carbon and energy. When both organisms possess the ability to interact with each other, this mutualistic relationship can promote high levels of plant growth. Overall, the interaction between arbuscular mycorrhizal fungi and plants is an excellent example of how a beneficial relationship between microorganisms and plants can result in significant benefits to both parties involved.

This book examines different aspects of symbiosis, such as biofertilizer, organic nitrogen in agricultural systems, nitrogen fertilization for sustainable crop production, and more. It is designed for researchers, students, and other interested readers.

I would like to thank my wife Fernanda, my daughter Maria Eduarda, and my son João Henrique for making my life happier. I also thank my deceased father Angelo Sidney Rigobelo, who lives in my mind and in my heart every day of my life.

Everlon Cid Rigobelo
Department of Plant Production,
Universidade Estadual Paulista (São Paulo State University),
Jaboticabal Campus, São Paulo, Brazil

Section 1

Plant Growth Rhizobacteria

Chapter 1

Plant Growth Promoting Rhizobacterial Consortium: A Sustainable Crop Production Strategy

Shayesta Islam, Malik A. Aziz, Zaffar M. Dar and Amjad Masood

Abstract

The prime concern for sustainable production is linked with biotic and abiotic pressures in environment as it impedes yield by producing ROS, which damage cell organelles and other biomolecules. Also the population is increasing at an alarming rate along with the climate change thereby leading to food insecurity. The only alternative to food security is adoption of Plant growth-promoting rhizobacteria (PGPR), as it provides an environmental-friendly and green substitute to chemical substance and traditional agricultural practices to achieve sustainable agriculture by enhancing plant growth and resistance to various pressures. The functions carried out by these microbes in agriculture include nutrient uptake, resistance of host plant to various animate and inanimate pressures. These surround the roots and affect the growth and development through various direct and indirect ways. Furthermore, they have the ability to combat harmful influence of pressures like salinity, drought, heavy metals, floods, and other stresses on plants by inducing the production of antioxidant enzymes such as catalase, peroxidase, and superoxide dismutase. To meet the increasing demand for food, and to evade environmental degradation, the utilization of PGPR consortium is a sustainable and ecofriendly technique to ameliorate the effectiveness of resource utilization and enhancing production under extreme climatic conditions and under increasing population.

Keywords: plant growth promoting rhizobacteria, sustainable, ecofriendly, microbial consortium, pathogens, growth promotion

1. Introduction

PGPR can be defined as the critical component of root zone bearing microorganisms that promote the development of host plants when they grow in association with host plants. Owing to great adjustment in diverse ecosystems, rapid growth, biochemical flexibility to break down numerous natural and synthetic substances, these have became prosperous root zone inhibiting bacteria sustaining in the soil environment [1]. These are regarded as considerable constituent in the regulation of agricultural practices with inherent genetic potential [2]. To qualify for PGPR the microbial strains should at minimum satisfy two out of three standards viz., aggressive colonization, promotion of plant growth and biological control [3, 4]. The basic types of interactions that occur between rhizospheric bacteria and host plants are neutral, positive and negative [5]. The interaction can be commensalism, in which bacteria establishes edible relationship with the host plant, thus no noticeable impact on the growth and development of host plants is exhibited [6]. In negative relationship, the growth and metabolism of host plants is impeded due to release of noxious substances like HCN, ethylene by Rhizobacteria, however some PGPRs have positive influence on the host plants through various direct and indirect processes like solubilization of nutrients, nitrogen fixation, formation of growth managers, encouraging development of mycorrhizae, elimination of pathogens or phytotoxic substances [7]. Furthermore, based on the extent of relationship between the roots of plants and microorganisms, these PGPRs are categorized into two types viz. extracellular plant growth promoting rhizobacteria (ePGPR) and intracellular plant growth promoting rhizobacteria (iPGPR) [8]. The extracellular plant growth promoting rhizobacteria can be found in the rhizosphere, on the rhizoplane or in the intercellular spaces in cortex. The typical examples of ePGPRs include Agrobacterium, Arthrobacter, Azotobacter, Azospirillum, Bacillus, Burkholderia, Caulobacter, Chromobacterium, Erwinia, Flavobacterium, Micrococcous, Pseudomonas and Serratia [9], while as intracellular plant growth promoting rhizobacteria (iPGPRs) occur in the nodules of roots. The typical iPGPRs are endophytes (Allorhizobium, Azorhizobium, Bradyrhizobium, Mesorhizobium and Rhizobium) and Frankia species, both having ability to fix nitrogen in symbiotic association with higher plants [10]. Endophytes colonize the roots of plants and promote growth ***(Wang and Martinez-Romero 2000). Plant growth-promoting rhizobacteria (PGPR) have the potential to increase development of plants through numerous processes viz. solubilization of phosphorus, synthesis of iron chelating compounds(Siderophores), biological fixation of nitrogen, rhizosphere engineering, generation of 1-Aminocyclopropane-1-carboxylate deaminase (ACC) and phytohormones, quorum sensing (QS) signal intervention and hampering of biofilm development, posses antifungal properties, produce volatile organic compounds (VOCs), induce systemic resistance, encourage useful interaction between plants and microbes, impede noxious substance production with pathogens etc. [11].

2. Why PGPR consortium

Microbial consortium is defined as the collection of microorganisms working together in a synergistic manner and is considerably more competent in comparison to single strain of organisms with varying potential. The various reasons that necessitate the utilization of PGPR consortium include meeting the demands of increasing population and increasing production under harsh climatic conditions. Since the global population is increasing at an alarming rate, it is estimated that by 2050, the global population will reach to 9 billion, so to meet the demands of growing population the production rate need to increase by 50% [12]. The food demand for increasing population necessitated the over utilization of synthetic fertilizers [13] since the advent of green revolution as well as over exploitation of arable land [14], thereby resulted in escalation of emission of greenhouse gas (GHG) [15] and change in climate [16]. Over last few decades the uncontrolled utilization of synthetic chemicals has resulted in decline in crop productivity globally due to deterioration in physicochemical and biological health of soil [17–19]. In addition to decline in crop production, climate alteration stem by GHG emissions, resulted in hike in price of agricultural products, which is intended to escalate the chance of food insecurity for 77 million people by 2050 [20]. Few plant species have the tendency to grow competently under adverse environmental conditions due to evolution of plasticity to combat alterations, but most plants cannot resist the adverse environmental conditions leading to diminish in productivity. It has been reported that maize and wheat has undergone a percentage loss in productivity by 3.8 and 5.5% due to climate alteration [21, 22]. To meet the increasing demand for food, and to evade the environmental degradation, while securing the productivity and biological health of soil, the only solution is to promote sustainable agriculture with progressive decline in the utilization of synthetic chemicals and outstanding use of bio-based products, as well as harnessing the biological and genetic potential of crops and associated microbes [23–25]. Thus the utilization of PGPR consortium is the sustainable and ecofriendly technique to ameliorate the effectiveness of resource utilization and enhancing the production under extreme climatic conditions [26]and under increasing population. Furthermore in addition to boost in crop production, these microbes have the capability to withstand various biotic and abiotic pressures [27, 28]. The microbe based products are safe, ecofriendly as well as means of growth stimulation and disease control. The harnessing of effective plant growth promoting rhizobacteria (PGPR) as biofertilizers and biological control agents is considered as a viable option for reducing the utilization of synthetic agrochemicals in crop production [4, 29–31].

3. PGPR for sustainable crop production

The utilization of PGPR in agriculture is progressing gradually, as it furnishes alternative to synthetic fertilizers, pesticides and other agrochemicals. The rhizosphere inhibiting microbes release significant amount of growth promoting substances that obliquely affect morphology of plants. The PGPRs have paramount importance in agriculture as it helps in welfare of crops by fixing atmospheric nitrogen, phosphate solubilization, decrease in concentration of heavy metals, releasing growth promoting hormones, degradation of crop refuse and soil organic matter, and checking plant pathogens [32, 33]. These PGPRs perform action through directly as well as indirectly as shown in **Figure 1**.

4. Growth promotion of plants by PGPRs through direct actions

PGPRs have the ability to promote growth of plants by assisting in the acquisition of essential nutrients viz., N, P and other micronutrients. Example, PGPRs assist in the transformation of atmospheric nitrogen into plant available forms, through a process called biological nitrogen fixation in presence of nitrogenase enzyme [34, 35]. These organisms either live in symbiotic association or non symbiotic association with the host plants [36]. Microbes like Rhizobium, Bradyrhizobium, Sinorhizobium, and Mesorhizobium live in symbiotic association with the roots of leguminous plants, on the other hand Frankia forms association with the non-leguminous plants [37]. Futher more

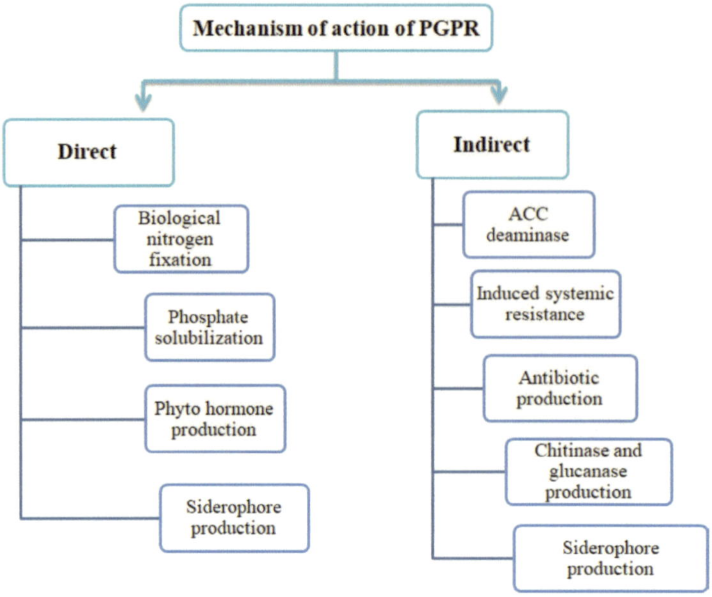

Figure 1.
Flow chart showing mechanism of action of PGPRs.

microbes like cyanobacteria, Azospirillum, Azotobacter, Burkholderia, Enterobacter, Gluconacetobacter, and Pseudomonas are non symbiotic in nature [11, 38, 39]. Hence inoculating these organisms with seeds, or seedlings or soil promotes the growth of plants, improves quality of soil and concentration of nitrogen [40]. Phosphate solubilizing microorganisms inhibiting root zone of plants have the tendency to mineralize, insoluble phosphate from soil into plant available forms (monobasic or dibasic ions) by producing different low molecular weight acids like gluconic acid and citric acid as 90% of phosphorus in soil is in insoluble forms [4, 41]. The phosphate solubilizing microbes are Arthrobacter, Bacillus, Beijerinckia, Burkholderia, Enterobacter, Microbacterium, Pseudomonas, Erwinia, Rhizobium, Mesorhizobium, Flavobacterium, Rhodococcus, and Serratia are phosphate solubilizers [42]. Another element is Iron which is critical for growth of plants, but it is commonly found in insoluble forms and thus not easily available to plants [43]. The PGPR strains viz. Pseudomonas, Bacillus, Rhizobium, Azotobactor, Enterobacter, and Serratia have the tendency to produce low molecular weight compounds called Siderophores which chelate iron under deficient conditions making it available to plants [44]. In addition to making iron available to plants, these Siderophores reduce the stress caused by heavy metals to plants [45–47]. Some Siderophores have the ability to control attack of pathogens to plants like inhibition of plants pathogens Fusarium, Pythium, and Aspergillus species by Siderophores from pseudomonads [48, 49]. It has been reported that potato wild disease by Fusarium oxysporum have been controlled by Siderophore (Pyoverdine) released by pseudomonads [50]. Another important function of PGPRs is the generation of various phytohormones, which regulate the growth, and activate defense [40]. The examples of bacteria producing phytohormones include Rhizobium, Bradyrhizobium, Mesorhizobium, Bacillus, Pantoea, Arthrobacter Pseudomonas, Enterobacter, and Burkholderia [51].

5. Growth promotion of plants by PGPRs through indirect actions

Growth promotion by indirect ways involve evading the attack of pathogens to the plants by generating various enzymes which cause break down of pathogens, anti microbial substances and induce systemic resistance [52]. A number

PGPR/Consortium	Effects
Rosa et al. [81] reported that effect of phosphate solubilizing consortia comprising of A. brasilense and *B. subtilis* under field conditions	Escalation in yield, dry matter, phosphorus accumulation, and 75% reduction in fertilization
Santos et al. [82] reported generation of IAA and enzymes (endoglucanases and xylanases) by *B. pumilus* under pot conditions	Increase in dry matter and number and diameter of tillers
Chandra et al. [83] reported solubilization of phosphorus and generation of siderophores, IAA, ammonia, and HCN under field conditions by *B. subtilis* (BSSC11) and *Bacillus megaterium* (BMSE7)	Enhancement in root and shoot length and total dry matter
Li et al. [84] reported nitrogen fixing capacity, phytohormones synthesis, and biocontrol by *P. koreensis* and *P. entomophila* under growth chamber	Enhancement of plant growth and development
Patel et al. [85] reported antagonism to phytopathogens, IAA production, P solubilization, and biological nitrogen fixation under green house conditions	Amplification in plant height, stem diameter, and number of leaves
Muthukumarasamy et al. [86] reported P solubilization by *Burkholderia gladioli* TNCSF 021 under pot conditions	Escalation in chlorophyll and N content of leaves and total biomass
Liu et al. [87] reported biological control by *Bacillus altitudinis* and *Bacillus velezensis* under green house conditions	Enhancement in dry weight, surface area, and total root length
Xia et al. [88] reported production of siderophores, IAA, amylase, pectinase, cellulase, chitinase, protease, and ACC deaminase and phosphate solubilization by Bacillus xiamenensis PM14 under green house conditions	Escalation in height, length, fresh weight, and root diameter and length
Ahmad et al. [89] reported production of IAA, siderophores and hydrogen cyanide; phosphate solubilization; and antifungal activity by Azotobacter sp. (AZS3), *P. fluorescens* (Ps5), and Bacillus sp. (Bc1) under pot conditions	Enhancement of dry weight of roots and shoots and shoot height
Viswanathan and Samiyappan [90] reported antifungal activity and induced systemic resistance of *P. fluorescens* under field conditions	Enhancement in germination and production
Lopes et al. [91] reported Nitrogenase activity of *A. brasilense* under field conditions	Increase in length, diameter, and Brix value
Hassan et al. [92] reported production of IAA, phosphate solubilization, and antifungal activity by *B. subtilis* NH-160 under green house conditions	Inhibition of red rot infection
Moutia et al. [93] reported capability of Azospirillum spp. against water stress under pot conditions	Increase in root dry matter

Table 1.
PGPR or their consortia for growth promotion of sugar cane under field as well as controlled conditions.

PGPR/Consortium	Effects
Breedt et al. [94] reported biological nitrogen fixation and IAA production by *Lysinibacillus sphaericus* (T19) under field conditions	Escalation in productivity
Cassan et al. [95] reported Phytohormone production by *A. brasilense* Az39, *Bradyrhizobium japonicum* E109 (individual experiments and consortia) in Growth chamber	seed germination promotion and early seedling development (use of isolated or combined species)
Kuan et al. [96] reported biological nitrogen fixation by *B. pumilus* S1r1 under Greenhouse conditions	Increase in corncob productivity (up to 30.9%)
Di Salvo et al. [97] reported IAA production and phosphate solubilization *A. brasilense* and *P. fluorescens* under Field conditions	Grain yield increment
Rocha et al. [98] reported nutrient addition by *P. fluorescens* F113 under greenhouse conditions	Addition of nutrients like N, K, Ca, Mg, and Mn with approximate percentage of 40, 49, 60, 100, and 141%, respectively, in the shoots
Danish et al. [99] reported under green house conditions ACC deaminase production by Enterobactercloacae	Escalation in grain production by 60%, photosynthetic rate by 73%, stomatal conductance by 43%, chlorophyll A by 69%, total chlorophyll by 76% and carotenoids by 42%
Pereira et al. [100] reported Phosphate solubilization potential by *B. subtilis* and *A. brasilense* under field conditions	Higher grain yield
Youseif [101] reported that Chryseobacterium sp. NGB-29 and Flavobacterium sp. O NGB-31 have BNF and production of large amounts of IAA under greenhouse conditions	Increment in all growth parameters
Moreira et al. [102] reported bioavailability of Zn in the soil under greenhouse conditions by *Ralstonia eutropha* 1C2 and *Chryseobacterium humi* ECP37 Zn bioavailability in the soil	Increased biomass and Zn accumulation and availability in plants
Rosas et al. [103] reported generation of phytohormones, antibiotics, and siderophores by *Pseudomonas aurantiaca* SR1 under field conditions	Increased productivity, length, and shoot and root dry weight
Lobo et al. [104] reported Phosphate solubilization and phytohormone production by *B. subtilis* 320 under field	Increase in productivity and P in the shoots
Zhao et al. [105] reported biocontrol properties and phosphate solubilization capacity of *Burkholderia cepacia* under greenhouse	Increase in leaf area, length, and shoot and root dry weight
Viruel et al. [106] reported Phosphate solubilization by *Pseudomonas tolaasii* IEXb under field conditions	Increase in seedling emergence, shoot length, grain yield, 1000-grain weight, total dry biomass, and P content in plants
Alori et al. [107] reported Phosphate solubilization and biocontrol ability of *Pseudomonas kilonensis* F113 and Pseudomonas protegens CHA0 under field conditions	Increase in leaf yield, height, and length
Verma et al. [108] reported Phosphate solubilization by *Enterobacter cloacae* PGLO9 under greenhouse conditions	Longer root length, shoot length, and increased shoot and root biomass

Table 2.
PGPR or their consortia for growth promotion of maize under field as well as controlled conditions.

of compounds generated by PGPR, which are non-volatile in nature like pyrrolnitrin, pyoluteorin, phenazines, phloroglucinols, and cyclic lipopeptides (CPLs) have potential to impede growth of pathogens [53]. It has been reported that damage to various vital crop plants has been controlled by PGPRs of genera Bacillus, Pseudomonas, and Streptomyces [54–56]. Pseudomonas, Burkholderia, Brevibacterium, and Streptomyces, have the ability to impede the growth of disease causing fungi and nematodes by releasing polycyclic nitrogenous substances like Phenazines [57–60]. The fungal and bacterial disease causing agents like Gaeumannomyces graminis, Pythium sp., Polyporus sp., Rhizoctonia solani, Actinomyces viscosus, Bacillus subtilis, and Erwinia amylovora has been known to be impeded by phenazine-1-carboxylic acid released by various species of Fluorescent pseudomonads [61]. Few metabolic substances like pyrrolnitrin with extraordinary action against fungal pathogens like Rhizoctonia solani, Fusarium graminearum, and Phytophthora capsici have been reported to be released by both fluorescent and non-fluorescent strains of Pseudomonas [62–64]. In addition to this Pseudomonas and Bacillus sp. boost the safeguard mechanism of host plants, promote root development and antimicrobial capacity by the release of numerous cyclic lipopeptides [65, 66]. Bacterial namely Pseudomonas, Bacillus, Burkholderia, Agrobacterium, Paenibacillus polymyxa, and Xanthomonas also produce numerous volatile substances swapping from aliphatic (dimethyl disulfide), aromatic (indole), ketones, alkanes, or alkenes (1-undecene), and terpenes (e.g., geosmin) [67–69], which activate signaling pathways of auxin, gibberellins, cytokinins, salicylic acid, and brassinosteroids [70–73]. Thus PGPRs have the ability to improve growth and development, fruit and seed development, germination of seeds as well as act as virulence-modulating factors, thereby reducing biotic and abiotic pressures [71, 74, 75]. Few PGPR strains viz., Pseudomonas, Bacillus, Serratia, Azospirillum, and Trichoderma have the potential to boost the defense system of plants without changing the genetic makeup called induced systemic resistance to fight against disease causing agents, in addition to this ethylene and jasmonic acid also activate signaling pathway to induce systemic resistance in plants [76, 77]. Abiotic pressures also affect growth and development of plants, the PGPR enhance resistance against abiotic pressures either directly or indirectly. It has been reported that Pseudomonas, Bacillus, Pantoea, Burkholderia, and Rhizobium increase resistance against abiotic pressures like drought, salinity, heat stress, and chilling injury [78–80]. A number of reports have revealed growth and development promotion of Sugar cane and maize by application of PGPR and their consortia as given in **Tables 1** and **2** respectively.

6. Conclusion

The PGPR research is of considerable interest worldwide, owing to increasing demand of organic food with increase in human population under changing climatic conditions. Adoption of PGPR is sustainable substitute to chemical fertilizers and pesticides. Among all possible biological means of sustainable agriculture, PGPR based formulations have triggered massive attention as they provide number of useful favors to plants. These ameliorate soil fertility, crop productivity, resistance to pathogens. A diverse group of microorganisms like Pseudomonas, Azospirillum, Azotobacter, Klebsiella, Enterobacter, Alcaligenes, Arthrobacter, Burkholderia,

Bacillus, and Serratia escalate growth and development of plants. Different Pseudomonas sp. cause prominent increment in germination, seedling growth and yield in different agricultural crops, including wheat. Development of beneficial PGPR formulation requires microbes possessing properties high ability to compete in the root zone, tremendous competitive saprophytic potential, capability to promote growth and development, easily produced at large scale, vide range of activities, should not pose any threat to environment and compatibility with other partnering organisms.

Author details

Shayesta Islam[1], Malik A. Aziz[2*], Zaffar M. Dar[3] and Amjad Masood[4]

1 Division of Environmental Sciences, FOH, Shalimar, SKUAST-Kashmir, India

2 Division of Basic Sciences and Humanities, FOA, Wadura, SKUAST-Kashmir, India

3 Division of Basic Sciences and Humanities, FOH, Shalimar, SKUAST-Kashmir, India

4 Division of Agronomy, FOA, Wadura, SKUAST-Kashmir, India

*Address all correspondence to: asifmalik@skuastkashmir.ac.in

References

[1] Kloepper JW, Schroth MN. Relationship of in vitro antibiosis of plant growth promoting rhizobacteria to plant growth and the displacement of root microflora. Phytopathology. 1981;**71**:1020-1024

[2] Cook RJ. Advances in plant health management in the twentieth century. Annual Review in Phytopathology. 2002;**38**:95-116

[3] Weller DM, Raaijmakers JM, Gardener BB, Thomashow LS. Microbial populations responsible for specific soil suppressiveness to plant pathogens. Annual Review in Phytopathology. 2002;**40**:309-348

[4] Vessey JK. Plant growth promoting rhizobacteria as biofertilizers. Plant and Soil. 2003;**255**:571-586

[5] Whipps JM. Microbial interactions and biocontrol in the rhizosphere. Journal of Experimental Botany. 2001;**52**:487-511

[6] Beattie GA. Plant-associated bacteria: Survey, molecular phylogeny, genomics and recent advances. In: Gnanamanickam SS, editor. Plant-associated Bacteria. The Netherlands: Springer; 2006. pp. 1-56

[7] Bashan Y, de-Bashan LE. How the plant growth-promoting bacterium Azospirillum promotes plant growth—A critical assessment. Advances in Agronomy. 2010;**108**:77-136

[8] Martinez-Viveros O, Jorquera MA, Crowley DE, Gajardo G, Mora ML. Mechanisms and practical considerations involved in plant growth promotion by rhizobacteria. Journal of Soil Science and Plant Nutrition. 2010;**10**:293-319

[9] Gray EJ, Smith DL. Intracellular and extracellular PGPR: Commonalities and distinctions in the plant-bacterium signaling processes. Soil Biology and Biochemistry. 2005;**37**:395-412

[10] Verma JP, Yadav J, Lavakush KNT, Singh V. Impact of plant growth promoting rhizobacteria on crop production. International Journal of Agricultural Research. 2010;**5**:954-983

[11] Bhattacharyya PN, Jha DK. Plant growth-promoting rhizobacteria (PGPR): Emergence in agriculture. World Journal of Microbiology and Biotechnology. 2012;**28**:1327-1350

[12] Alexandratos N, Bruinsma J. World Agriculture Towards 2030/2050: The 2012 Revision. Rome: FAO; 2012

[13] Canfield DE, Glazer AN, Falkowski PG. The evolution and future of Earth's nitrogen cycle. Science. 2010;**330**:192-196

[14] Pastor A, Palazzo A, Havlik P, Biemans H, Wada Y, Obersteiner M. The global nexus of food–trade–water sustaining environmental flows by 2050. Nature Sustainability. 2019;**2**:499-507

[15] Smith P, Haberl H, Popp A, Erb KH, Lauk C, Harper R. How much land-based greenhouse gas mitigation can be achieved without compromising food security and environmental goals? Global Change Biology. 2013;**19**:2285-2302

[16] Richardson Y, Blin J, Julbe A. A short overview on purification and conditioning of syngas produced by biomass gasification: Catalytic strategies, process intensification and new concepts. Progress in Energy and Combustion Science. 2012;**38**:765-781

[17] Pingali PL. Green revolution: Impacts, limits, and the path ahead. Proceedings of the National Academy of Sciences, USA. 2012;**109**:12302-12,308

[18] Yang X, Fang S. Practices, perceptions, and implications of fertilizer use in East-Central China. Ambio. 2015;**44**:647-652

[19] Bishnoi U. Agriculture and the dark side of chemical fertilizers. Environmental Analyses and Ecological Studies. 2018;**3**:EAES.000552.2018

[20] Janssens C, Havlík P, Krisztin T, Baker J, Frank S, Hasegawa T. Global hunger and climate change adaptation through international trade. Nature Climate Change. 2020;**10**:829-835

[21] Lobell DB, Schlenker W, Costa-Roberts J. Climate trends and global crop production since 1980. Science. 2011;**333**:616-620

[22] Lipper L, Thornton P, Campbell BM, Baedeker T, Braimoh A, Bwalya M. Climate-smart agriculture for food security. Nature Climate Change. 2014;**4**:1068-1072

[23] Fascella G, Montoneri E, Ginepro M, Francavilla M. Effect of urban biowaste derived soluble substances on growth, photosynthesis and ornamental value of Euphorbia × lomi. Scientia Horticulturae. 2015;**197**:90-98

[24] Fascella G, Montoneri E, Francavilla M. Biowaste versus fossil sourced auxiliaries for plant cultivation: The Lantana case study. Journal of Cleaner Production. 2018;**185**:322-330

[25] Liu J, Ma K, Ciais P, Polasky S. Reducing human nitrogen use for food production. Scientific Reports. 2016;**6**:30-104

[26] Pareek A, Dhankher OP, Foyer CH. Mitigating the Impact of Climate Change on Plant Productivity and Ecosystem Sustainability. Oxford: Oxford University Press UK; 2020

[27] Backer R, Rokem JS, Ilangumaran G, Lamont J, Praslickova D, Ricci E. Plant growth-promoting rhizobacteria: Context, mechanisms of action, and roadmap to commercialization of biostimulants for sustainable agriculture. Frontiers in Plant Science. 2018;**9**:1473

[28] Lyu D, Backer R, Subramanian S, Smith DL. Phytomicrobiome coordination signals hold potential for climate change-resilient agriculture. Frontiers in Plant Science. 2020;**11**:634

[29] Anli M, Baslam M, Tahiri A, Raklami A, Symanczik S, Boutasknit A, et al. Biofertilizers as strategies to improve photosynthetic apparatus, growth, and drought stress tolerance in the date palm. Frontiers in Plant Science. 2020;**11**:516-818

[30] Dong L, Li Y, Xu J, Yang J, Wei G, Shen L, et al. Biofertilizers regulate the soil microbial community and enhance *Panax ginseng* yields. Chinese Medical Journal. 2019;**14**:20

[31] Atieno M, Herrmann L, Nguyen HT, Phan HT, Nguyen NK, Srean P, et al. Assessment of biofertilizer use for sustainable agriculture in the Great Mekong Region. Journal of Environmental Management. 2020;**275**:111-300

[32] Etesami H, Maheshwari DK. Use of plant growth promoting rhizobacteria (PGPRs) with multiple plant growth promoting traits in stress agriculture: Action mechanisms and future prospects. Ecotoxicology and Environmental Safety. 2018;**156**:225-246

[33] He Y, Pantigoso HA, Wu Z, Vivanco JM. Co-inoculation of Bacillus sp. and *Pseudomonas putida* at different development stages acts as a biostimulant to promote growth, yield and nutrient uptake of tomato. Journal of Applied Microbiology. 2019;**127**:196-207

[34] Tairo EV, Ndakidemi PA. Possible benefits of rhizobial inoculation and phosphorus supplementation on nutrition, growth and economic sustainability in grain legumes. American Journal of Research Communication. 2013;**1**:532-556

[35] Smith BE, Richards RL, Newton WE. Catalysts for Nitrogen Fixation: Nitrogenases, Relevant Chemical Models and Commercial Processes. Vol. 2013. Berlin, Germany: Springer Science & Business Media. p. 340

[36] Ahemad M, Khan MS. Evaluation of plant growth promoting activities of rhizobacterium *P. putida* under herbicide stress. Annals of Microbiology. 2012;**62**:1531-1540

[37] Zahran HH. Rhizobia from wild legumes: Diversity, taxonomy, ecology, nitrogen fixation and biotechnology. Journal of Biotechnology. 2001;**91**:143-153

[38] Chittora D, Meena M, Barupal T, Swapnil P, Sharma K. Cyanobacteria as a source of biofertilizers for sustainable agriculture. Biochemistry and Biophysics Reports. 2020;**22**:100737

[39] Meena M, Zehra A, Swapnil P, Harish Marwal A, Yadav G, Sonigra P. Endophytic nanotechnology: An approach to study scope and potential applications. Frontiers in Nanoscience and Nanotechnology. 2021;**9**:613343

[40] Damam M, Kaloori K, Gaddam B, Kausar R. Plant growth promoting substances (phytohormones) produced by rhizobacterial strains isolated from the rhizosphere of medicinal plants. International Journal of Pharmaceutical Sciences Review and Research. 2016;**37**:130-136

[41] Sharma S, Chen C, Navathe S, Chand R, Pandey SPA. halotolerant growth promoting rhizobacteria triggers induced systemic resistance in plants and defends against fungal infection. Scientific Reports. 2019;**9**:40-54

[42] Oteino N, Lally RD, Kiwanuka S, Lloyd A, Ryan D, Germaine KJ, et al. Plant growth promotion induced by phosphate solubilizing endophytic Pseudomonas isolates. Frontiers in Microbiology. 2015;**6**:745

[43] Rajkumar M, Ae N, Prasad MNV, Freitas H. Potential of siderophore-producing bacteria for improving heavy metal phytoextraction. Trends in Biotechnology. 2010;**28**:142-149

[44] Ansari RA, Mahmood I, Rizvi R, Sumbul A Siderophores: Augmentation of soil health and crop productivity. In Probiotics in Agroecosystem, 1st ed.; Kumar V, Kumar M, Sharma S, Prasad R, Eds.; Springer Nature: Singapore; 2017. pp. 291-312

[45] Glick BR. Plant growth-promoting bacteria: Mechanisms and applications. Scientifica. 2012;**963**:401

[46] Hider RC, Kong X. Chemistry and biology of siderophores. Natural Product Reports. 2010;**27**:637-657

[47] Ahemad M, Khan MS. Assessment of plant growth promoting activities of rhizobacterium *P. putida* under insecticide stress. Journal of Microbiology. 2011;**1**:54-64

[48] Trapet P, Avoscan L, Klinguer A, Pateyron S, Citerne S, Chervin C, et al.

The *Pseudomonas fluorescens* siderophore pyoverdine weakens *Arabidopsis thaliana* defense in favor of growth in iron-deficient conditions. Plant Physiology. 2016;**171**:675-693

[49] Ali MA, Ren H, Ahmed T, Luo J, An Q, Qi X, et al. Antifungal effects of rhizospheric Bacillus species against bayberry twig blight pathogen Pestalotiopsis versicolor. Agronomy. 2020;**10**:1811

[50] Schippers B, Bakker AW, Bakker PAH. Interactions of deleterious and beneficial rhizosphere microorganisms and the effect of cropping practices. Annual Review of Phytopathology. 1987;**25**:339-358

[51] Egamberdieva D, Wirth SJ, Alqarawi AA, Abd Allah EF, Hashem A. Phytohormones and beneficial microbes: Essential components for plants to balance stress and fitness. Frontiers in Microbiology. 2017;**8**:2104

[52] Gouda S, Kerry RG, Das G, Paramithiotis S, Shin HS, Patra JK. Revitalization of plant growth promoting rhizobacteria for sustainable development in agriculture. Microbiology Research. 2018;**206**:131-140

[53] Vacheron J, Desbrosses G, Bouffaud ML, Touraine B, Moenne-Loccoz Y, Muller D, et al. Plant growth-promoting rhizobacteria and root system functioning. Frontiers in Plant Science. 2013;**4**:356

[54] Meena M, Swapnil P, Zehra A, Dubey MK, Aamir M, Patel CB, et al. Virulence factors and their associated genes in microbes. In: Singh HB, Gupta VK, Jogaiah S, editors. New and Future Developments in Microbial Biotechnology and Bioengineering: Microbial Genes Biochemistry and Applications. 1st ed. Hoboken, NJ, USA: Elsevier; 2019. pp. 181-208

[55] Almoneafy AA, Moustafa-Farag M, Mohamed HI. The auspicious role of plant growth-promoting rhizobacteria in the sustainable management of plant diseases. In: Mohamed HI, El-Beltagi HEDS, Abd-Elsalam KA, editors. Plant Growth-Promoting Microbes for Sustainable Biotic and Abiotic Stress Management. 1st ed. Cham, Switzerland: Springer Nature; 2021. pp. 251-283

[56] Ngalimat MS, Mohd Hata E, Zulperi D, Ismail SI, Ismail MR, Mohd Zainudin NAI, et al. Plant growth-promoting bacteria as an emerging tool to manage bacterial rice pathogens. Microorganisms. 2021;**9**:682

[57] Zhou D, Feng H, Schuelke T, De Santiago A, Zhang Q, Zhang J, et al. Rhizosphere microbiomes from root-knot nematode non-infested plants suppress nematode infection. Microbial Ecology. 2019;**78**:470-481

[58] Muller T, Ruppel S, Behrendt U, Lentzsch P, Muller MEH. Antagonistic potential of fluorescent pseudomonads colonizing wheat heads against mycotoxin producing Alternaria and Fusaria. Frontiers in Microbiology. 2018;**9**:2124

[59] Chen S, Zou J, Hu Z, Chen H, Lu Y. Global annual soil respiration in relation to climate, soil properties and vegetation characteristics: Summary of available data. Agricultural and Forest Meteorology. 2014;**198**:335-346

[60] Dasgupta D, Kumar A, Mukhopadhyay B, Sengupta TK. Isolation of phenazine 1,6-di-carboxylic acid from *Pseudomonas aeruginosa* strain HRW.1-S3 and its role in biofilm-mediated crude oil degradation and cytotoxicity against

bacterial and cancer cells. Applied Microbiology and Biotechnology. 2015;**99**:8653-8665

[61] Saraf M, Pandya U, Thakkar A. Role of allelochemicals in plant growth promoting rhizobacteria for biocontrol of phytopathogens. Microbiology Research. 2014;**169**:18-29

[62] Jung BK, Hong SJ, Park GS, Kim MC, Shin JH. Isolation of *Burkholderia cepacia* JBK9 with plant growth-promoting activity while producing pyrrolnitrin antagonistic to plant fungal diseases. Applied Biological Chemistry. 2018;**61**:173-180

[63] Pawar S, Chaudhari A, Prabha R, Shukla R, Singh DP. Microbial pyrrolnitrin: Natural metabolite with immense practical utility. Biomolecules. 2019;**9**:443

[64] Zhang M, Yang L, Hao R, Bai X, Wang Y, Yu X. Drought-tolerant plant growth-promoting rhizobacteria isolated from jujube (*Ziziphus jujuba*) and their potential to enhance drought tolerance. Plant and Soil. 2020;**452**:423-440

[65] Raaijmakers JM, De Bruijn I, Nybroe O, Ongena M. Natural functions of lipopeptides from Bacillus and Pseudomonas: More than surfactants and antibiotics. FEMS Microbiology Reviews. 2010;**34**:1037-1062

[66] Malviya D, Sahu PK, Singh UB, Paul S, Gupta A, Gupta AR, et al. Lesson from ecotoxicity: Revisiting the microbial lipopeptides for the management of emerging diseases for crop protection. International Journal of Environmental Research and Public Health. 2020;**17**:1434

[67] Effmert U, Kalderas J, Warnke R, Piechulla B. Volatile mediated interactions between bacteria and fungi in the soil. Journal of Chemical Ecology. 2012;**38**:665-703

[68] Penuelas J, Asensio D, Tholl D, Wenke K, Rosenkranz M, Piechulla B, et al. Biogenic volatile emissions from the soil. Plant, Cell & Environment. 2014;**37**:1866-1891

[69] Sharifi R, Ryu CM. Revisiting bacterial volatile-mediated plant growth promotion: Lessons from the past and objectives for the future. Annals of Botany. 2018;**122**:349-358

[70] Ryu CM, Farag MA, Hu CH, Reddy MS, Wei HX, Pare PW, et al. Bacterial volatiles promote growth in Arabidopsis. Proceedings of the National Academy of Sciences USA. 2003;**100**:4927-4932

[71] Zhang H, Kim MS, Krishnamachari V, Payton P, Sun Y, Grimson M, et al. Rhizobacterial volatile emissions regulate auxin homeostasis and cell expansion in Arabidopsis. Planta. 2007;**226**:839-851

[72] Zhang H, Xie X, Kim MS, Kornyeyev DA, Holaday S, Pare PW. Soil bacteria augment Arabidopsis photosynthesis by decreasing glucose sensing and abscisic acid levels in planta. The Plant Journal. 2008;**56**:264-273

[73] Meena M, Swapnil P, Zehra A, Dubey MK, Upadhyay RS. Antagonistic assessment of Trichoderma spp. by producing volatile and non-volatile compounds against different fungal pathogens. Archives of Phytopathology and Plant Protection. 2017;**50**:629-648

[74] Ossowicki A, Jafra S, Garbeva P. The antimicrobial volatile power of the rhizospheric isolate Pseudomonas donghuensis P482. PLoS One. 2017;**12**:e0174362

[75] Sharifi R, Lee SM, Ryu CM. Microbe-induced plant volatiles. The New Phytologist. 2018;**220**:684-691

[76] Choudhary DK, Prakash A, Johri BN. Induced systemic resistance (ISR) in plants: Mechanism of action. Indian Journal of Microbiology Research. 2007;**47**:289-297

[77] Pieterse CMJ, Zamioudis C, Berendsen RL, Weller DM, Van Wees SCM, Bakker PAHM. Induced systemic resistance by beneficial microbes. Annual Review of Phytopathology. 2014;**52**:347-375

[78] Egamberdieva D, Wirth S, Bellingrath-Kimura SD, Mishra J, Arora NK. Salt-tolerant plant growth promoting rhizobacteria for enhancing crop productivity of saline soils. Frontiers in Microbiology. 2019;**10**:2791

[79] Jha Y, Subramanian RB. PGPR regulate caspase-like activity, programmed cell death, and antioxidant enzyme activity in paddy under salinity. Physiology and Molecular Biology of Plants. 2014;**20**:201-207

[80] Noorieh B, Arzanesh MH, Mahlegha G, Maryam S. The effect of plant growth promoting rhizobacteria on growth parameters, antioxidant enzymes and microelements of canola under salt stress. Journal of Applied Environmental and Biological Sciences. 2013;**3**:17-27

[81] Rosa PAL, Mortinho ES, Jalal A, Galindo FS, Buzetti S, Fernandes GC. Inoculation with growth-promoting bacteria associated with the reduction of phosphate fertilization in sugarcane. Frontiers in Environmental Science. 2020;**8**:32

[82] Santos RM, Kandasamy S, Rigobelo EC. Sugarcane growth and nutrition levels are differentially affected by the application of PGPR and cane waste. Microbiology. 2018;**7**:e00617

[83] Chandra P, Tripathi P, Chandra A. Isolation and molecular characterization of plant growth-promoting Bacillus spp. and their impact on sugarcane (Saccharum spp. hybrids) growth and tolerance towards drought stress. Acta Physiologiae Plantarum. 2018;**40**:199

[84] Li HB, Singh RK, Singh P, Song QQ, Xing YX, Yang LT. Genetic diversity of nitrogen-fixing and plant growth promoting pseudomonas species isolated from sugarcane rhizosphere. Frontiers in Microbiology. 2017;**8**:20

[85] Patel P, Shah R, Joshi B, Ramar K, Natarajan A. Molecular identification and biocontrol activity of sugarcane rhizosphere bacteria against red rot pathogen Colletotrichum falcatum. Biotechnology Reports. 2019;**21**:e00317

[86] Muthukumarasamy R, Revathi G, Vadivelu M, Aruri K. Isolation of bacterial strains possessing nitrogen-fixation, phosphate and potassiumsolubilization and their inoculation effects on sugarcane. Indian Journal of Experimental Biology. 2017;**55**:161-170

[87] Liu K, McInroy JA, Hu CH, Kloepper JW. Mixtures of plantgrowth-promoting rhizobacteria enhance biological control of multiple plant diseases and plant-growth promotion in the presence of pathogens. Plant Disease. 2018;**102**:67-72

[88] Xia Y, Farooq MA, Javed MT, Kamran MA, Mukhtar T, Ali J. Multi-stress tolerant PGPR Bacillus xiamenensis PM14 activating sugarcane (*Saccharum officinarum* L.) red rot disease resistance. Plant Physiology and Biochemistry. 2020;**151**:640-649

[89] Ahmad F, Ahmad I, Aqil F, Ahmed Wani A, Sousche YS. Plant growth promoting potential of free-living diazotrophs and other rhizobacteria isolated from Northern Indian soil. Biotechnology Journal. 2016;**1**:1112-1123

[90] Viswanathan R, Samiyappan R. Induced systemic resistance by fluorescent pseudomonads against red rot disease of sugarcane caused by Colletotrichum falcatum. Crop Protection. 2002;**21**:1-10

[91] Lopes VR, Bespalhok-Filho JC, Araujo LM, Rodrigues FV, Daros E, Oliveira RA. The selection of sugarcane families that display better associations with plant growth promoting rhizobacteria. Agronomy Journal. 2012;**11**:43-52

[92] Hassan MN, Afghan S, Hafeez FY. Suppression of red rot caused by Colletotrichum falcatum on sugarcane plants using plant growth-promoting rhizobacteria. BioControl. 2010;**55**:695

[93] Moutia JFY, Saumtally S, Spaepen S, Vanderleyden J. Plant growth promotion by Azospirillum sp. in sugarcane is influenced by genotype and drought stress. Plant and Soil. 2010;**337**:233-242

[94] Breedt G, Labuschagne N, Coutinho TA. Seed treatment with selected plant growth-promoting rhizobacteria increases maize yield in the field. The Annals of Applied Biology. 2017;**171**:229-236

[95] Cassan F, Perrig D, Sgroy V, Masciarelli O, Penna C, Luna V. Azospirillum brasilense Az39 and *Bradyrhizobium japonicum* E109, inoculated singly or in combination, promote seed germination and early seedling growth in corn (*Zea mays* L.) and soybean (Glycine max L.). European Journal of Soil Biology. 2009;**45**:28-35

[96] Kaur H, Kaur J, Gera R. Plant growth promoting rhizobacteria: A boon to agriculture. International Journal of Cell Science and Bitechnology. 2016;**5**:17-22

[97] Di Salvo LP, Cellucci GC, Carlino ME, de Salamone IEG. Plant growth-promoting rhizobacteria inoculation and nitrogen fertilization increase maize (*Z. mays* L.) grain yield and modified rhizosphere microbial communities. Applied Soil Ecology. 2018;**126**:113-120

[98] Rocha I, Ma Y, Carvalho MF, Magalhaes C, Janouskova M, Vosatka M. Seed coating with inocula of arbuscular mycorrhizal fungi and plant growth promoting rhizobacteria for nutritional enhancement of maize under different fertilization regimes. Archives of Agronomy and Soil Science. 2019;**65**:31-43

[99] Danish S, Zafar-ul-Hye M, Mohsin F, Hussain M. ACCdeaminase producing plant growth promoting rhizobacteria and biochar mitigate adverse effects of drought stress on maize growth. PLoS One. 2020;**15**:e0230615

[100] Pereira NCM, Galindo FS, Gazola RPD, Dupas E, Rosa PAL, Mortinho ES. Corn yield and phosphorus use efficiency response to phosphorus rates associated with plant growth promoting bacteria. Frontiers in Environmental Science. 2020;**8**:40

[101] Youseif SH. Genetic diversity of plant growth promoting rhizobacteria and their effects on the growth of maize plants under greenhouse conditions. Annals of Agricultural Science. 2018;**63**:25-35

[102] Moreira H, Pereira SIA, Marques A, Rangel A, Castro PML. Effects of soil sterilization and metal spiking in plant growth promoting rhizobacteria

selection for phytotechnology purposes. Geoderma. 2019;**334**:72-81

[103] Rosas SB, Avanzini G, Carlier E, Pasluosta C, Pastor N, Rovera M. Root colonization and growth promotion of wheat and maize by *Pseudomonas aurantiaca* SR1. Soil Biology and Biochemistry. 2009;**41**:1802-1806

[104] Lobo LLB, dos Santos RM, Rigobelo EC. Promotion of maize growth using endophytic bacteria under greenhouse and field condition. Australian Journal of Crop Science. 2019;**13**:2067-2074

[105] Zhao K, Penttinen P, Zhang XP, Ao XL, Liu MK, Yu XM. Maize rhizosphere in Sichuan, China, hosts plant growth promoting *B. cepacia* with phosphate solubilizing and antifungal abilities. Microbiology Research. 2014;**169**:76-82

[106] Viruel E, Erazzu LE, Calsina LM, Ferrero MA, Lucca ME, Sineriz F. Inoculation of maize with phosphate solubilizing bacteria: Effect on plant growth and yield. Journal of Soil Science and Plant Nutrition. 2014;**14**:819-831

[107] Alori ET, Babalola OO, Prigent-Combaret C. Impacts of microbial inoculants on the growth and yield of maize plant. Open Agriculture Journal. 2019;**13**:1-8

[108] Verma P, Agrawal N, Shahi SK. Enterobacter cloacae strain PGLO9: Potential source of maize growth promoting rhizobacteria. International Journal of Botany Studies. 2018;**3**:172-175

Chapter 2

Plants and Their Microbes

Olaoluwa Oluwafunto Ajayi and M. Dianda

Abstract

The soil harbors billions of Microbes supporting the growth of several plants, resulting in the constant symbiotic or non-symbiotic interactions between the plants and these microorganisms known as plant-microbe interactions. Plants need nutrients (available or available but inaccessible) in the soil to grow while microbes need shelter and nutrients supplied by plants and also promote plant-growth. Microbes are constantly trading with plants either buying or selling nutrients (the nutrients are considered the main currencies for trading and the product being bought in the soil rhizosphere). Plants including legumes inoculated with Rhizobia and microbes in experimental studies show that they are able to solubilizing phosphate and metals, fixing nitrogen fixers, producing IAA, cytokinins, Gibberellins and Ethylene. Microbiomes are therefore essential for plant growth and health as they govern most soil functions affecting plant-growth. For higher crop-yields and increased soil- fertility using ecofriendly manner, researches focused on the new concepts of exotic biomolecules, hormones, enzymes and metabolites which create a suitable environment for the interaction between plant and microbes using molecular and biotechnological approaches are necessary to increase our knowledge of rhizosphere biology and to achieve an integrated management of soil microbial populations that can ultimately enhance the health of plants.

Keywords: plants-microbes, symbiosis, plant-health, soil-fertility, plant growth

1. Introduction

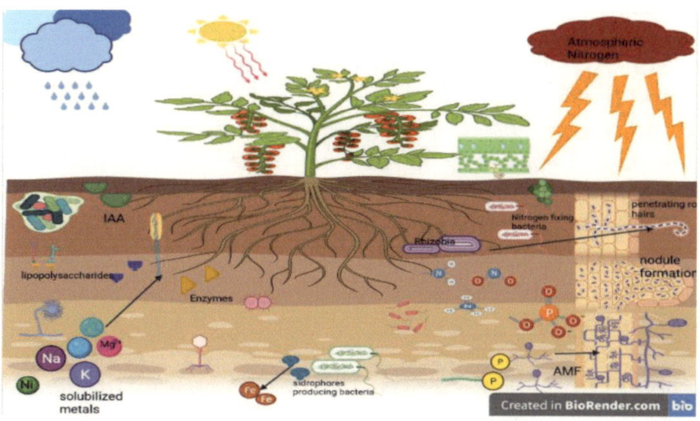

The soil is the habitat for billions of bacteria and the support for the growth of all or most plants. There is therefore bound to be constant interactions between the plants and these microorganism hence the study of plant-microbe interactions (these interactions can be symbiotic or non-symbiotic). Plants are constantly in need of nutrients in the soil (such as zinc, nitrogen, phosphorus, calcium, cobalt, molybdenum, iron, etc.) to grow and these nutrients may be available or sometimes available but inaccessible for the plants. While bacteria are constantly in need of shelter form harsh environmental conditions, and also in need of nutrients that the plants can supply to them. Hence they are constantly in contact with the roots of the plant trying to either buy or sell nutrients with the plant (we may say that nutrients are the main currencies for trading and the product being bought in the soil rhizosphere). The rhizosphere is the narrow region or (zone around the root in which the microbial biomass is impacted by the presence of the root of soil) substrate that is directly influenced by root secretions and associated soil microorganisms known as the root microbiome. The rhizosphere of each plant the roots of each plant is has been shown to be colonized by a specific set of bacteria that are useful to it or benefiting from it and this microbes are able to interact with plant tissues and cells with different degrees of dependence these interactions can be beneficial or harmful especially in the case of pathogenic microbes that attack the plants making them fall sick. These bacteria are referred to as Plant Growth Promoting Rhizobacteria (PGPR). In essence plants and their microbiomes need each other but can they survive without each other in the soil? Most likely not.

2. Common microbiomes associated with plant the good and the bad

Microbiomes commonly associated with plants are largely dependent on the environmental conditions of the soil such as the pH, the temperature, the moisture content, the soil texture, availability of nutrients etc. For example, in mountainous regions (such as Indian Himalayan Region and Tropical Andes), the climate is mainly characterized by low soil and atmospheric temperatures especially during the winter months, and the soils are generally acidic and severely lack nutrients. On the other hand in the desert regions, the climate is mainly characterized by high temperature, low moisture content, poor soil structure, high salinity and is mostly unfavorable for micrognanisms. These two regions face several major challenges in agriculture while in temperate regions the weather and environmental conditions are more favorable to microorganisms although it does not come without its own challenges. Several of the PGPR found in the soil belong to four phyla namely Proteobacteria, Firmicutes, Actinobacteria and Bacteroidetes respectively. And the most common genera found in the soil includes *Pseudomonas* spp., *Bacillus* spp., *Stenotrophomonas* spp., *Methylobacterium* spp., *Arthrobacter* spp., *Pantoea* spp., *Achromobacter* spp., *Acinetobacter* spp., *Exiguobacterium* spp., *Staphylococcus* spp., *Enterobacter* spp., *Providencia* spp., *Klebsiella* spp., *Leclercia* spp., *Brevundimonas* spp., *Flavobacterium* spp., *Kocuria* spp., *Kluyvera* spp., *Rhizobia* spp., *Serratia* spp., and *Planococcus* spp.

2.1 Phosphate solubilizers

Plants are able to access phosphorus in two forms which are the monobasic and dibasic. It can be found in the soil in the inorganic forms, such as apatite, or as organic forms such as inositol phosphate, phosphomonoesters, and phosphotriesters, but it is mostly insoluble. Phosphate solubilizing bacteria are able to solubilize the insoluble forms of phosphorus thereby making it available for plants to use. They inculde bacteria such as *Achromobacter* spp., *Agrobacterium* spp., *Bacillus* spp., *Enterobacter* spp., *Erwinia* spp., *Flavobacterium* spp., *Gluconacetobacter* spp., *Mycobacterium* spp., *Pseudomonas* spp., *Rhizobium* spp., *Serratia* spp.

2.2 Nitrogen fixers

They are able to fix nitrogen which is generally unavailable to plants although it is an essentially important element for developing and producing of fruits and seeds in plants. Leguminous plants in particular are able to form symbiotic associations with *Rhizobia* sp., which include the genera Bradyrhizobium, Sinorhizobium, Mesorhizobium, Rhizobium, Azorhizobium, Neorhizobium, and Pararhizobium. They enter into the plant tissue of the roots by exchanging chemical signals with the plant forming nodules where they are able to fixing atmospheric nitrogen and converting it into ammonia a form of nitrogen that plants can use. Some other bacteria (non-nitrogen-fixing bacteria), for example Pseudomonas, can stimulate symbiosis between legumes and their associated rhizobia while helping to increase levels of nitrogen fixation, ultimately helping to improve the nutrition and growth of the plants. For example, when the activity level of the enzyme 1-aminocyclopropane-1-carboxylate (ACC) deaminase is high in Pseudomonas it essentially helps to enhance the process of nodulation in rhizobia, an indication that interactions among bacteria are also beneficial to plants.

2.3 Pathogenic microrganisms

These are bacteria or fungi that produce substances, enzymes or metabolites that affect the plant they also feed on the plant or its fruits causing disease, infection and yield losses they impair the normal state of the plant and interrupt the normal functioning in the plant. Such microbes include fungi, viruses, bacterium.

2.4 Metal solubilizers bioremediators

PGPR are able to stimulate solubilization of metals, such as zinc, potassium, Iron and manganase are generally essential for the fertility of the soil and improvement of crops. Iron is naturally highly abundant, but for it to be available to plant, there is a need for the formation of Fe-siderophore complexes. Bacteria, can secrete siderophores as iron-chelating compounds that are able to reduce iron (Fe^{3+} to Fe^{2+}) intra and intercellularly which are then made available to plants and endophytes. Siderophores are therefore highly have a relevant as they are able to perform the function of making Fe available when it is scarce in the rhizosphere.

2.5 IAA producer

Most PGPR can produce auxins that have very strong effects on the growth of plant root and structure and they are therefore involved in the plant-microbe interactions examples include IAA (Indole-3-acetic acid) produce cytokinins and gibberellins. They are able to give rise to transcription changes in hormones, defense-related and cell wall related genes, resulting in longer roots, root biomass increase and stomata size and density decrease and activate auxin response genes that enhance plant growth.

2.6 Cytokinins and gibberellins producers

PGPR are able to produce cytokinins and gibberellins, which are able to enhance plant shoot growth. Gibberellins interact with auxins to alter the structure of the root favorably while cytokinins act by enhancing root exudate production in plants this helps to encourage the presence of PGPR associated with the plant.

2.7 Ethylene producers

Ethylene gas hormone is produced in response to stress (abiotic and biotic) conditions and is very active in extremely low concentrations (0.05 mL L – 1). It helps plant in tolerating stress and senescence by providing more growth stimulation under drought stress, heat stress, salinity, unfavorable pH and other types of stress. Ethylene plays an important role for improving plant stress tolerance for some PGPR.

2.8 Abscisic acid (ABA) producers

Abscisic acid is a plant hormone that regulates numerous aspects of plant growth, development (seed and bud dormancy, control of organ size, germination, cell development, synthesis of protein etc.) and stress responses (abiotic), modulation of synthesis and inhibition of cytokinins. Their functions are conserved in plants. They are usually produced in small quantities in the root as well as terminal buds.

2.9 ACC (1-aminocyclopropane-1-carboxylic acid producers)

1-aminocyclopropane-1-carboxylic acid is a non- protein amino acid acting as the direct precursor of ethylene (which is a plant hormone that regulates a wide range of vegetative and developmental processes). ACC is the central molecule of ethylene biosynthesis.

2.10 Fungal PGRP

Fungi such as *Trichoderma* spp. (promote the growth of plants), *Beauveria bassiana* (used to control pests), Mychorriza (enhance water utilization and plant growth), have been found to be very useful in plant.

3. Response mechanism of plants to their microbiomes friends or foes

Plants and bacteria respond to the each other by producing hormones known phytohormones which are key requirements for regulating plant growth and development. They do these mostly to facilitate uptake of nutrients are in defense against pathogens or to enhance and stimulate proper growth and development in the plants. They also function as molecular signals in response to environmental factors that otherwise limit plant growth or become lethal when uncontrolled they can have either positive, neutral or negative effects on plants.

PGPR are able to produce acid phosphatases, to mineralize organic phosphorus in the soil and organic acids such as gluconic acid and citric acid to solubilize phosphorus. They also produce organic and inorganic acids such as citrate, oxalate, acetate, sulfuric acid, carbonic acid and nitric acids to stimulate the solubilization of other elements, such as zinc and potassium, which are essential for soil fertility and crop improvement. Siderophores such as pyochelin, pseudobactin, and pyoverdine are produced by bacteria to reduce iron (Fe^{3+} to Fe^{2+}) intra and intercellularly and are very useful to both the plant by (attacking pathogens by making it unavailable to them) and the PGRP. They increase plant growth through the improvement in iron acquisition, since this element has several important biological functions for the cell. Others like Rhizobia are able to enter the root plant tissue through the exchange of chemical signals and form nodules. Other non-nitrogen-fixing bacteria, such as Pseudomonas are able to enhance nodulation process by producing enzymes like 1-aminocyclopropane-1-carboxylate (ACC) deaminase. They also produce auxins, such as volatile organic compounds (VOC) which stimulate plant growth and are required in several physiological processes such as seed germination, formation of lateral roots, gravidity, and photosynthesis; it affects photosynthesis and the production of metabolites and other relevant compounds involved in the development of the plant.

PGPR also act as antagonist against plant pathogens, they do so by restricting their growth or killing the pathogens promoting plant growth and health. PGPR containing a collection of compounds and enzymes are usually able to restrict or entirely eliminate the activities of pathogens. Bacteria of the genus *Pseudomonas* are able to produce siderophores that chelate Fe available in the rhizosphere, to restrict pathogens. Another mechanism widely used by bacterial endophytes with antifungal activity is the production of enzymes, such as chitinases, cellulases, and β-1, 3-glucanases that degrade or attack the fungal cell wall, chitin (component of the fungal cell walls). They also produce enzymes with antibiotic activity including peptides of ribosomal origin (polyketide synthases), such as subtilin, subtilosin A, TasA, and sublancin and peptides of non-ribosomal origin such as bacillaene, bacilysin, chlorotetain, difficidin, mycobacillin, and some rhizocticins and antibiotics, fluorescent pigment or by cyanide production they are therefore able to act as bio controls against pathogen.

4. Utilizing the useful microbes

New biotechnologies are being developed using the plant defense response, plant beneficial bacteria to control diseases in plants (biopesticides), to control pathogens (biopathogens), as biocontrol agents and for plant growth promotion (biofertilizer) replacing chemical fertilizers and pesticides. They can be in the solid (e.g., powder) or liquid forms and can be either inoculated in the plants in the case of biofertilizers, biopathogens and biocontrol agents or sprayed on the plants biopesticides and biopathogens. They can be applied in the soil or on the aerial parts of the plant depending on the target. If it is intended for attacking soil fungi or oomycetes, it is more suitable to apply them in the soil, near the root while if the potential infection by pathogens is in the aerial part of the plants application in liquid or powder at the foliar parts is more suitable. They can be applied pre-field, on-field or post-field depending on when their action is required.

New biotechnologies are being developed using the plant defense response, plant beneficial bacteria to control diseases in plants (bio-pesticides) [1], to control pathogens (bio-pathogens), as biocontrol agents and for plant growth promotion (bio-fertilizer) replacing chemical fertilizers and pesticides. They can be in the solid (e.g., powder) or liquid forms and can be either inoculated in the plants in the case of biofertilizers, biopathogens and biocontrol agents or sprayed on the plants biopesticides and biopathogens [2, 3]. They can be applied in the soil or on the aerial parts of the plant depending on the target.

5. Symbiosis in legumes: biological nitrogen fixation

Legumes are crops which are known to fix nitrogen in their roots with the help of nitrogen fixing bacteria known as Rhizobia [4, 5]. Although the roles of rhizobia is well known in legumes we know that several other bacteria are present and associated with legumes in the soil play important roles to facilitate the health and growth of legumes [4]. Plants and bacteria respond to the each other by producing hormones known phytohormones which are key requirements for regulating plant growth and development [6]. They also function as molecular signals in response to environmental factors that otherwise limit plant growth or become lethal when uncontrolled they can have either positive, neutral or negative effects on plants. Plant growth promoting bacteria (PGPR) are able to produce acid phosphatases, to mineralize organic phosphorus in the soil and organic acids such as gluconic acid and citric acid to solubilize phosphorus [7, 8]. They also produce organic and inorganic acids such as citrate, oxalate, acetate, sulfuric acid, carbonic acid and nitric acids to stimulate the solubilization of other elements, such as zinc and potassium, which are essential for soil fertility and crop improvement [9].

6. Other plant growth promoting bacteria and their contribution/roles in legume health

The microbiomes associated with legumes showed abilities to improve %N_2 fixed, biomass of the plants, leaf chlorophyll and seed emergence as shown in the findings of [2, 4]. It is suggested that they are able to do this by able to stimulate solubilization of metals, produce Ethylene gas hormone is produced in response to stress (abiotic and biotic) conditions, and organic acids such as gluconic acid and citric acid to solubilize phosphorus and also produce IAA (Indole-3-acetic acid) produce cytokinins and gibberellins produce acid phosphatases, to mineralize organic phosphorus in the soil and organic acids such as gluconic acid and citric acid to solubilize phosphorus as shown in the findings of [2, 3]. They also produce organic and inorganic acids such as citrate, oxalate, acetate, sulfuric acid, carbonic acid and nitric acids to stimulate the solubilization of other elements, such as zinc and potassium, which are essential for soil fertility and crop improvement as explained in the findings of Andreote and Pereira e Silva [9] and Wille et al. [1].

6.1 Plant biomass

The plant biomass shoot, root, nodule and seed weight for both the wet and dry matter has been found to be enhanced by Plant growth promoting bacteria [10, 11]. The nodule weight and dry weight of plants receiving *Serratia proteamaculans* increased by 33 and 25%, respectively at 15°C RZT, while for plants receiving *Aeromonas hydrophila* P73 they were increased by 72 and 31% at 17 ± 5°C RZT. Rhizobia has also been shown to significantly increase the yield of legumes and influence the plant biomass positively, ultimately enhancing the plants growth and also the nodule biomass of legumes [10, 12–14].

6.2 Nutrient uptake

The uptake of nutrients such as nitrogen, phosphorus, macronutrient and micronutrients has been shown to be significantly increased [15] in plants by the presence of plant growth promoting bacteria. The bacteria make them available by solubilizing them thereby releasing them into the rhizosphere of the plant and making them assessable for the roots of the plants to absorb for the plant to use for various biochemical processes [10, 16]. In legumes the percentage nitrogen fixed in the plant is also significantly increased [10, 12, 16]. *Herbaspirillum sp*. is another gram-negative which can be found associated with most agricultural crops like rice (*Oryza sativa L.*) [17, 18], maize (*Zea mays L.*) [19], sorghum (*Sorghum bicolor L.*) [20], and sugarcane [17]. Studies using electron and fluorescence microscopy were able to revealed that the *Herbaspirillum sp*. strain B501 is able to colonize the intercellular space in the leaves of wild rice (Oryza officinalis), and fix N_2 in vivo [21]. It also is able to express nif genes in the shoot of wild rice [22]. *Herbaspirillum sp* can also colonize the intercellular space in the root and stem tissue of sugarcane plant, they are therefore non-specific to host plants [23]. When plants of barley (*Hordeum vulgare L.*) and Miscanthus are inoculations with *Herbaspirillum frisingense* strains colonization occurs showing that it is a true plant endophyte [24].

6.3 Flowering, height and maturation (Podding and yield)

Flowering and maturation (Podding and yield) have been shown to be enhanced by the activities of Plant growth promoting bacteria. In cow pea plants inoculated with PGPRs (*Rhizobial* spp., *Agrobacterium* spp., *Pseudomonas sp*. *Paenibacillus sp*., *Azotobacter*), the first signs of flowering commenced by the 6th week while plants to which PGPRs were added commenced flowering ahead of plants with N^+ (fertilizer application) and in non-sterile soils, an indication of faster maturity than the control plants. By the end of the 7th week, flowering had occurred in all plants expect for the plants in non-sterile soils (showed no signs of flowering), and by the 8th week no signs of flowering could still be seen [25]. In 82% of the plants to which PGPRs were applied, podding rate and size was enhanced with visible cowpea pods before harvested at the 8th week, while neither controls i.e. N^+ and plants in non-sterile soil showed signs of pod formation. Flowering and podding were also enhanced on co-inoculation of *Rhizobial* spp. alongside a consortium of *Agrobacterium* spp., *Pseudomonas* sp. *Paenibacillus* sp. and a consortium of *Azotobacter*. Co-innoculation of *Bradhyrizobium* with PGPRs increased seed yield in soybean plants [26], this may vary from plant to plant and strain type as rhizobial inoculation were shown to induce late flowering and pod formation in soybean [27].

Much study has been done on PGPRs related to genus Azospirillum largely because of their efficiency to promote the growth of plants with of different agronomical interest. The work of Garcia de Salamone et al. [28], shows that *Azospirillum sp*. is a major contributor to plant fitness through BNF. The genus Burkholderia also includes bacteria species that are able to fix N_2. *Burkholderia vietnamiensis*, a human pathogenic species, was efficient in colonizing rice roots and fixing N_2 [29]. Other bacteria genera asides Burkholderia, such as *Pantoea*, *Bacillus* and *Klebsiella*, have the potentials for BNF and endophytic colonization in different maize genotypes [30]. Another extensively studied endophyte is *Gluconacetobacter diazotrophicus* [31–34]. Plant height was also significantly enhanced on inoculation with PGPRs, where the plants with PGPR were taller than their counterpart containing only rhizobia, with only 22–33% of the plants, being shorter than the N^+ (nitrogen fertilizer) plants. Also increased availability of nutrient with inoculation of PGPRs in the rhizosphere enhanced to plant height.

Co-innoculation of Bradhyrhizobium with PGPRs increased soybean plant vigor [35, 36]. Plant height, flowering and podding were enhanced by co-inoculation of *Rhizobial* spp. with a consortium of *Agrobacterium* spp., *Pseudomonas* sp. *Paenibacillus* sp. while a consortium of *Azotobacte*r spp. showed no ability to improve plant height [25]. The plant growth-promoting rhizobacteria (or PGPR) generally belong to heterogeneously beneficial groups of microorganisms residing in the plant rhizosphere, either on associated or on the surface of the root surface with the capacity to enhance plant-growth and provide protection for the plants from disease and other abiotic stresses [37–39].

6.4 Phosphorus uptake

Phosphorus (P) is required for several molecular and biochemical plant processes, especially in acquisition, storage and utilization of energy. When phosphorous supply is deficient, it results in severe limitations in the nitrogen fixing and symbiotic interacting processes. It's requirement maybe more for the roots or shoots of the host plant. There are definite differences in Rhizobial and plant requirements for P [40]. Usually, the slow- growers are more tolerant to low P than the fast-growing Rhizobia. Nodules are also strong sinks for P and nodulation and N_2 fixation are greatly affected by P availability [40]. In cases where leguminous plants that depend on symbiotic nitrogen get insufficient supplies of phosphorus, they also suffer nitrogen deficiency. The phosphorus content in a dry weight unit of a plant is usually greater in nodules than in the roots and shoots, especially when there are low supplies of phosphorus. It has been discovered that Nitrogen fixing plants require more P than plants getting nitrogen directly from fertilizer, this is likely because it is needed for nodule development and signal transductions, and for P-lipids in the very numerous bacterioids [40].

6.5 Nitrogen fixation

Biological nitrogen fixation (BNF) is the process by which Rhizobia (which are rod shape, Gram negative soil bacteria) that relate symbiotically with legumes, play the crucial role of replenishing the soi1 nitrogen by converting atmospheric nitrogen to ammonia [41]. They have the potential to fix up to 300 kg/ha of nitrogen. The biological fixation of atmospheric nitrogen by *Rhizobium*/legumes association has been estimated to be 50% of all nitrogen fixed on the earth [42, 43].

The common *nodABC* genes is found in all *Azorhizobium*, *Rhizobium*, and *Bradyrhizobium* isolates which have been studied so far are structurally conserved and functionally interchangeable between *Rhizobia* spp. [44, 45]. They encode enzymes involved in the synthesis of the lipo-oligosaccharide core. And their inactivation of abolishes the ability to elicit any symbiotic reaction in the plant [44, 46]. *nodZ*A and *nodZ*B are two genes that have been shown to have specificity for soybean associated rhizobia and most *Rhizobia* spp. that have ability to nodulate soybean [47, 48] as they play a major role in the nodulation of soybean.

Both *nodA* and *nodB* were found to be present in all the strains confirming that they are rhizobia strains as these two nod genes are found in all rhizobia and are functionally similar in *Rhizobia* spp. and was similar to the findings of Toolarood et al. [49] and Reeve et al. [50] where both genes have been found to be conserved in *Rhizobia* spp. *nodZ*A and *nodZ*B may also be found in a Bambara symbiotic

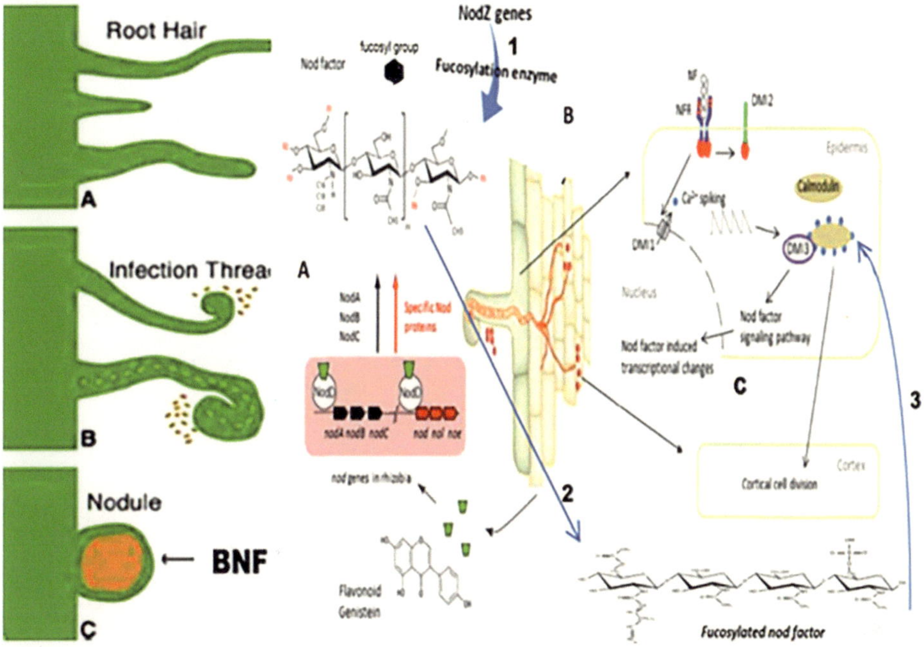

Figure 1.
(A) Legume secrete flavonoid eg genistein into the rhizosphere as signal and are perceived by rhizobial nodD regulatory proteins which binds to appropriate flavonoid, which are transcription factors that activate transcriptones of various nod genes that induce synthesis of rhizobial nod factors (NF). (B) Nod factors NFs are perceived by the Nod factor receptors (NFR) at the legume root. (C) DM12 causes root hair curling. DM11 encodes an ion channel that is localized to the nuclear membrane. DM13 encodes a calcium-calmodulin-dependent protein kinase which is required for the induction of transcriptional changes done to Nod factors binding as well as cortical cell division coined from Oldroyd and Downie [51] and Masson-Bovin et al. [52]. 1—The product of the nodZ genes are responsible for the addition of a fucosyl group to Nod factors; 2—Fucosylation of nod factors occurs; and 3—Fucosylation results in sharp increase of cytosolic ca^{2+}.

strains [15], implying that these strains may have the ability to form nodules in soybean plants as these two genes are shown to be specifically found in soybean symbiotic rhizobia because their products is the fucosyl group which a common product for the symbiont of soybean and most *Rhizobia* spp. that have ability to nodulate soybean as confirmed by [47, 48] (**Figure 1**).

The importance of bacteria endophytes with N_2-fixing ability have been an interest of study in non-leguminous plants like sugarcane (*Saccharum officinarum L.*), [53] and several other plants. Other studies suggest that bradyrhizobia is able to colonize the root nodules in legume plants and also the roots of tubers such as sweet potatoes (*Ipomoea batatas L.*), expressing nifH, and also acting as diazotrophic endophytes [54]. In conditions of N deficiency, the *G. diazotrophicus* strain Pal5 isolated from sugarcane is able to increase the N content when compared with plants inoculated with a Nif – mutant or uninoculated plants [55].

6.6 ACC (1-aminocyclopropane-1-carboxylic acid producers)

1-aminocyclopropane-1-carboxylic acid (ACC) is also a fairly simple molecule and is the biochemical precursor of ethylene. It's amino acid has a triangularly shape with several more features in addition to being a precursor for the lead-role player ethylene. For instance, ACC could be conjugated into three types of derivatives, although

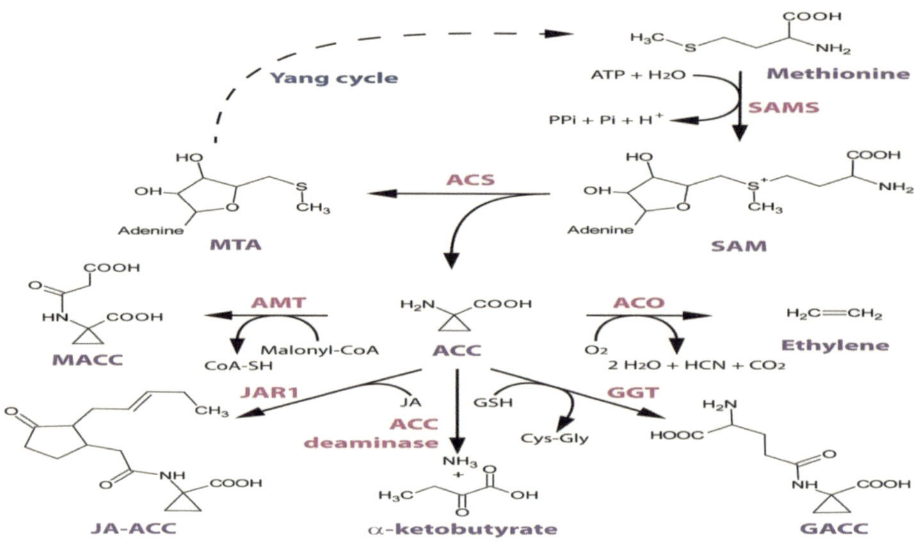

Figure 2.
Structural scheme of ethylene biosynthesis and 1-aminocyclopropane-1-carboxylic acid (ACC) conjugation/metabolism.

their role biological is still uncertain. ACC favors plant growth and lowers susceptibility to stress when they are metabolized by bacteria using the enzyme ACC-deaminase. ACC is also subjected to a sophisticated transport mechanism to ensure local and long-distance ethylene responses. ACC has been reported to function as a signal itself, independently from ethylene. Plants possess several mechanisms to control their pool of ACC, for example by converting it to ethylene or to conjugates like MACC, GACC, or JA-ACC. ACC can also be metabolized by deamination (**Figure 2**). The ACC deaminase enzyme was discovered in bacteria first with PGPRs having the ability to process plant-borne ACC and convert it to ammonia and α-ketobutyrate by metabolizing it with the ACC deaminase enzyme [56]. ACC deaminase enzyme has been retrieved in bacteria such as *Pseudomonas sp*. strain ACP [56], *Pseudomonas putida* GR12-2 [57] Pseudomonas chloroaphis 6G5 [58] and *Pseudomonas putida* UW4 [59].

The ACC deaminase enzyme produced by bacterial is PLP-dependent having low affinity for ACC (Km value is 1.5–15 mM have been reported [60];). Nonetheless, expressions of the ACC deaminase gene (acdS) can be induced under low relative concentrations of ACC (100 nM), much-like other amino acids including DL-alanine, L-alanine, and DL-valine [57] (**Figure 2**). The expression of acdS is also regulated by the nitrogen fixation (nif) promotors of most Rhizobia strains, with links ACC deaminase with the process of nodule formation [61, 62].

The amino acid methionine is converted to S-adenosyl-L-methionine (SAM) by SAM-synthetase (SAMS) with the requirement of ATP. The general precursor SAM is then converted to ACC by ACC-synthase (ACS). This reaction also involves the cleavage of 5′-methylthioadenosine (MTA), which is recycled back to methionine by the Yang cycle (dotted line indicates multiple enzymatic steps). ACC can be converted to ethylene by ACC-oxidase (ACO) in the presence of oxygen. ACC can also be converted to its major conjugate 1-malonyl-ACC (MACC) by the yet uncharacterized ACC-N-malonyl transferase (AMT) with the requirement of malonyl-Coenzyme-A. A second derivate of ACC is γ-glutamyl-ACC (GACC) which is formed by γ-glutamyl-transpeptidase (GGT) with the requirement of glutathione (GSH). Another novel derivate of ACC is jasmonyl-ACC (JA-ACC), which is formed by jasmonic acid resistance 1 (JAR1). ACC can also be metabolized by the bacterial (and plant) ACC deaminase into ammonium and α-ketobutyrate [63].

6.7 Soil viral infection

Legumes (form symbiotic relationships with nitrogen fixing *Rhizobia* spp) are of major economic importance worldwide but are damaged by viral diseases causing significant economic losses as a result of reduced grain production, seed-quality and costs incurred during disease-control. Most viruses infecting legumes are transferred by insect-vectors, or via seed. Viral infections in plants are currently incurable due to unavailability of antiviral compounds, only the use of efficient measures of control or agents that can suppress viral-activity can be employed to prevent or reduce their effects. Symptoms of infection in Legumes: flowering, leaves, podding, seedling stages, nodule formation, reduced yield and productivity (**Figure 3**).

Recent researches show that virus-carrying-aphids induced salicylic-acid in plants, affecting the nitrogen-fixing-ability of rhizobia in the plants resulting in fewer root nodules and nodule biomass. But fewer essential amino-acids were obtained in the aphids that were collected on *P. sativum* plants inoculated with rhizobia than aphids collected from *P. sativum* plants grown without rhizobia inoculation. Also, in faba-bean plants previously inoculated with rhizobia there was a reduction in disease-occurrence in plants infected with BYMV (bean yellow mosaic virus) when studied using ELISA reads but the volatile emissions from the plant host were suppressed (**Figure 3**).

Rhizobia may be useful as plant-pathogen fighters providing defense against viruses and protect leguminous plants, ultimately reducing susceptibility and damage-effects in plants. In addition, the manipulation of the host by the virus may be dependent on the physiological condition of the host. These mechanisms are not yet well understood and more work needs to be done to discover the interaction between Rhizobia and soil virome in leguminous plants resistance to viral infections.

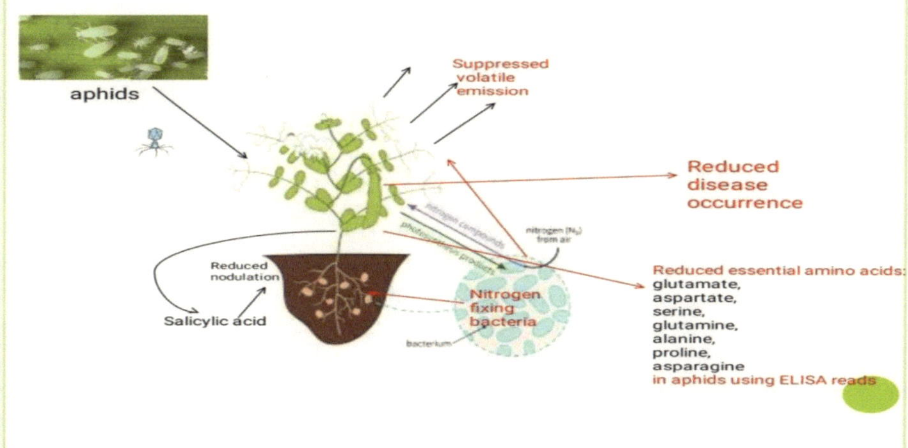

Figure 3.
Soil- plant- virus interaction.

6.8 Other factors

Other factors that are affected by PGPRs are seedling health, fight against diseases, drought tolerance, germination rate. Bacterial inoculation contributes substantially to increasing the agronomic efficiency by reduced production costs and pollution to the environment, reduced or complete elimination of use of chemical fertilizers can be accomplished with inoculant efficiency. To achieve success with the use of bacterial inoculants in improved plant growth and production in plants, many processes which ultimately influence the efficiency of inoculation are involved, these includes the release of plant root exudates, the root colonization by bacteria and soil health. This review gives an overview of how important soil-plant-microbe interactions are for developing efficient bacterial inoculants, and the extensively studied PGPRs that represents a large diversity group of easily accessible beneficial bacteria [37–39].

7. Plant analysis to check the activities of plant growth promoting bacteria in the soil and their effect on the plant

7.1 Determination of chlorophyll

Leaf chlorophyll content is determined determined using a chlorophyll meter SPAD- 502 this can be done weekly starting from the 2nd–3rd week [64]. The measurments are recorded in SPAD units. This is done by placing the leaf in-between the teeth of the meter which has a sensor the measures the chlorophyll based on the color and transpiration rate. Three leaves per plant including the youngest the oldest and the matured leaves [4, 15]. chlorophyll content can also be measured traditionally in the laboratory by extracting the chlorophyll itself from the leaf sample using acetone before calculating the chlorophyll concentration can be measured invitro by using a spectrophotometer at absorption of 663 nm and 645 nm [65, 66]. Its unit are recorded as nmol/cm, mg/L, mSPU (m specific pigment unit), μmol m-2 (micromoles per square meter), mg m-2 (milligrams per square meter).

7.2 Biomass determination harvesting

To determine the biomass of the plant, the plants are harvested between 4 and 8 weeks after planting (WAP) (depending on the plant type), by cutting at the base with a secateurs. The shoots and roots are separated, collected and put inside labeled paper bags before being placed in an oven were they were dried at 68°C for 72 hours or until constant weights are obtained [4].

7.3 Determination of nitrogen

Plant samples (shoot, root) are collected and dried in an oven at 68°C for 2 days or brown dry. The samples are weighed and then grinded to 1 mm using a grinder after which they are packed in small white envelops and transferred to the laboratory and stored under dry conditions. Approximately 0.1 g of the shoot samples is weighed into digest tubes. To digest the samples the following steps are carried out addition of 2.5 mL of acid digest solution to the samples while stirring the solution of the acid digest. Samples are placed in a fume cupboard and 3 mL of H_2O_2 is added to it. The digest tubes are arranged in a digester and the temperature is set to 360°C and heated for 3 hours.

Samples are checked to see if they had turned white (those that were not white can be left for 1 hr. more) and then removed. The digested samples are allowed to cool and 50 mL of distilled water is added. The sample solution is then poured into centrifuge tubes. 10 mL of the digested sample is taken and added to a 50 mL standard volumetric flask the volume was made up to 50 mL. It was allowed to stay for 1 hour and the readings were taken using a spectrophotometer at 650 nm [67].

7.4 Determination of nutrients in plant shoots

The total Nitrogen in plant shoots can be determined using the micro Kjeldahl method [15, 67] while phosphorus is determined by the molybdenum blue method as described by [68]. The concentrations of Ca, K, Na and Mg in plant shoot and root can be determined using the method as described by [15, 69]. Un-inoculated plants or plant with control PGRP inoculation e.g., N^+ treatment can be used as reference plants for comparison. Dried plant samples are digested (as described above) after which they can be used for detections of metals such as sodium and calcium using a flame photometer while the detection of nutrients such as potassium and magnesium are done using an atomic spectrometer from the diluted digested samples.

7.5 Determination of phosphate solubilization

Pakovskaya medium is used for the determination of the ability of the selected strains to solubilize calcium phosphate ($Ca_3 (PO_4)_2$). Pakovskaya's agar [70] was composed (Glucose-10 g, Yeast extract-0.5 g, Ammonium sulphate-0.5 g, Magnesium Sulphate 0.1 g, Calcium Phosphate-5 g, KCl-0.2 g, MnSO4.7H_2O-0.0001 g, $FeSO_4$.7H2O-0.0001 g, Agar powder-20 g). The medium is autoclaved at 121°C for 15 min allowed to cool to 45°C, poured into petri dishes and allowed to set. Broth (10 μL) of bacteria with concentrations of $\leq 10^7$cfu/ml are spotted on the plate in triplicates and incubated at 30°C for 7 days. Isolates with visible clear halos around the colonies are considered to be able to solubilize calcium phosphate. The phosphate solubilization index (PSI) can be calculated by dividing the halo diameter formed by the bacteria (colony diameter + diameter of the clear zone around the colony, B) by the colony diameter (b) of the bacteria itself as shown in **Figure 4**. Tests are done using three–four replicates.

7.6 Indole-acetic acid (IAA) production

The production of IAA production can be determined using the method of Gordon and Weber [71] with three replicates. The isolated bacteria strains are cultured in 50 ml Laria Broth (LB) for 2 days in a rotary shaker. The broth is then collected and centrifuged at 10,000 × g for 15 min and the clear supernatant is collected. The supernatant (1 ml) is mixed with 2 ml of Salkovski reagent (1 ml of 0.5 M ferric chloride solution added to 50 ml of 35% perchloric acid (HClO4)), and 10 μl (a drop) of the ortho-phosphoric acid is added. The solution is incubated for 30 min. Pink color change is used to determine positive IAA production. The absorbance of the pink (positive) samples is measured at 530 nm using a spectrophotometer to evaluate the amount of IAA produced. A standard curve is prepared from pure IAA with concentration points ranging from 5 to 50 $\mu g.ml^{-1}$ (or ppm) from the prepared standard concentrations. A standard graph is prepared by plotting the IAA concentrations against their respective optical density (OD530). The linear equation

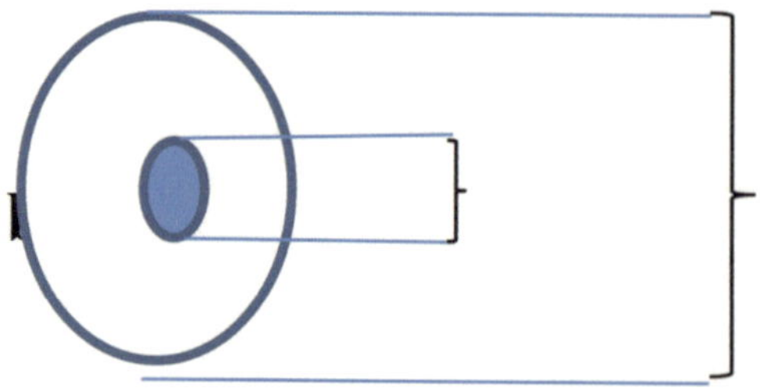

Figure 4.
Phosphorus-solubilizing and parameters for the determination of bacteria PSI (right).

obtained can be used to determine the concentration of IAA in samples from their optic density at 530 nm. IAA sample (μg/ml) = (OD530-(−0.0052)/0.0016). IAA can also be determined using HPLC (high performance liquid chromatography), liquid chromatography, mass spectrometry and GC (gas chromatography). IAA can also be analyzed by gas chromatography-mass spectrometry (GC-MS) using selected-ion monitoring (SIM).

7.7 Nitrogen fixation assay

The amount of nitrogen fixed can be determined using Uriede analysis. Stems and stem segments of the plant is harvested and leaves are collected and placed in properly labeled bags and then dried at 65–80°C in an air oven for 2 days. The dried plant tissue are grounded, passed through a 60-mesh (1.0 mm) screen and then stored in a dry place until extraction. Approximately 0.5 g of the dry grounded material is weighed and transferred into 100 mL beakers or Erlenmeyer flasks to which 25 mL of distilled water is added and boiled for one to 2 minutes. The extract is then filtered while it was hot through a 15 cm Whatman No. 4 filter paper folded into a funnel inserted into a 50 mL volumetric flask and the residue is washed into the filter and rinsed with a little distilled water. The residue is allowed to cool, and the volume made up to 50 mL with distilled water. The extract are then stored in a freezer in small vials until analysis of N solutes.

Into the extract an amount of 0.5 mL of 0.5 N sodium hydroxide (NaOH) is added (forcefully that additional mixing is unnecessary) using a Dispenser and placed in a rack of tubes in a boiling water bath for 10 mins after which 1.0 mL HCl/phenylhydrazine mix is added using a second Dispenser and placed in the boiling water bath for exactly 2 minutes. It is then removed and immediately plunged into an ice bath and left for 15 minutes after which 2.5 mL of cold $HCl/K_3Fe(CN)_6$ is added using a Dispenser, thoroughly mixed and left on the bench for 10 mins after which the optical density is read at 525 nm on a spectrophotometer. A standard curve which covers the 0–0.10 mM allantoin range can be constructed from time to time as part of the standard laboratory's quality control using 2.5 mL of each of the five different concentrations in test tubes using 2 replicates corresponding to 0, 0.025, 0.05, 0.1, and 0.25 μmol of Ureide [72].

7.8 Determination of abscisic acid (ABA) and ethylene production

ABA can be isolated from mature seeds and from plant leaves of at different stages of growth. The seeds or leaves are dried and grinded into powder and 10 g (using three replicates) are extracted and homogenizied using 40 mL of 80% aqueous methanol for 30 min at 4°C. The mixture is filtered using a conical flask (for each sample) using the filter paper (size No. 1. Whatman). The filtrate is then subjected to vaccum evaporation in a lyophilizer and the dried vacuumed residue is re-dissolved into 10 mL of 0.5 M phosphate buffer (set to pH 8) afterwhich it is stirried for 30 min. The resulting suspension is washed with 20 mL light petroleum sprit and pH re-adjusted to 2.8 (with dil. HCl), and the extraction is done four times using ethyl acetate (4 × 10 mL). The extract is then lyophilized and re-dissolved using 5 mL buffer of 0.5 M phosphate (pH 8). The sample can then be purified by passing it through a sephadex G10 column. To prepare the Sephadex column, Sephdex (1 g) is allowed to swell in 10 mL of distilled water (double) over night before it is packed into a glass column (1 cm diameter and 5 cm length) up to 2 cm length and the column should be equilibrated using a phosphate buffer. The eluted solution can then be lyophilized again and the resulting residue dissolved using 1 mL of acetonitrile and analyzed by HPLC [73]. ABA can also be analyzed by gas chromatography-mass spectrometry (GC-MS) using selected-ion monitoring (SIM). Three main categories of methods commonly used (or having the potential to be used) for ethylene detection in plants: (1) GC detection [74, 75], (2) electrochemical sensing [76] and (3) optical detection [77].

8. Conclusion

Microbiomes are a very essential part of the plant growth and health and they govern most soil functions affecting the growth of plant directly or indirectly. There is a need for higher crop yield production in our current world as well as fertility of soil in an ecofriendly manner and this requires research that is focused on the new concepts of the exotic biomolecules, hormones, enzymes and metabolites which create a suitable environment for the interaction between plant and microbes using molecular and biotechnological approaches to increase our knowledge of rhizosphere biology and to achieve an integrated management of soil microbial populations.

Author details

Olaoluwa Oluwafunto Ajayi[1*] and M. Dianda[2]

1 Soil Microbiology Unit, International Institute for Tropical Agriculture, Ibadan, Nigeria

2 Laboratoire de Microbiologie Forestière (INERA/DEF), Ouagadougou, Burkina Faso

*Address all correspondence to: oluwafuntoajayi@yahoo.com

References

[1] Wille L, Messmer MM, Studer B, Hohmann P. Insights to plant–microbe interactions provide opportunities to improve resistance breeding against root diseases in grain legumes. Plant, Cell & Environment. 2019;**42**:20-40. DOI: 10.1111/pce.13214

[2] Müller Pérez-Jaramillo JE, Carrión VJ, Bosse M, Ferrão LFV, De Hollander M, Garcia AAF, et al. Linking rhizosphere microbiome composition of wild and domesticated *Phaseolus vulgaris* to genotypic and root phenotypic traits. ISME Journal. 2017;**11**:2244-2257

[3] Müller DB, Vogel C, Bai Y, Vorholt JA. The plant microbiota: Systems-level insights and perspectives. Annual Review of Genetics. 2016;**50**:211-234

[4] Ajayi OO, Dianda M, Fagade OE, Nwadike B. Effect of co-inoculation of non- Rhizobial microorganisms (NRMs) with Rhizobia on nodulation and leaf chlorophyll content on cowpea Vigna Ungigulata [(L.) Walp.] (Tvx3236). International Journal of Research and Innovation in Applied Science (IJRIAS). 2019;**IV**(XII). ISSN 2454-6194

[5] Angelini J, Castro S, Fabra A. Alterations in root colonization and nodC gene induction and inpeanut-rhizobia interaction under acidic conditions. Plant Physiology and Biochemistry. 2003;**41**:289-294

[6] Sy A, Giraud E, Jourand P, Garcia N, Willems A, De Lajudie P, et al. *Methylotrophic Methylobacterium* bacteria nodulate and fix, nitrogen in symbiosis with legumes. The Journal of Bacteriology. 2001;**183**:214-220

[7] Cheng Z, McConkey BJ, Glick BR. Proteomic studies of plant- bacterial interactions. Soil Biology and Biochemistry. 2010;**42**:1673-1684

[8] Walker V, Cédric B, Floriam B, Yvan M, René B, Gilles C. Host plant secondary metabolite profiling shows a complex, strain-dependent response of maize to plant growth-promoting rhizobacteria of the genus Azospirillum. New Phytologist. 2011;**189**:494-506. www.newphytologist.com

[9] Andreote FD, de Pereira e Silva MC. Microbial communities associated with plants: Learning from nature to apply it in agriculture. Current Opinion in Microbiology. 2017;**37**:29-34

[10] Ojo CT, Heerwaaeden JV, Dianda M, Geurts R, Giller K. Cowpea Can Benefit from Rhizobia Inoculation: Does Indigenous Status Matter. Ibadan: IITA research for development conference (R4D) N2Africa. Poster presentation. 2017

[11] Zhu GS, Yu ZN, Gui Y, Liu ZY. A Novel Technique for Isolating Orchid; 2008

[12] Egbe OM, Bar-nyam CU. Nodulation, nitrogen fixation and harvest index of extra-short- and short-duration cowpea varieties intercropped with maize at Otobi, Benue state, Nigeria. Journal of Animal & Plant Sciences. 2011;**10**:1315-1324

[13] Egbe OM. Assessment of biological nitrogen fixing potentials of pigeonpea genotypes intercropped with sorghum in Southern Guinea Savanna of Nigeria. Agro-Science. 2007;**6**:33-45

[14] Kaleem F. Biological nitrogen fixation in a soybean / sorghum intercropping system. In: Karanja N, Kahindi HP, editors. Challenges and Imperatives for Biological Nitrogen Fixation Researchand

Applications in Africa for the 21st Century. Nairobi, Kenya: Proceedings of the nineth congress of the African Association for Biological Nitrogen Fixation; 2010, 2000. pp. 50-62

[15] Ajayi OO, Dianda M, Fagade OE, Nwadike B. Plant growth promoting bacteria play a major role in nutrient uptake of Cowpea plants. International Journal of Research and Innovation in Applied Science (IJRIAS). 2020, 2020;**V**(II). ISSN 2454-6194

[16] Dianda M, Boukar O, Ibrahim H, Muaz WA. Indigenous bradyrhizobium and commercial inoculants in growth of soybean [Glycine max (L.) MERR.] in northern Nigeria. In: Proceedings of the 1st International Conference on Drylands. Kano, Nigeria. 2014

[17] Pedrosa FO, Monteiro RA, Wassem R, Cruz LM, Ayub RA, Colauto NB, et al. Genome of Herbaspirillum seropedicae strain SmR1, a specialized diazotrophic endophyte of tropical grasses. PLoS Genetics. 2011;7

[18] Souza R, Beneduzi A, Ambrosini A, Costa PB, Meyer J, Vargas LK, et al. The effect of plant growth-promoting rhizobacteria on the growth of rice (Oryza sativa L.) cropped in southern Brazilian fields. Plant and Soil. 2013;**366**:585-603

[19] Monteiro RA, Schmidt MA, de Baura VA, Balsanelli E, Wassem R, Yates MG, et al. Early colonization pattern of maize (Zea mays L. Poales, Poaceae) roots by Herbaspirillum seropedicae (Burkholderiales, Oxalobacteraceae). Genetics and Molecular Biology. 2008;**31**:932-937

[20] James EK, Olivares FL, Baldani JI, Dobereiner J. Herbaspirillum, an endophytic diazotroph colonizing vascular tissue of Sorghum bicolor L. Moench. Journal of Expremental Botany. 1997;**48**:785-798

[21] Elbeltagy A, Nishioka K, Sato T, Suzuki H, Ye B, Hamada T, et al. Endophytic colonization and in plant nitrogen fixation by a sHerbaspirillum sp isolated from wild rice species. Applied and Environmental Microbiology. 2001;**67**:5285-5293

[22] You M, Nishiguchi T, Saito A, Isawa T, Mitsui H, Minamisawa K. Expression of the nifH geneof a Herbaspirillum endophyte in wild rice species: daily rhythm during the light-dark cycle. Applied Environment Microbiology. 2005;**71**(12):8183-8190. DOI: 10.1128/AEM.71.12.8183-8190.2005. PMID: 16332801; PMCID: PMC1317309

[23] Njoloma J, Tanaka K, Shimizu T, Nishiguchi T, Zakria M, Akashi R, et al. Infection and colonization of aseptically micropropagated sugarcane seedlings by nitrogen-fixing endophytic bacterium, Herbaspirillum sp. B501gfp1. Biology and Fertility of Soils. 2006;**43**:137-143

[24] Rothballer M, Eckert B, Schmid M, Fekete A, Schloter M, Lehner A, et al. Endophytic root colonization of gramineous plants by Herbaspirillum frisingense. FEMS Microbiology Ecology. 2008;**66**:85-95

[25] Ajayi OO et al. Plant growth promoting rhizobacteria and how they affect the physical properties of cowpea Vigna ungigulata [(L.) Walp.]. In: Isteams Conference Paper Preceeding. Accra, Ghana. 2018

[26] Li DM, Alexxandrer A. Co-innoculaton with antibiotic-producing bacteria to increase colonization and nodulation by Rhizobia. Plant and Soil. 1988;**108**:211-219

[27] Eutropia VT, Patrick AN. Bradhyrhizobium japonicum inoculation and phosphorous supplementation on growth and chlorophyll accumulation

in soybean (Glycine max L.). American Journal of Plant Sciences. 2013;**4**:2281-2289

[28] Garcia de Salamone IE, Dobereiner J, Urquiaga S, Boddey RM. Biological nitrogen fixation in Azospirillum strain–maize genotype associations as evaluated by the 15N isotope dilution technique. Biology Fertilizer Soils. 1997;**23**:249-256

[29] Govindarajan M, Balandreau J, Kwon SW, Weon HY, Lakshminarasimhan C. Effects of the inoculation of Burkholderia vietnamensis and related endophytic diazotrophic bacteria on grain yield of rice. Microbial Ecology. 2008;**55**:21-37

[30] Ikeda AG, Bassani LL, Adamoski D, Stringari D, Cordeiro VK, Glienke C, et al. Morphological and genetic characterization of endophytic bacteria isolated from roots of different maize genotypes. Microbial Ecology. 2013;**65**:154-160

[31] Baldani JI, Caruso L, Baldani VLD, Goi SR, Döbereiner J. Recent advances in BNF with non-legume plants. Soil Biology and Biochemistry. 1997;**29**:911-922

[32] Bertalan M, Albano R, de Pádua V, Rouws L, Rojas C, Hemerly A, et al. Complete genome sequence of the sugarcane nitrogen-fixing endophyte Gluconacetobacter diazotrophicus PAl5. BMC Genomics. 2009;**10**

[33] Muthukumarasamy R, Cleenwerk I, Revathi G, Vadivelu M, Janssens D, Hoste B, et al. Natural association of Gluconacetobacter diazotrophicus and diazotrophic Acetobacter peroxydans with wetland rice. Systematic and Applied Microbiology. 2005;**28**:277-286

[34] Oliveira ALM, Urquiaga S, Döbereiner J, Baldani JI. The effect of inoculating endophytic N2 fixing bacteria on micro propagated sugarcane plants. Plant and Soil. 2002;**2**:205-215

[35] Vermar DPS, Fortin MG, Stanley VP, Mauro S, Purohit S, Morrison N. Nodulins and nodulin genes of Glycine max. A Perspective Plant Molecular Biology. 1986;**7**:51-61

[36] Yaholom E, Okon Y, Dovrat A. Azospirillium effects on susceptibility to Rhizobium nodulaton and nitrogen fixation of several forage legumes. Canadian Journal of Microbiology. 1987;**33**:510-514

[37] Dimkpa C, Weinand T, Asch F. Plant-rhizobacteria interactions alleviate abiotic stress conditions. Plant, Cell & Environment. 2009;**32**:1682-1694

[38] Glick B. Plant growth-promoting bacteria: Mechanisms and applications. Scientifica. 2012;**2012**:1-15

[39] Grover M, Ali SKZ, Sandhya V, Rasul A, Venkateswarlu B. Role of microorganisms in adaptation of agriculture crops to abiotic stresses. World Journal of Microbiology and Biotechnology. 2011;**27**:1231-1240

[40] Weria W, Yaghoub R, Kaveh HA. Role of some of mineral nutrients in biological nitrogen fixation. Bulletin of Environment, Pharmacology and Life Sciences. 2013;**2**:77-84

[41] Gage JD. Infection and invasion of roots by symbiotic, nitrogen-fixing Rhizobia during nodulation of temperate legumes. Microbiological and Molecular Biology Review. 2017;**68**:203

[42] Ravinder KG, Autar KM, Maria AS. Rhizobial–host interactions and symbiotic nitrogen fixation in legume crops toward agriculture sustainability. Frontiers in Microbiology. 2021;**12**:1290

[43] Food and Agriculture Organization. Fertilizer Use Under Multiple Cropping

Systems. FAO Fertilizer and Plant Nutrition Bull 5. Rome: Food and Agriculture Organization; 1983

[44] Long SR. Rhizobium genetics. Annual Reviews of Genetics. 1989;**23**(483-506):13

[45] Stacey, G. 1990. Workshop summary: Compilation of the nod, fix and nif genes of Rhizobia and information concerning their function. pp. 239-244

[46] Martinez E, Romero D, Palacios R. The Rhizobium genome. Critical Reviews in Plant Sciences. 1990;**9**:59-93

[47] Branden JF, Conley SP, Ané J-M. Enumeration of soybean-associated Rhizobia with quantitative real-time polymerase chain reaction. Crop Science. 2010;**50**:2591-2596

[48] Stacey G, Luka S, Sanjuan J, Banfalvi Z, Nieuwkoop AJ, Chun JY, et al. NodZ, a unique host-specifi c nodulation gene, is involved in the fucosylation of the lipooligosaccharide nodulation signal of Bradyrhizobium japonicum. Journal of Bacteriology. 1996;**176**:620-633

[49] Toolarood AAS, Alikhani HA, Salehi G, Asadi-Rahmani H, Khavazi K, Poorbabaee AA, et al. Genetic diversity and phylogeny of alfalfa nodulating rhizobia assessed by nifH and nodA genes. International Research Journal of Applied and Basic Sciences. 2012;**3**:1470-1476

[50] Reeve W, Ardley J, Tian R, Eshragi L, Yoon JW, Ngamwisetkun P, et al. A genomic encyclopedia of the root nodule bacteria: Assessing genetic diversity through a systematic biogeographic survey. Standard Genomic Science. 2015;**10**:14

[51] Oldroyd GE, Downie JA. Coordinating nodule morphogenesis with rhizobial infection in legumes. Annual Review of Plant Biology. 2008;**59**:519-546. DOI: 10.1146/annurev.arplant.59.032607.092839. PMID: 18444906

[52] Masson-Boivin C, Giraud E, Perret X, Batut J. Establishing nitrogen-fixing symbiosis with legumes: How many rhizobium recipes? Trends Microbiology. 2009;**17**:458-466

[53] Thaweenut N, Hachisuka Y, Ando S, Yanagisawa S, Yoneyama T. Two seasons' study on nifH gene expression and nitrogen fixation by diazotrophic endophytes in sugarcane (Saccharum spp. hybrids): Expression of nifH genes similar to those of rhizobia. Plant and Soil. 2011;**338**:435-449

[54] Terakado-Tonooka J, Ohwaki Y, Yamakawa H, Tanaka F, Yoneyama T, Fujihara S. Expresses nifH genes of endophytic bacteria detected in field-growth sweet potatoes (Ipomoea batata L.). Microbes and Environments. 2008;**23**:89-93

[55] Sevilla M, Burris R, Gunapala N, Kennedy C. "Comparison of benefit to sugarcane plant growth and 15N2 incorporation following inoculation of sterile plants with Acetobacter diazotrophicus wild-type and Nif-mutant strains" Molecular Plant-Microbe Interactions. 2001;**14**(3):358-366. DOI: 10.1094/MPMI.2001.14.3.358

[56] Honma M, Shimomura T. Metabolism of 1-aminocyclopropane-1-carboxylic acid. Agricultural and Biological Chemistry. 1978;**43**:1825-1831. DOI: 10.1271/bbb1961.42.1825

[57] Jacobson CB, Pasternak JJ, Glick BR. Partial purification and characterization of ACC deaminase from the plant growth-promoting rhizobacterium pseudomonas putida GR12-2. Canadian Journal of Microbiology. 1994;**40**:1019-1025. DOI: 10.1139/m94-162

[58] Klee HJ, Hayford MB, Kretzmer KA, Barry GF, Kishmore GM. Control of

ethylene synthesis by expression of a bacterial enzyme in transgenic tomato plants. The Plant Cell. 1991;**3**:1187-1193. DOI: 10.1105/tpc.3.11.1187

[59] Hontzeas N, Zoidakis J, Glick BR, Abu-Omar MM. Expression and characterization of 1-aminocyclopropane-1-carboxylate deaminase from the rhizobacterium pseudomonas putida UW4: A key enzyme in bacterial plant growth promotion. Biochimica et Biophysica Acta. 2004;**1703**:11-19. DOI: 10.1016/j.bbapap.2004.09.015

[60] Glick BR, Cheng Z, Czarny J, Duan J. Promotion of plant growth by ACC deaminase-producing soil bacteria. European Journal of Plant Pathology. 2007;**119**:329-339. DOI: 10.1007/s10658-007-9162-4

[61] Nascimento FX, Brígido C, Glick BR, Oliveira S, Alho L. Mesorhizobium ciceri LMS-1 expressing an exogenous 1-aminocyclopropane-1-carboxylate (ACC) deaminase increases its nodulation abilities and chickpea plant resistance to soil constraints. Letters in Applied Microbiology. 2012;**55**:15-21. DOI: 10.1111/j.1472-765X.2012.03251.x

[62] Nukui N, Minamisawa K, Ayabe S-I, Aoki T. Expression of the 1-aminocyclopropane-1-carboxylic acid deaminase gene requires symbiotic nitrogen-fixing regulator gene nifA2 in Mesorhizobium loti MAFF303099. Applied and Environmental Microbiology. 2006;**72**:4964-4969. DOI: 10.1128/AEM.02745-05

[63] Van de Poel B, Van Der Straeten D. 1-aminocyclopropane-1-carboxylic acid (ACC) in plants: More than just the precursor of ethylene! Frontiers in Plant Science. 2014;**5**:640. DOI: 10.3389/fpls.2014.00640

[64] Uddling J, Gelang-Alfredsson J, Piiki K, Pleijel H. Evaluating the relationship between leaf chlorophyll concentration and SPAD-502chlorophyll meter readings. Photosynthesis Research. 2007;**91**:37-46

[65] Cassol D, Silva FSP, Falqueto AR, Bacarin MA. An evaluation of non-destructive methods to estimate total chlorophyll content. Photosynthetica. 2008;**46**:634-636. DOI: 10.1007/s11099-008-0109-6

[66] Kalaji HM, Schansker G, Brestič M, et al. Frequently asked questions about chlorophyll fluorescence, the sequel. – Photosynthesis Research. 2017;**132**:13-66

[67] Kjeldahl J. New method for the determination of nitrogen in organic substances. Zeitschrift fur analytische chemie. 1883;**22**:366-383

[68] Murphy J, Rilley JP. A modified single solution method for the determination of phosphate in natural water. Analytical Chemistry Acta. 1962;**27**:31-36

[69] Hesse PR. A Textbook of the Soil Chemistry Analysis. London: John Murray ltd; 1971. pp. 120-307

[70] Paul D, Sinha SN. Isolation and Characterization of Phosphate Solubilizing Bacterium Pseudomonas aeruginosa KUPSB12 with Antibacterial Potential from River Ganga, India. Annals of Agricultural Sciences. 2017;**15**:130

[71] Gordon SA, Weber RP. Colorimetric estimation of indoleacetic acid Plant Physiology. 1951;**26**(1):192-195. DOI: 10.1104/pp.26.1.192

[72] Herridge DF, Peoples MB. Ureide assay for measuring nitrogen fixation by nodulated soybean calibrated by 15N methods. Plant Physiology. 1990;**93**:495-503

[73] Thompson M, Ellison SLR, Wood R. Harmonized guidelines for

singlelaboratory validation of methods of analysis. Pure and Applied Chemistry. 2002;**74**:835-855

[74] Schomburg G. Gas Chromatography. Weinheim: Wiley VCH; 1990

[75] James AT, Martin AJ. Gas-liquid partition chromatography: The separation and micro-estimation of volatile fatty acids from formic acid to dodecanoic acid. Biochemical Journal. 1952;**50**(5):679-690. DOI: 10.1042/bj0500679. PMID: 14934673; PMCID: PMC1197726

[76] Bard AJ, Faulkner L. Electrochemical Methods. New York, NY: Wiley; 1980

[77] Phillips CSG. Organic and biochemical: The chromatography of gases and vapours. Discussions of the Faraday Society. 1949;7:241-248

Chapter 3

Bacterial Symbiotic Signaling in Modulating Plant-Rhizobacterial Interactions

Aqsa Tariq and Ambreen Ahmed

Abstract

Rhizosphere is the hub for microbial activities where microbes and plants interact with complex signaling mechanisms. Plants release various metabolites in response to environmental factors which are significant in shaping rhizospheric microbial communities. These microbes develop symbiotic relation with plants by quorum sensing signals and regulate various microbial activities including biofilm formation. Biofilms are important in inhabiting rhizosphere and provide platform for cell-to-cell microbial interactions. Biofilm- forming rhizobacteria can successfully colonize plant roots and establish symbiotic relations with host. During this association, rhizobacteria are flourished by using plant root exudates, while the bacteria benefit the plants by synthesizing phytohormones, locking soil minerals for plant, protecting them from pathogenic invasions and enhancing plant immunity by improving plant tolerance against various environmental conditions. Indole is an effector molecule in regulating bacterial gene expression related to biofilm production. These interactions are coordinated by bacterially released phytohormones mainly auxin which act as key factor in regulating plant-microbe symbiotic interactions. It is characterized as inter- kingdom signaling molecule that coordinates various plant and rhizobacterial activities. Thus, understanding the nature and interacting behaviors of these molecules would lead to the exploitation of plant growth-promoting rhizobacteria for better plant growth in agricultural fields.

Keywords: quorum sensing, auxin, biofilms, rhizosphere, bacterial signaling, biofertilizers, PGPR

1. Introduction

Rhizosphere is the most complex ecosystem where plants interact with billions of microbes that colonize plant roots. Symbiotic interactions between host and its associated microbiota are ecologically significant in maintaining ecosystem stability. The associated rhizobacteria is also referred to as the second genome as it allows the host metabolism to trigger various genomic and functional repertoire to deal with a wide range of ecological challenges [1]. Rhizosphere is a hotspot for various plant-rhizobacterial activities derived mainly through chemical signals. Plants can preferentially interact with tiny soil creatures recruiting them according to their own ecological demands. Plants can filter rhizospheric bacteria by secreting certain chemical messengers that disproportionally increase relative abundance of mutualistic rhizobacteria and decrease antagonistic species in rhizosphere [2, 3]. These chemical messengers can either be plant primary metabolites such as carbohydrates, organic acids, lipids, amino acids, nucleic acids, or secondary metabolites such as alkaloids, lignin, phenolics, terpenoids, sterols, steroids, essential oils, etc. These metabolites are released in the form of root exudates and are a source of nitrogen and carbon for rhizobacteria [4]. Crosstalk among microbes and pathogens is an important mechanism to ensure successful root colonization and development of symbiotic relationship with their photobiont partner. Generally, rhizospheric communications are identified as inter- and intra-species microbial signaling, plant to microbes signaling and vice versa [5]. Quorum sensing (QS) is the fundamental means of rhizobacterial crosstalk. These sensing signals rely on inter- and intra-species microbial information that coordinate and control bacterial cell-to-cell activities in microbial communities. QS directs various bacterial functions involving biofilm formation [6]. Some phytohormones released by rhizobacteria also act as signaling molecules. Majority of rhizobacteria, that is, 80% are able to produce auxins (IAA) by utilizing tryptophan precursor present in plant root exudates. IAA is an effector molecule in bacterial gene expression related to indole signaling molecules that regulate biofilm production [7]. These signals are the chief elements affecting the structure and physical heterogeneity of soil microbial communities which in turn influence plant resilience and plasticity. These plant-microbial cross communications are dynamic and interdependent on both partners as well as rhizospheric ecological parameters [8]. Microbial processes have great impact on ecological parameters by regulating various biogeochemical cycles to establish a compatible niche for healthier plants. Understanding these soil interactions occurring among microorganisms and their host is critical for the manipulation of untapped potential of microbiome and their usability in fields of environmental and sustainable agriculture [9]. Deciphering rhizospheric signals could enable scientists to exploit rhizobacteria to promote plant growth and production. The overall plant growth depends on the soil microbiota and its association with host plant [10]. Hence, rhizospheric engineering can provide a healthier environment for plant growth but rhizospheric engineering require understanding of these signals. The current chapter unentangles the complexity of rhizospheric signals among microbes and their photobionts and the significance of these signalings in the development of symbiotic associations.

2. Decrypting phyto-microbiome signals

As described earlier, three major categories of rhizospheric signals are plant-to-microbe signaling which is regulated by low molecular weight molecules secreted in the form of root exudates and are responsible for establishing plant-microbes interactions, inter- and intra-microbial signaling regulated by QS which is responsible for diversity and contemporizing microbial behavior, and lastly, microbe to plant signals regulated by microbially produced substances responsible for altering plant physiology and enhancing immunity to cope with environmental fluctuations [11]. Consequently, plants and microbes develop cooperative relationship which improves plant growth and health.

2.1 Microbial intra-kingdom crosstalk

Quorum sensing is bacterial signaling process which is initiated through detection of exogenous chemical signals. Bacterial cells synthesize and release chemical signals known as autoinducers (AIs) which in turn regulate the synthesis of various microbial functions such as biofilm formation, virulence, adhesion, motility, metabolism, and symbiotic association [12]. QS signals provide bacteria to survive competitive environment, increase microbial survival under various habitats and assist microbial rhizospheric colonization to develop plant microbial associations. Various quorum sensing signaling networks have been identified among rhizobacterial species such as diketopiperazines (DKPs), acylated homoserine lactones (AHLs), diffusible signaling factor (DSF), phytohormones, antibiotics and secondary metabolites, etc. [13, 14]. Quorum sensing signals are categorized as acyl-homoserine lactone (AHLs) which are commonly produced by Gram-negative bacteria and autoinducer peptides (AIPs) which are commonly produced by Gram-positive bacteria, out of AIPs autoinducer type 2 have properties of both AHLs and AIPs, hence, can be produced by both Gram-negative and Gram-positive bacteria. AHL-regulated QS signaling is known as the LuxI/Rlike QS system [15].

Rhizobacteria recognize specific signals within their vicinity and stimulate the synthesis of QS that control transcription of various microbial mechanisms [16]. Plant-beneficial bacteria also protect plants by releasing certain signals such as acylase and lactonase, which degrade pathogenic AHLs or release VOCs to compete with rhizospheric phytopathogens. Besides this, most of the rhizobacteria have ability to produce antimicrobial compounds such as cyclic-lipodepsipeptides and polycyclic tetramate macrolactams that hinder the phytopathogenic abundance and colonization [17]. A complex communal signaling network mechanism initiates as a result of interactions of microbial QS with plant signals. These signal exchanges occur both outside and inside plants [18].

2.2 Microbial inter-kingdom crosstalk

Among these QS, AHLs are commonly produced signaling molecules in rhizobacteria which also act as interkingdom signaling molecules as it also regulates expression of various genes in plants [19, 20]. Some AHLs such as cyclodipepetides and their derivatives also modulate auxin-responsive genes expression in plant roots [21]. Moreover, AHLs also regulate plant auxin to cytokinin ratio by triggering genes responsible for upregulation of auxin and downregulation of cytokinin [22]. AHLs, that is, N-butanoyl homoserine lactone and N-hexanoyl homoserine lactone, greatly impact plant physiology. Short-chain AHLs regulate plant growth especially root elongation [23]. QS also promotes signals for nodulations. These signals are responsible for regulating plant defense systems related to systemic acquired resistance (SAR) and induced systemic resistance (ISR) [24]. There is a molecular mechanism known as microbe-associated molecular patterns (MAMPs) that allows recognition based on highly conserved signature molecular patterns of microbes that are unique to the whole class. This is similar to pattern recognition receptors (PRRs) present in plants that recognize microbial compounds within root vicinity [25, 26]. Identification of unique MAMPs by phytoreceptors trigger signals from roots that modulate signaling pathways of various phytohormones such as salicylic acid, jasmonic acid, ethylene, etc. which trigger the expression of genes related to plant secondary metabolites and defensive proteins to activate resistance within plants [15]. Microbial QS interact with VOCs within rhizosphere. VOCs are significant for long-distance interactions with rhizobacteria and responsible for managing microbial populations and plant-microbe associations. VOCs serve as inter- and intra-species signaling molecules by regulating gene expression of various bacterial functions including biofilm formation. VOCs also modulate plant physiology, that is, root system architecture and plant hormonal signaling [27, 28]. Microbial VOCs trigger and interfere with auxin-signaling pathway in host plants [29]. Microbial diffusible signaling factor is also responsible for regulating plant immunity by trigging various plant defensive signaling. Similarly, various Gram-positive bacteria use peptides as QS to colonize plant roots which is also considered to regulate defensive signaling within plants [30, 31]. Moreover, QS also regulates nodulation in leguminous plants. Flavonoids present in root exudates enhance microbial gene expression for AHL synthesis [32]. Plants possess quorum quenching (QQ) mechanisms to QS signaling. Plants have adapted various mechanisms such as enzymatic degradation, biosynthesis and secretion of anti-QS compounds, disruption of binding, receptor, and regulatory sites [33, 34]. Sometimes, plants produce QS-intimate molecules, that is, p-coumaric acid, isothiocyanate sulforaphane, curcumin, patulin, etc. [35–38]. Among these molecules, alkamides andN-acylethanolamines (NAEs) are excellent contenders regarding their chemistry and structure to act as AHL [38]. These AHL analogs interfere with pathogenic QS regulatory pathways inhibiting their gene expression [39].

2.3 Rhizospheric phyto-signaling

The chemicals released by plants directly influence the structure of microbial communities by influencing abundance and diversity of rhizobacteria. Root exudates are important for microbial root colonization. Metabolically active bacterial species are attracted toward root exudates [40]. Root exudation occurs through active and passive mechanisms. Secretion of low molecular weight signals is a passive process whereas other metabolites such as mucopolysaccharides are released actively. ATP-binding cassette transporters such as ABC transporters are involved in exudation phenomena [41]. The microbes within root vicinity use the substances present in root exudates. These substances act as plant pheromones that either stimulate or antagonize other microbes. Moreover, the composition of root exudates also modifies microbial dynamics which favors abundance of plant growth-promoting rhizobacteria (PGPR) and prevents pathogenic microbial growth [42]. Hence, root exudate mediates a tripartite signal among PGPR, pathogens, and the host plant [43]. Root exudates can either be low molecular weight compounds such as sugars, amino acids, organic acids, phenolics, and other secondary metabolites or high molecular weight compounds, that is, mucopolysaccharides [44]. The quantity and quality of exudates depend on various plant physiological parameters such as plant species, nutritional status, growth stage, etc. These compounds act as sources of nitrogen and carbon for microbes. The presence of these catabolic pathways is significant in root colonization and disease suppression [45]. Amino acids and carbohydrates act as microbial pheromones whereas protein molecules are essential for antagonistic signaling [46, 47]. For instance, lectin proteins released as root exudates assist in both symbiosis and defense responses [48]. Moreover, arabinogalactans (mucopolysaccharides) exuded from roots attracts PGPR and repel rhizospheric pathogens [49]. Besides amino acids, proteins, and carbohydrates, a considerable amount of phytohormones are also present in root exudates. Salicylic and jasmonic acids are involved in shaping rhizospheric microbial community. Microbial abundance is relatively high in the presence of these phytohormones and provoke plant immunity [50]. Although volatile organic compounds (VOCs) are quite significant aboveground signaling molecules, however, minute quantity of VOCs is also released in the form of root exudates. These compounds attract and kill root-colonizing rhizospheric pathogens [51].

Transcriptional profiling showed that expression of various microbial genes is regulated by root exudates. Phytochemicals direct various bacterial processes involved in rhizosphere colonization such as chemotaxis, motility, phase variation, membrane integrity, environmental resistance, and nutrient sequestration [45]. Root exudates also trigger microbial biofilm-related pathways. Biofilm formation is significant in rhizospheric colonization, adhesion to seeds, and roots [52]. Microbial motility in response to phytochemicals is crucial for recruiting rhizobacteria within plant vicinity [53]. Rhizospheric microbial activities determine the amount of root exudates which in turn affect patterns of root distribution toward rhizospheric nutrient pools. These exudates induce microbial chemotaxis by trigging the expression of various proteins related to methyl-accepting chemotaxis (MCP) such as Mcp (A, B, C), Tlp (A, B, C), YvaQ, YoaH, HemAT, and YfmS. These proteins adhere to the chemoreceptors when concentration of root exudated reaches to threshold. The binding of ligand to the ligand binding domain of chemoreceptors induces autophosphorylation of protein followed by transfer of phosphoryl group to response regulatory protein which interacts with flagellar motor and controls microbial movement responsible for chemotaxis [54, 55].

2.4 Auxins mediated cross-kingdom signal transmission

Auxins (IAA) are master phyto-hormone that control majority of plant functions. Besides plant, rhizobacteria also have ability to produce auxins as a result of their defensive mechanism. Tryptophan is most significant amino acid exuded by plants. Microbes in root vicinity detoxify tryptophan and produce auxins [56]. PIN proteins mediated auxin flux allows initiation of cell-to-cell communicating network [57]. Auxins act as signaling molecules for communication between microbes to coordinate their activities. Auxins also trigger diverse plant metabolic functions. Modification of root system architecture is one of the most promising features regulated by auxin producing PGPR [58]. Auxins also act as phytochemicals that regulate rhizobacterial chemotaxis. IAA signals received by bacterial methyl accepting proteins present in chemoreceptors initiate flagellar movement in rhizobacteria. Moreover, IAA also serves as nutrient reservoir in rhizosphere for microbes [59, 60]. IAA also enhances competence within rhizosphere. IAA-producing bacteria are more competent and successful in colonizing rhizosphere. This indicates that IAA might have some evolutionary significance that enhances bacterial survival under various environmental conditions. It is considered that IAA provides protection to rhizobacteria under biotic and abiotic stress conditions ensuring their survival [61]. IAA also interacts with QS molecules and regulates secretion of extracellular polymeric substances, lipopolysaccharides, and formation of biofilms [7]. Hence, rhizobacterial colonization is regulated by the interaction of IAA and QS signals and their high survival is also linked with biofilm production. IAA plays important role in modulating gene expression of bacterial cells and making them more competent [56]. Transcriptome analysis revealed that IAA signaling causes downregulation and upregulation of various respiration affecting genes in rhizobacteria [62].

Auxin signaling is significant in root morphological processes such as root hair formation and development of root tips. Plant phenotypic plasticity relases on auxin signaling and transport. It induces hyperpolarization that causes physicochemical alterations in plasma membrane [63]. Plant root morphological alterations depends on auxin concentration. Auxin primarily induces formation of primary and lateral roots under optimum concentrations and adventitious root formation under higher concentrations [64]. Moreover, auxins also initiate the nodulation process in leguminous plants. Flavonoids interact with auxin-signaling pathway and inhibit auxin biosynthesis and initiate the nodulation process. Auxin signaling controls various processes of nodulation, that is, formation of bacteroids, vascular bundles, number of nodules, etc. [65]. Auxins also regulates signaling pathways related to other phytohormones. It regulates transcription of 1-aminocyclopropane-l-carboxylic acid (ACC) that regulate ACC deaminase production which in the presence of low ethylene converts to α-ketobutyrate and ammonia. Hence, auxins interact with ethylene and reduce effects of environmental stress on plants [66, 67]. Coordination of auxins with cytokinins regulates cell division and differentiation at root meristem resulting in root growth. Auxins initiate formation of lateral roots while interactions of auxins with ethylene are significant for the development of root hair and elongation. Overall mechanism of root growth is regulated by auxin-derived phytohormonal coordination. Similarly, brassinosteroids, gibberellins, strigolactones act synergistically with auxins to regulate plant physiological processes [68, 69]. These underlying signal coordinations are responsive for plant root adjustments to external stimuli through inducing primary root elongation, lateral roots, and root hair formation [70, 71]. The overall mechanism of signaling crosstalk is illustrated in **Figure 1**.

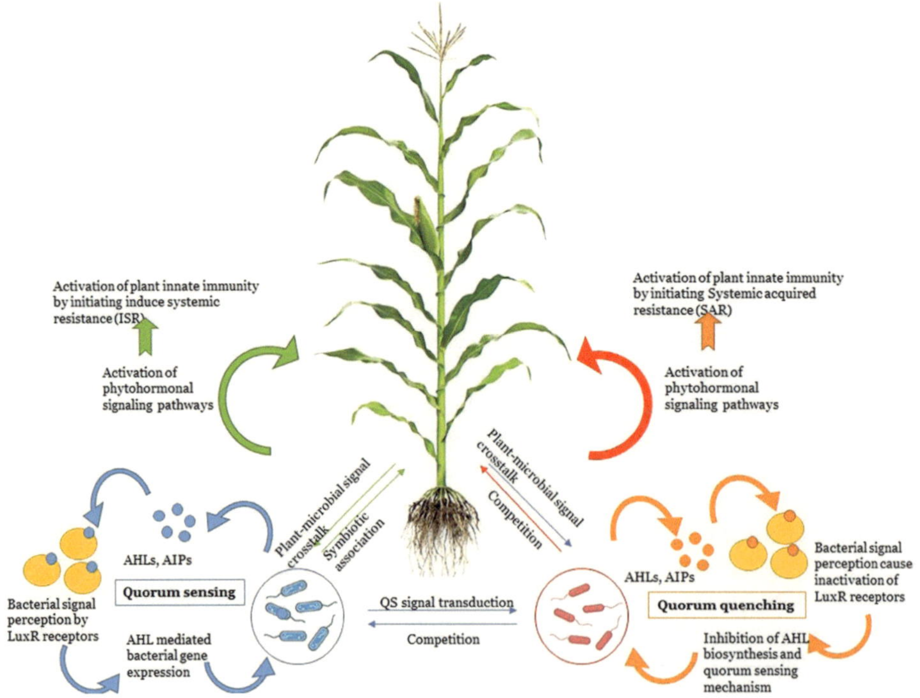

Figure 1.
Signal crosstalk among plant-Rhizobacteria and phytopathogens.

3. Signal-mediated phyto-microbial symbiotic responses

These signaling networks provide benefits to both partners in regulating symbiotic associations.

3.1 Rhizospheric colonization by biofilm formation

Microbial quorum sensing signals are involved in biofilm production that facilitates rhizobacteria to colonize plant roots and rhizosphere. Rhizobacteria have ability to produce multicellular communities embedded within extracellular polymeric substances regulated by QS. This phenomenon is regulated by complex signaling network. QS-mediated bacterial chemotaxis is important in regulating early stages of biofilms. QS such as cyclic dimeric guanosine 3′,5′-monophosphate (c-di-GMP), cyclic dimeric adenosine 3′,5′-monophosphate (c-di-AMP), and cyclic adenosine 3′,5′-monophosphate (cAMP) are the frontline signaling molecules in the regulation of biofilm development [72]. Biofilms are bacterial protecting shells under unpleasant environmental conditions to ensure their survivability and colonization in rhizosphere and to plant roots. Biofilms provide protection by secreting extracellular polymer. Extracellular polymer of biofilms is composed of exopolysaccharides (EPS), nucleic acids, proteins, and lipids which help in sequestration of toxic substances by elevating catabolic gene expression. Higher bacterial cell density within biofilms provides larger gene pool that facilitates adhesion and other metabolic activities [73]. This synchronization is controlled by QS mediated cell-to cell contact. The internal structures are interconnected for communication and transfer of nutrients. EPS primarily contains carbohydrates which provide foundation for the adhesion of various substances (proteins, lipids, nucleic acids) within biofilm microenvironment [74]. The release of QS depends on bacterial cell density. When population size reaches the threshold value QS initiates the formation of biofilms. In *Pseudomonas aeruginosa*, there are two AHL signaling systems, that is, LasI/R and RhlI/R system that controls microbial activities. Binding of LasI-mediated protein to LasR activates LasR-AHL complex. Similarly, RhlI mediated signal activates RhlR-AHL complex. After its activation, Las system can take over Rhl system to produce Pel polysaccharides and pyocyanin which are important for the formation of complete and mature biofilms. LasI/R system also regulates biofilms by activating PelD via controlling the expression of diguanylate cyclase TpbB [75, 76]. Production of rhamnolipids by RhlI/R system also regulates bacterial motility which influences and determines initial attachment, stability, and integrity of biofilm structure. AHLs trigger the expression of lectins (LecA and LecB) responsible for binding the bacterial communities to EPS. These proteins are responsible for the formation of cell clusters for EPS retention and stability of biofilm structure. LecB protein is also involved in the formation of biofilm matrix [77]. In addition to this, AHLs also stimulate the secretion of EPS through a positive feedback mechanism [78]. luxS also contributes to biofilm formation mechanism. It regulates the expression of iron transporter to increase the intracellular Fe concentrations which stimulate biofilm formation. Similarly, it is also involved in the release of extracellular nucleic acids into biofilm matrix. Besides these, AIPs and DSF also contribute to biofilm formation and bacterial colonization [74]. In addition to QS, indole (IAA) also acts as signaling molecule in regulating microbial communities. Rhizobacteria produce indole during stationary phase of bacterial growth [79]. Microbes utilize root-exuded tryptophane to produce indole via tryptophanase. Indole concentration is crucial for biofilm formation. It acts as high-level pulse and low-level persistent signal for bacterial persistence and biofilm formation [80]. Indoles induce DksA binding to RNA polymerase and activation of VpsR protein that trigger the upregulation of cdgA gene responsible for biofilm formation [81]. Besides, trigging biofilm formation, under certain conditions, indole inhibits its formation. For instance, in the presence of antistress protein (YcfR), indole synthesis inhibit biofilm formation in *Escherichia coli* [82].

3.2 Development of root nodules

Rhizobia-legume symbiotic interactions are widely known for their unique character, that is, nodule formation on plant roots in which rhizobia provide nitrogen to plants and in return get dicarboxylates. This process is initiated by the phytochemical signals, that is, flavonoid. Flavonoids interact with bacterial QS to activate nodulation genes (nod, noe, and nol) which then synthesize nodulation factor (Nod), that is, lipo-chitooligosaccharides (LCOs) [83]. These signals are received by receptors on roots epidermis to initiate nodulation process. Besides flavonoids, some other molecules present in root exudates such as vanillin, chalcones, aldonic acid, betaines, and jasmonate also activate nod genes. These signals are specific to the plant species to ensure accurate host-microbe relationship [84]. Flavonoid signals determine auxin transport patterns in plant roots. Alterations in auxin transport patterns are important in the nodulation mechanism. During initial stages of nodule organogenesis, auxin transport is inhibited and later during initiation and differentiation of nodular cells, auxin transport increases, and auxin accumulates in nodule primordium. This auxin influx and efflux is controlled by PIN and AUX1 genes present in plant roots. Flavonoids itself can regulate PIN activation and localization for auxin transport. Plant signals trigger reduction in auxin efflux in response to rhizobial encounter leading to auxin accumulation in pericycle and inner cortex. Moreover, exogenous auxin (bacterially produced) also disrupts plant phytohormonal homeostasis to adjust auxin at permissive levels for nodulation [85, 86]. Rhizobia produce auxin that enhances plant root growth, cell division, differentiation, and nodule formation. It also increased the number of nodules which is associated with biological nitrogen fixation ability [87]. During nodulation, auxins also interact with other phytohormones (such as abscisic acid) that inhibit lateral root formation and stimulate nodulation.

3.3 Root architecture system

The impact of rhizobacteria on plant root system architecture is significant in plant growth induction. PGPR modify plant root system architecture by their phytohormone production ability which interact with plant hormonal balance. An enhanced lateral root with more branching pattern is one of the more distinctive impacts of PGPRs on plants [88]. The overall root architecture is the dynamic response of roots toward soil localized nutrient resources through meristematic activity. Root architecture is linked with nutrient acquisition, thus, roots with higher surface area can interact more with soil particles, soil water, and microorganisms. PGPR-associated roots displayed increased rigidity with increased length and diameter of the roots as well as the cortex and aerenchyma space, increased protoxylem poles, and metaxylem vessel components with enhanced flavonoid and phenolic contents. PGPR releases phytohormones which manipulates plant hormonal homeostasis. Impact of IAA on the architecture and growth of roots is very much significant [89]. Auxin to cytokinin ratio is important in shaping plant root system. IAA regulates a wide range of processes involved in plant growth and development, low concentrations of IAA can promote primary root elongation, whereas high concentrations of IAA promote the growth of lateral roots, shorten primary root length, and promote root hair formation [90]. The concentration of auxin in the plants changes according to its uptake and excretion from tissues, its production from tryptophan, and the generation of IAA conjugates [91]. Auxins encourage root development to its full potential. Moreover, its increased concentrations reduce root growth [92].

3.4 Regulation of plant defense responses

PGPR-associated plants configure high tolerance to unfavorable environmental conditions. PGPR induces innate immune response in plants against phytopathogens and also PGPR regulates defense responses in plants by activating defense-related signaling pathways and prepares plants for accurate response. Plants recognize phytopathogens through pattern recognition receptors (PRRs) by recognizing pathogen-associated molecular patterns (PAMPs or MAMPs) to initiate signal cascades for the regulation of specific transcription factors to activate multiple intracellular defense responses [20]. This induction of signal cascade is either hormone dependent or independent. Rhizobacteria-produced phytohormones such as auxins, abscisic acid, gibberellins, and cytokinins also contribute toward these signal transductions. This stimulates intracellular defense responses [93]. AHLs and DSFs act as signaling molecule in plant-pathogenic interactions. This signal is perceived by plants and initiate the phytohormone-mediated biosynthesis of phytoanticipins or phytoalexins [94]. PGPRs can also recognize specific phytopathogens. PGPR-produced salicylic acid (SA) regulate plant defense response

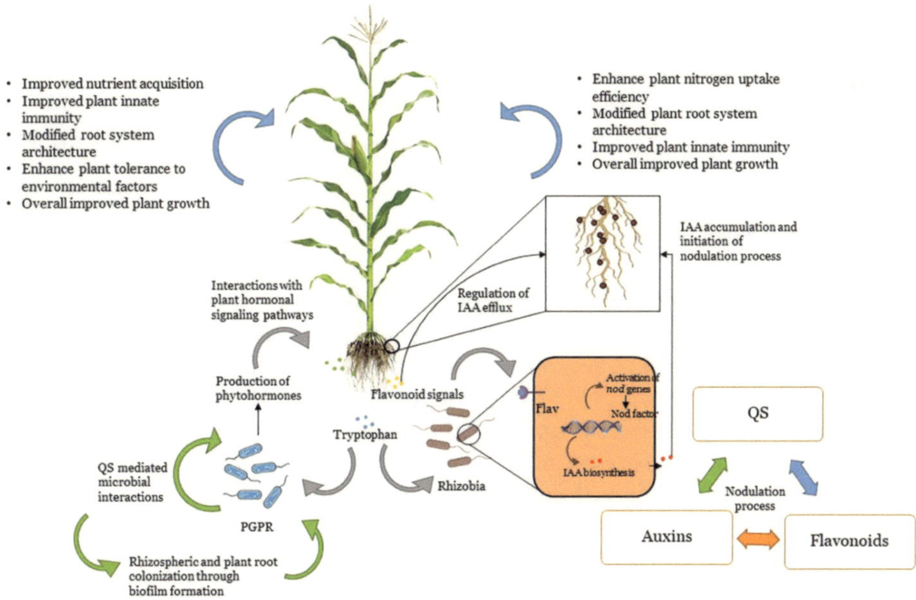

Figure 2.
Development of symbiotic associations between plants and beneficial bacteria by regulating plant phytohormonal pathways.

known as systemic acquired resistance (SAR). It binds with non-repressor of pathogenesis-related genes (NPR1). SA pathway is also involved in regulating flavonoid signaling which inhibits pathogenic proliferations [95]. SA-independent PGPR-mediated plant defense system is referred to as induced systemic resistance (ISR). This pathway includes jasmonic acid and ethylene signals. Invasion of PGPR trigger MAMPs recognition leading to the activation of ISR mechanism [95]. The development of symbiotic associations between plants and microbes has been illustrated in **Figure 2**.

4. Agronomic efficiencies of symbiotic signaling

Plant microbial signaling mechanisms are responsible for the development of plant-microbial associations which is crucial for sustainable agricultural. Nutrient acquisition is a significant consequence of these associations. A cascade of signals regulates the continuous flow of nutrient exchange between both partners.

4.1 PGPR as biofertilizers

Various commercially available PGPR bioformulations are used as biofertilizers to increase crop productivity [96]. PGPR can restore the nutrient cycle between the soil, plant roots, and existing microorganisms [97]. Fertilizers containing nitrogen drive up manufacturing costs. Nitrogen biofixation is the first method of accelerating plant development which regulates nitrogenase activity and helps the plant regain its nitrogen equilibrium. In order to break the bond between the nitrogen atoms, the N2-fixing process carried out by microbes requires energy, which is supplied by readily available organic carbon. The root mutualistic bacteria may fix up to 30 kg/ha of pure nitrogen each year. The amount of nitrogen fixed by these bacteria can be useful for maintaining soil fertility over the long run. Biological N2-fixing PGPR is injected into crops and crop fields to promote growth, control disease, and maintain the nitrogen content in agricultural soil [98]. They are reliable, eco-friendly sources of the renewable nutrients needed to sustain the biology and health of the soil [99]. On the basis of their capacity to draw nutrients from the soil, fix atmospheric N2, promote the solubilization of nutrients, a variety of microbial taxa have been utilized commercially as effective biofertilizers [100].

4.2 PGPR as biocontrol agent

The utilization of these rhizobacteria by farmers in the field is currently insufficient. To address variability in the efficacy of the biocontrol system, combined activity of several bacterial strains is necessary. These rhizobacteria mixes can be employed as a seed treatment, which might help to cut down on the amount of bacterial inoculum needed. This will promote both the hostile activity of the bacteria on the root surface during the early root infection by the pathogens and their systemic dissemination over the surface of the growing root system [101].

4.3 PGPR in phytoremediation

Use of PGPR biofilms in phytoremediation technologies plays a significant role. PGPR enhances the qualities of plants employed in phytoremediation, such as biomass production, low-level pollutant absorption, plant nutrition, and health, however, it is crucial to pick PGPR that can last and be effective in phytoremediation procedures. Metal-tolerant PGPR hinders direct uptake of heavy metals to plants by converting them to less toxic forms as a result plant growth is improved [102]. When PGPR are applied to a polluted site, they increase the capacity of the plants that grow there to detoxify synthetic chemicals and preserve soil structure. Additionally, PGPR reduces the toxicity of heavy metals by changing their bioavailability in plants [103].

5. Conclusion and future prospects

Microbial interactions with plants are carried out through an interconnected signaling network that work inter- and intra-specifically. These signaling mechanisms decipher the microbiome communications related to plant productivity. Various plant metabolites released through roots in the form of root exudates attract rhizobacteria that interact by producing quorum-sensing signals. These signals are responsible for various bacterial activities in rhizosphere. Digging into their molecular pattern is necessary to take their full advantage. Manipulating these signaling pathways can help in rhizospheric engineering which will be beneficial for sustainable agriculture. Expanding research related to metagenomic analysis of rhizomicrobiome is important to understand whole infrastructure of symbiotic associations. Moreover, use of appropriate PGPR and their commercialization is necessary as they are more economic and environmentally safe compared to chemical fertilizers. Thereby, utilization of PGPR base biofertilizers are much appreciated.

Conflict of interest

Authors declare no conflict of interest.

Author details

Aqsa Tariq and Ambreen Ahmed*
Institute of Botany, University of the Punjab, Quaid-e-Azam Campus, Lahore, Pakistan

*Address all correspondence to: ambreenahmed1@hotmail.com

References

[1] Klein M, Stewart JD, Porter SS, Weedon JT, Kiers ET. Evolution of manipulative microbial behaviors in the rhizosphere. Evolutionary Applications. 2022;**15**(10):1521-1536. DOI: 10.1111/eva.13333

[2] Bakker PA, Berendsen RL, Van Pelt JA, Vismans G, Yu K, Li E, et al. The soil-borne identity and microbiome-assisted agriculture: Looking back to the future. Molecular Plant. 2020;**13**(10):1394-1401. DOI: 10.1016/j.molp.2020.09.017

[3] Li E, de Jonge R, Liu C, Jiang H, Friman VP, Pieterse CM, et al. Rapid evolution of bacterial mutualism in the plant rhizosphere. Nature Communications. 2021;**12**(1):1-3. DOI: 10.1038/s41467-021-24005-y

[4] Olanrewaju OS, Ayangbenro AS, Glick BR, Babalola OO. Plant health: Feedback effect of root exudates-rhizobiome interactions. Applied Microbiology and Biotechnology. 2019;**103**(3):1155-1166. DOI: 10.1007/s00253-018-9556-6

[5] Venturi V, Keel C. Signaling in the rhizosphere. Trends in Plant Science. 2016;**21**(3):187-198. DOI: 10.1111/1462-2920.13571

[6] Zhou L, Zhang Y, Ge Y, Zhu X, Pan J. Regulatory mechanisms and promising applications of quorum sensing-inhibiting agents in control of bacterial biofilm formation. Frontiers in Microbiology. 2020;**11**:589640. DOI: 10.3389/fmicb.2020.589640

[7] Spaepen S, Vanderleyden J. Auxin and plant-microbe interactions. Cold Spring Harbor Perspectives in Biology. 2011;**3**(4):1-13. DOI: 10.1101/cshperspect.a001438

[8] Hassan MK, McInroy JA, Kloepper JW. The interactions of rhizodeposits with plant growth-promoting rhizobacteria in the rhizosphere: A review. Agriculture. 2019;**9**(7):142. DOI: 0.3390/agriculture9070142

[9] Hakim S, Naqqash T, Nawaz MS, Laraib I, Siddique MJ, Zia R, et al. Rhizosphere engineering with plant growth-promoting microorganisms for agriculture and ecological sustainability. Frontiers in Sustainable Food Systems. 2021;**5**:617157. DOI: 10.3389/fsufs.2021.617157

[10] Masood S, Zhao XQ, Shen RF. *Bacillus pumilus* promotes the growth and nitrogen uptake of tomato plants under nitrogen fertilization. Scientia Horticulturae. 2020;**272**:109581. DOI: 10.1016/j.scienta.2020.109581

[11] Altaf M, Khan M, Ahmad S, Abulreesh HH, Ahmad I. Quorum sensing in plant growth-promoting rhizobacteria and its impact on plant-microbe interaction. In: Dhananjaya Pratap Singh DP, Singh HB, Prabha P, editors. Plant-Microbe Interactions in AgroEcological Perspectives. Singapore: Springer; 2017. pp. 311-331. DOI: 10.1007/978-981-10-5813-4_16

[12] Checcucci A, Marchetti M. The rhizosphere talk show: The rhizobia on stage. Frontiers in Agronomy. 2020;**2**:591494. DOI: 10.3389/fagro.2020.59149

[13] Cortez M, Handy D, Headlee A, Montanez C, Pryor S, Cutshaw K, et al. Quorum sensing in the rhizosphere. In: Horwitz BA, Mukherjee PK, editors. Microbial Cross-Talk in the Rhizosphere. Singapore: Springer; 2022. pp. 99-134. DOI: 10.1007/978-981-16-9507-0_5

[14] Mondal S, Baksi S. Signalling of rhizosphere microbiomes: Benign and malign borders. In: Arora NK, Bouizgarne B, editors. Microbial BioTechnology for Sustainable Agriculture. Vol. 1. Singapore: Springer; 2022. pp. 237-260. DOI: 10.1007/978-981-16-4843-4_7

[15] Jamil F, Mukhtar H, Fouillaud M, Dufossé L. Rhizosphere signaling: Insights into plant–rhizomicrobiome interactions for sustainable agronomy. Microorganisms. 2022;**10**(5):899. DOI: 10.3390/microorganisms10050899

[16] Bramhachari PV. Implication of quorum sensing system in biofilm formation and virulence. Springer; Singapore. 2019. p. 28. DOI: 10.1007/978-981-13-2429-1

[17] Brescia F, Marchetti-Deschmann M, Musetti R, Perazzolli M, Pertot I, Puopolo G. The rhizosphere signature on the cell motility, biofilm formation and secondary metabolite production of a plant-associated Lysobacter strain. Microbiological Research. 2020;**234**:126424. DOI: 10.1016/j.micres.2020.126424

[18] van der Burgh AM, Joosten MH. Plant immunity: Thinking outside and inside the box. Trends in Plant Science. 2019;**24**(7):587-601. DOI: 10.1016/j.tplants.2019.04.009

[19] Singh A, Chauhan PS. N-acyl homoserine lactone mediated quorum sensing exhibiting plant growth-promoting and abiotic stress tolerant bacteria demonstrates drought stress amelioration. Journal of Pure and Applied Microbiology. 2022;**1**:669-684. DOI: 10.22207/JPAM.16.1.69

[20] Bukhat S, Imran A, Javaid S, Shahid M, Majeed A, Naqqash T. Communication of plants with microbial world: Exploring the regulatory networks for PGPR mediated defense signaling. Microbiological Research. 2020;**238**:126486. DOI: 10.1016/j.micres.2020.126486

[21] Ortiz-Castro R, Campos-García J, López-Bucio J. *Pseudomonas putida* and *Pseudomonas fluorescens* influence Arabidopsis root system architecture through an auxin response mediated by bioactive cyclodipeptides. Journal of Plant Growth Regulation. 2020;**39**(1):254-265. DOI: 10.1007/s00344-019-09979-w

[22] Nadarajah K, Abdul Rahman NS. Plant–microbe interaction: Aboveground to belowground, from the good to the bad. International Journal of Molecular Sciences. 2021;**22**(19):10388. DOI: 10.3390/ijms221910388

[23] Phour M, Sehrawat A, Sindhu SS, Glick BR. Interkingdom signaling in plant-rhizomicrobiome interactions for sustainable agriculture. Microbiological Research. 2020;**241**:126589. DOI: 10.1016/j.micres.2020.126589

[24] Shrestha A, Grimm M, Ojiro I, Krumwiede J, Schikora A. Impact of quorum sensing molecules on plant growth and immune system. Frontiers in Microbiology. 2020;**11**:1545. DOI: 10.3389/fmicb.2020.01545

[25] Pantigoso HA, Newberger D, Vivanco JM. The rhizosphere microbiome: Plant–microbial interactions for resource acquisition. Journal of Applied Microbiology. 2022;**133**(5):2864-2876. DOI: 10.1111/jam.15686

[26] Yu J, Tu X, Huang AC. Functions and biosynthesis of plant signaling metabolites mediating plant–microbe interactions. Natural Product Reports. 2022;**39**(7):1393-1422. DOI: 10.1039/D2NP00010E

[27] Poveda J. Beneficial effects of microbial volatile organic compounds (MVOCs) in plants. Applied Soil Ecology. 2021;**168**:104118. DOI: 10.1016/j.apsoil.2021.104118

[28] Weisskopf L, Schulz S, Garbeva P. Microbial volatile organic compounds in intra-kingdom and inter-kingdom interactions. Nature Reviews. Microbiology. 2021;**19**(6):391-404. DOI: 10.1038/s41579-020-00508-1

[29] Tyagi S, Mulla SI, Lee KJ, Chae JC, Shukla P. VOCs-mediated hormonal signaling and crosstalk with plant growth promoting microbes. Critical Reviews in Biotechnology. 2018;**38**(8):1277-1296. DOI: 10.1080/07388551.2018.1472551

[30] Monnet V, Juillard V, Gardan R. Peptide conversations in Gram-positive bacteria. Critical Reviews in Microbiology. 2016;**42**(3):339-351. DOI: 10.3109/1040841X.2014.948804

[31] He YW, Deng Y, Miao Y, Chatterjee S, Tran TM, Tian J, et al. DSF-family quorum sensing signal-mediated intraspecies, interspecies, and inter-kingdom communication. Trends in Microbiology. 2022;**5**:36-50. DOI: 10.1016/j.tim.2022.07.006

[32] Babenko LM, Romanenko KO, Iungin OS, Kosakovska IV. Acyl-homoserine lactones for crop production and stress tolerance of agricultural plants. Agricultural Biology [Sel'skokhozyaistvennaya Biologiya]. 2021;**56**(1):3-19. DOI: 10.15389/agrobiology.2021.1.3eng

[33] Deryabin D, Galadzhieva A, Kosyan D, Duskaev G. Plant-derived inhibitors of AHL-mediated quorum sensing in bacteria: Modes of action. International Journal of Molecular Sciences. 2019;**20**(22):5588. DOI: 10.3390/ijms20225588

[34] Billot R, Plener L, Jacquet P, Elias M, Chabrière E, Daudé D. Engineering acyl-homoserine lactone-interfering enzymes toward bacterial control. Journal of Biological Chemistry. 2020;**295**(37):12993-13007. DOI: 10.1074/jbc.REV120.013531

[35] Jakobsen TH, van Gennip M, Phipps RK, Shanmugham MS, Christensen LD, Alhede M, et al. Ajoene, a sulfur-rich molecule from garlic, inhibits genes controlled by quorum sensing. Antimicrobial Agents and Chemotherapy. 2012;**56**(5):2314-2325. DOI: 10.1128/AAC.05919-11

[36] Ganin H, Rayo J, Amara N, Levy N, Krief P, Meijler MM. Sulforaphane and erucin, natural isothiocyanates from broccoli, inhibit bacterial quorum sensing. Medicinal Chemistry Communications. 2013;**4**(1):175-179. DOI: 10.1039/C2MD20196H

[37] Hartmann A, Rothballer M, Hense BA, Schröder P. Bacterial quorum sensing compounds are important modulators of microbe-plant interactions. Frontiers in Plant Science. 2014;**5**:131. DOI: 10.3389/fpls.2014.00131

[38] Aziz M, Chapman KD. Fatty acid amide hydrolases: An expanded capacity for chemical communication? Trends in Plant Science. 2020;**25**(3):236-249. DOI: 10.1016/j.tplants.2019.11.002

[39] Kamath A, Shukla A, Patel D. Quorum sensing and quorum quenching: Two sides of the same coin. Physiological and Molecular Plant Pathology. 2022;**123**:101927. DOI: 10.1016/j.pmpp.2022.101927

[40] Wang NQ, Kong CH, Wang P, Meiners SJ. Root exudate signals

in plant–plant interactions. Plant, Cell & Environment. 2021;**44**(4):1044-1058. DOI: 10.1111/pce.13892

[41] Vives-Peris V, de Ollas C, Gómez-Cadenas A, Pérez-Clemente RM. Root exudates: From plant to rhizosphere and beyond. Plant Cell Reports. 2020;**39**(1):3-17. DOI: 10.1007/s00299-019-02447-5

[42] Yu Y, Gui Y, Li Z, Jiang C, Guo J, Niu D. Induced systemic resistance for improving plant immunity by beneficial microbes. Plants. 2022;**11**(3):386. DOI: 10.3390/plants11030386

[43] Liu Y, Zhang N, Qiu M, Feng H, Vivanco JM, Shen Q, et al. Enhanced rhizosphere colonization of beneficial *Bacillus amyloliquefaciens* SQR9 by pathogen infection. FEMS Microbiology Letters. 2014;**353**(1):49-56. DOI: 10.1111/1574-6968.12406

[44] Zhalnina K, Louie KB, Hao Z, Mansoori N, da Rocha UN, Shi S, et al. Dynamic root exudate chemistry and microbial substrate preferences drive patterns in rhizosphere microbial community assembly. Nature Microbiology. 2018;**3**(4):470-480. DOI: 10.1038/s41564-018-0129-3

[45] Mavrodi OV, McWilliams JR, Peter JO, Berim A, Hassan KA, Elbourne LD, et al. Root exudates alter the expression of diverse metabolic, transport, regulatory, and stress response genes in rhizosphere *Pseudomonas*. Frontiers in Microbiology. 2021;**12**:698. DOI: 10.3389/fmicb.2021.651282

[46] More SS, Shinde SE, Kasture MC. Root exudates a key factor for soil and plant: An overview. The Pharma Innovation Journal. 2020;**8**:449-459

[47] Pascale A, Proietti S, Pantelides IS, Stringlis IA. Modulation of the root microbiome by plant molecules: The basis for targeted disease suppression and plant growth promotion. Frontiers in Plant Science. 2020;**10**:1741. DOI: 10.3389/fpls.2019.01741

[48] Reyes-Montaño EA, Vega-Castro NA. Plant lectins with insecticidal and insect static activities. In: Begum G, editors. Insecticides Agriculture and Toxicology. London, UK: IntechOpen; 2018. pp. 17-42. DOI: 10.5772/intechopen.74962

[49] Villa-Rivera MG, Cano-Camacho H, López-Romero E, Zavala-Páramo MG. The role of arabinogalactan type II degradation in plant-microbe interactions. Frontiers in Microbiology. 2021;**12**:1-14. DOI: 10.3389/fmicb.2021.730543

[50] Singh K, Chandra R, Purchase D. Unraveling the secrets of rhizobacteria signaling in rhizosphere. Rhizosphere. 2022;**21**:100484. DOI: 10.1016/j.rhisph.2022.100484

[51] Gfeller V, Huber M, Förster C, Huang W, Köllner TG, Erb M. Root volatiles in plant–plant interactions I: High root sesquiterpene release is associated with increased germination and growth of plant neighbours. Plant, Cell & Environment. 2019;**42**(6):1950-1963. DOI: 10.1111/pce.13532

[52] Zboralski A, Filion M. Genetic factors involved in rhizosphere colonization by phytobeneficial *Pseudomonas* spp. Computational and Structural Biotechnology Journal. 2020;**18**:3539-3554. DOI: 10.1016/j.csbj.2020.11.025

[53] Feng H, Fu R, Hou X, Lv Y, Zhang N, Liu Y, et al. Chemotaxis of beneficial rhizobacteria to root exudates: The first step towards root–microbe rhizosphere interactions. International Journal of Molecular Sciences. 2021;**22**(13):6655. DOI: 10.3390/ijms22136655

[54] López-Farfán D, Reyes-Darias JA, Matilla MA, Krell T. Concentration dependent effect of plant root exudates on the chemosensory systems of *Pseudomonas putida* KT2440. Frontiers in Microbiology. 2019;**10**:78. DOI: 10.3389/fmicb.2019.00078

[55] Colin R, Ni B, Laganenka L, Sourjik V. Multiple functions of flagellar motility and chemotaxis in bacterial physiology. FEMS Microbiology Reviews. 2021;**45**(6):fuab038. DOI: 10.1093/femsre/fuab038

[56] Tariq A, Ahmed A. Auxins interkingdom signaling molecules. In: Hano C, editors. Plant Hormones: Recent Advances, New Perspectives and Applications. Vol. 25. London, UK: IntechOpen; 2022. p. 3. DOI: 10.5772/intechopen.95608

[57] Saritha M, Kumar P, Panwar NR, Burman U. Intelligent plant–microbe interactions. Archives of Agronomy and Soil Science. 2022;**68**(7):1002-1018. DOI: 10.1080/03650340.2020.1870677

[58] Ahmed A, Hasnain S. Auxins as one of the factors of plant growth improvement by plant growth promoting rhizobacteria. Polish Journal of Microbiology. 2014;**63**(3):261

[59] Duca D, Lorv J, Patten CL, Rose D, Glick BR. Indole-3-acetic acid in plant–microbe interactions. Antonie Van Leeuwenhoek. 2014;**106**(1):85-125. DOI: 10.1007/s10482-013-0095-y

[60] Duca DR, Glick BR. Indole-3-acetic acid biosynthesis and its regulation in plant-associated bacteria. Applied Microbiology and Biotechnology. 2020;**104**(20):8607-8619. DOI: 10.1007/s00253-020-10869-5

[61] Zhang Y, Li Y, Hassan MJ, Li Z, Peng Y. Indole-3-acetic acid improves drought tolerance of white clover via activating auxin, abscisic acid and jasmonic acid related genes and inhibiting senescence genes. BMC Plant Biology. 2020;**20**(1):1-2. DOI: 10.1186/s12870-020-02354-y

[62] Kunkel BN, Johnson JM. Auxin plays multiple roles during plant–pathogen interactions. Cold Spring Harbor Perspectives in Biology. 2021;**13**(9):a040022. DOI: 10.1101/cshperspect.a040022

[63] Napier R. The story of auxin-binding protein 1 (ABP1). Cold Spring Harbor Perspectives in Biology. 2021;**13**(12):a039909. DOI: 10.1101/cshperspect.a039909

[64] Ahmed A, Tariq A, Habib S. Interactive biology of auxins and phenolics in plant environment. In: Lone R, Shuab R, Kamili AN, editors. Plant Phenolics in Sustainable Agriculture. Singapore: Springer; 2020. pp. 117-133. DOI: 10.1007/978-981-15-4890-1_5

[65] Grover M, Bodhankar S, Sharma A, Sharma P, Singh J, Nain L. PGPR mediated alterations in root traits: Way toward sustainable crop production. Frontiers in Sustainable Food Systems. 2021;**4**:618230. DOI: 10.3389/fsufs.2020.61823

[66] Zemlyanskaya EV, Omelyanchuk NA, Ubogoeva EV, Mironova VV. Deciphering auxin-ethylene crosstalk at a systems level. International Journal of Molecular Sciences. 2018;**19**(12):4060. DOI: 10.3390/ijms19124060

[67] Bhardwaj S, Sharma D, Jan S, Singh R, Bhardwaj R, Kapoor D. Crosstalk of ethylene and other phytohormones in the regulation of plant development. Ethylene in Plant Biology.

2022;**23**:17-31. DOI: 10.1002/9781119744719.ch2

[68] Vissenberg K, Claeijs N, Balcerowicz D, Schoenaers S. Hormonal regulation of root hair growth and responses to the environment in Arabidopsis. Journal of Experimental Botany. 2020;**71**(8):2412-2427. DOI: 10.1093/jxb/eraa048

[69] Mazzoni-Putman SM, Brumos J, Zhao C, Alonso JM, Stepanova AN. Auxin interactions with other hormones in plant development. Cold Spring Harbor Perspectives in Biology. 2021;**13**(10):a039990. DOI: 10.1101/cshperspect.a039990

[70] Eichmann R, Richards L, Schäfer P. Hormones as go-betweens in plant microbiome assembly. The Plant Journal. 2021;**105**(2):518-541. DOI: 10.1111/tpj.15135

[71] Nakano M, Omae N, Tsuda K. Inter-organismal phytohormone networks in plant-microbe interactions. Current Opinion in Plant Biology. 2022;**68**:102258. DOI: 10.1016/j.pbi.2022.102258

[72] Mahto KU, Kumari S, Das S. Unraveling the complex regulatory networks in biofilm formation in bacteria and relevance of biofilms in environmental remediation. Critical Reviews in Biochemistry and Molecular Biology. 2022;**57**(3):305-332. DOI: 10.1080/10409238.2021.2015747

[73] Arnaouteli S, Bamford NC, Stanley-Wall NR, Kovács ÁT. Bacillus subtilis biofilm formation and social interactions. Nature Reviews. Microbiology. 2021;**19**(9):600-614. DOI: 10.1038/s41579-021-00540-9

[74] Wang Y, Bian Z, Wang Y. Biofilm formation and inhibition mediated by bacterial quorum sensing. Applied Microbiology and Biotechnology. 2022;**12**:1-7. DOI: 10.1007/s00253-022-12150-3

[75] Oh MH, Han K. AbaR is a LuxR type regulator essential for motility and the formation of biofilm and pellicle in *Acinetobacter baumannii*. Genes & Genomics. 2020;**42**(11):1339-1346. DOI: 10.1007/s13258-020-01005-8

[76] Thi MT, Wibowo D, Rehm BH. *Pseudomonas aeruginosa* biofilms. International Journal of Molecular Sciences. 2020;**21**(22):8671. DOI: 10.3390/ijms21228671

[77] Passos da Silva D, Matwichuk ML, Townsend DO, Reichhardt C, Lamba D, Wozniak DJ, et al. The *Pseudomonas aeruginosa* lectin LecB binds to the exopolysaccharide Psl and stabilizes the biofilm matrix. Nature Communications. 2019;**10**(1):1-11. DOI: 10.1038/s41467-019-10201-4

[78] Jiang Q, Chen J, Yang C, Yin Y, Yao K. Quorum sensing: A prospective therapeutic target for bacterial diseases. BioMed Research International. 2019;**2019**:1-15. DOI: 10.1155/2019/2015978

[79] Li Y, Feng T, Wang Y. The role of bacterial signaling networks in antibiotics response and resistance regulation. Marine Life Science & Technology. 2022;**28**:1-6. DOI: 10.1007/s42995-022-00126-1

[80] Zarkan A, Liu J, Matuszewska M, Gaimster H, Summers DK. Local and universal action: The paradoxes of indole signalling in bacteria. Trends in Microbiology. 2020;**28**(7):566-577. DOI: 10.1016/j.tim.2020.02.007

[81] Lee JH, Lee J. Indole as an intercellular signal in microbial communities. FEMS Microbiology

Reviews. 2010;**34**(4):426-444. DOI: 10.1111/j.1574-6976.2009.00204.x

[82] Cui B, Chen X, Guo Q, Song S, Wang M, Liu J, et al. The cell–cell communication signal indole controls the physiology and interspecies communication of *Acinetobacter baumannii*. Microbiology Spectrum. 2022;**10**(4):691-696. DOI: 10.1111/1758-2229.13100

[83] Acosta-Jurado S, Fuentes-Romero F, Ruiz-Sainz JE, Janczarek M, Vinardell JM. Rhizobial exopolysaccharides: Genetic regulation of their synthesis and relevance in symbiosis with legumes. International Journal of Molecular Sciences. 2021;**22**(12):6233. DOI: 10.3390/ijms22126233

[84] Smith DL, Gravel V, Yergeau E. Signaling in the Phytomicrobiome. Frontiers in Plant Science. 2017;**8**:611. DOI: 10.3389/fpls.2017.00611

[85] Zhang P, Jin T, Kumar Sahu S, Xu J, Shi Q, et al. The distribution of tryptophan-dependent indole-3-acetic acid synthesis pathways in bacteria unraveled by large-scale genomic analysis. Molecules. 2019;**24**(7):1411. DOI: 10.3390/molecules24071411

[86] Bosse MA, da Silva MB, de Oliveira NG, de Araujo MA, Rodrigues C, de Azevedo JP, et al. Physiological impact of flavonoids on nodulation and ureide metabolism in legume plants. Plant Physiology and Biochemistry. 2021;**166**:512-521. DOI: 10.1016/j.plaphy.2021.06.007

[87] Kohlen W, Ng JL, Deinum EE, Mathesius U. Auxin transport, metabolism, and signalling during nodule initiation: Indeterminate and determinate nodules. Journal of Experimental Botany. 2018;**69**(2):229-244. DOI: 10.1093/jxb/erx308

[88] Erturk Y, Ercisli S, Haznedar A, Cakmakci R. Effects of plant growth promoting rhizobacteria (PGPR) on rooting and root growth of kiwifruit (*Actinidia deliciosa*) stem cuttings. Biological Research. 2010;**43**(1):91-98. DOI: 10.4067/S0716-97602010000100011

[89] Vacheron J, Desbrosses G, Bouffaud ML, Touraine B, Moënne-Loccoz Y, Muller D, et al. Plant growth-promoting rhizobacteria and root system functioning. Frontiers in Plant Science. 2013;**4**:356. DOI: 10.3389/fpls.2013.00356

[90] Remans R, Beebe S, Blair M, Manrique G, Tovar E, Rao I, et al. Physiological and genetic analysis of root responsiveness to auxin-producing plant growth-promoting bacteria in common bean (*Phaseolus vulgaris* L.). Plant and Soil. 2008;**302**(1):149-161. DOI: 10.1007/s11104-007-9462-7

[91] Lima JV, Tinôco RS, Olivares FL, Chia GS, de Melo Júnior JA, da Silva GB. Rhizobacteria modify root architecture and improve nutrient uptake in oil palm seedlings despite reduced fertilizer. Rhizosphere. 2021;**19**:100420. DOI: 10.1016/j.rhisph.2021.100420

[92] Facella P, Daddiego L, Giuliano G, Perrotta G. Gibberellin and auxin influence the diurnal transcription pattern of photoreceptor genes via CRY1a in tomato. PLoS One. 2012;**7**(1):e30121

[93] Vannier N, Agler M, Hacquard S. Microbiota-mediated disease resistance in plants. PLoS Pathogens. 2019;**15**(6):e1007740

[94] Nishad R, Ahmed T, Rahman VJ, Kareem A. Modulation of plant defense

system in response to microbial interactions. Frontiers in Microbiology. 2020;**11**:1298. DOI: 10.3389/fmicb.2020.01298

[95] Pršić J, Ongena M. Elicitors of plant immunity triggered by beneficial bacteria. Frontiers in Plant Science. 2020;**11**:594530. DOI: 10.3389/fpls.2020.594530

[96] García-Fraile P, Menéndez E, Celador-Lera L, Díez-Méndez A, Jiménez-Gómez A, Marcos-García M, et al. Bacterial probiotics: A truly green revolution. In: Kumar V, Kumar M, Sharma S, Prasad R, editors. Probiotics and Plant Health. Singapore: Springer; 2017. pp. 131-162. DOI: 10.1007/978-981-10- 3473-2_6

[97] Martínez-Viveros O, Jorquera MA, Crowley DE, Gajardo GM, Mora ML. Mechanisms and practical considerations involved in plant growth promotion by rhizobacteria. Journal of Soil Science and Plant Nutrition. 2010;**10**(3):293-319. DOI: 10.4067/S0718-95162010000100006

[98] Damam M, Kaloori K, Gaddam B, Kausar R. Plant growth promoting substances (phytohormones) produced by rhizobacterial strains isolated from the rhizosphere of medicinal plants. International Journal of Pharmaceutical Sciences Review and Research. 2016;**37**(1):130-136

[99] Sun B, Bai Z, Bao L, Xue L, Zhang S, Wei Y, et al. *Bacillus subtilis* biofertilizer mitigating agricultural ammonia emission and shifting soil nitrogen cycling microbiomes. Environment International. 2020;**144**:105989. DOI: 10.1016/j.envint.2020.105989

[100] Schütz L, Gattinger A, Meier M, Müller A, Boller T, Mäder P, et al. Improving crop yield and nutrient use efficiency via biofertilization—A global meta-analysis. Frontiers in Plant Science. 2018;**8**:2204. DOI: 10.3389/fpls.2017.02204

[101] Akhtar M, Siddiqui ZA. Role of plant growth promoting rhizobacteria in biocontrol of plant diseases and sustainable agriculture. In: Maheshwari DK, editors. Plant Growth and Health Promoting Bacteria. Berlin, Heidelberg: Springer; 2010. pp. 157-195. DOI: 10.1007/978-3-642-13612-2_7

[102] Fatima H, Ahmed A. Indole-3-acetic acid synthesizing chromium-resistant bacteria can mitigate chromium toxicity in *Helianthus annuus* L. Plant, Soil and Environment. 2020;**66**(5):216-221. DOI: 10.17221/581/2019-PSE

[103] Habib S, Ahmed A. Comparative analysis of pre-germination and post-germination inoculation treatments of *Zea mays* L. to mitigate chromium toxicity in Cr-contaminated soils. Polish Journal of Environmental Studies. 2018;**28**(2):597-607. DOI: 10.15244/pjoes/81570

Chapter 4

Symbiosis under Abiotic Stress and Its Challenges

Maria Daniela Artigas Ramírez and Jean Louise Cocson Damo

Abstract

Many abiotic factors have affected symbiosis effectiveness. However, the responses and interactions vary depending on the plant host, environmental factors, and symbiotic strains. The effect of various environmental factors on the competitiveness of rhizobial strains in host legumes has been examined, but many questions are still unresolved. For example, in the Rhizobia-legume symbiosis, the nitrogen fixation and nodulation processes are strongly related to the physiological state of the host plant. Therefore, a competitive and persistent rhizobial strain is not expected to express its total capacity for nitrogen fixation under limiting factors (e.g., salinity, unfavorable soil pH, nutrient deficiency, mineral toxicity, extreme temperatures, soil moisture problems, and inadequate photoperiods). Moreover, populations of rhizobial species vary in their tolerance to major environmental factors. Furthermore, this chapter emphasizes the studies on symbiosis under abiotic stress and its challenges. Additionally, this can help to understand and establish an effective biological process for improvement in agricultural productivity.

Keywords: metal-stress, temperature-stress, salinity tolerance, *Rhizobium*, acid soils

1. Introduction

In recent years, the effect of various environmental factors on the competitiveness of rhizobial strains for nodulation of host legumes has been examined. During the revision and search for an answer based on abiotic stress and symbiosis as the main topic, it was found that the progress of the studies has been advancing accordingly through the hard times for cultivation in different crops as time goes on. However, the type of manuscripts is based on 32.9% of published reviews of previous year's studies and 6.1% of books. The remaining published articles were 61%, which indicated that the effect of various environmental factors on the cultivation plant host and their symbionts and the competitiveness of rhizobial strains with their host legumes had been examined (**Figure 1**). However, many questions still need to be solved, which is indicated by the low amount of studies based on the percentage of reported studies included in other such metabolism or mechanisms, inoculants, and interactions (**Figure 1**).

Furthermore, this chapter emphasizes the symbiosis under abiotic stress studies and its challenges. Additionally, this will help to lead future studies and to understand

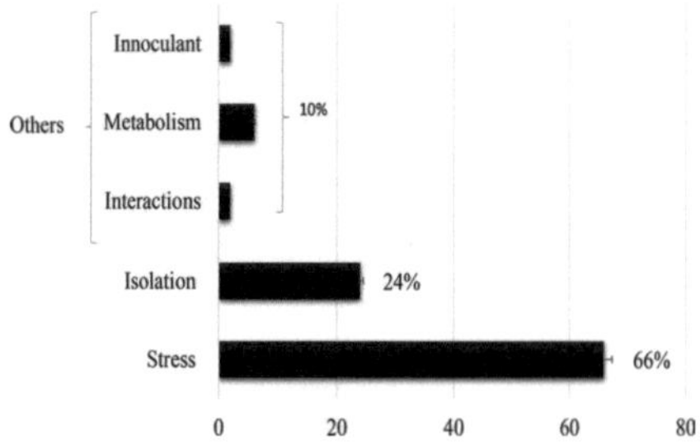

Figure 1.
Approximation of the total distribution of the main or critical topic of the studies during 2021 ~ 2022.

and establish an effective biological process for improving agricultural productivity or recovery of regions affected by contaminants, deforestations, or natural causes of losing their natural vegetation.

Foremost, the arid climate is characterized by hot, dry summers and cold winters, which limits the use of different species for soil revegetation [1–5]. Therefore, using native species for revegetation may be an interesting practice, especially in dry climatic conditions, where salinity and drought are often serious problems. About one-third of the world's land area comprises arid and semiarid climates [2]. In the *Rhizobium*-legume symbiosis, the nitrogen fixation process is strongly related to the physiological state of the host plant. Therefore, a competitive and persistent rhizobial strain is not expected to express its total capacity for nitrogen fixation under limiting factors, such as salinity, unfavorable soil pH, nutrient deficiency, mineral toxicity, temperature extremes, insufficient or excessive soil moisture, inadequate photosynthesis, presence of other microorganisms in the rhizosphere [2–5]. Several studies showed that the competition pattern between some strains might be affected by one or more factors mentioned above [2–4]. However, other factors, such as the host genotype, are also involved in determining the outcome of competition between strains of *Rhizobium* [3, 6] and, in recent years, between other rhizobia groups such as *Microvirga, Paraburkholderia* [7]. The most complex environments for rhizobia are marginal lands with low rainfall, extreme temperatures, acidic soils with low nutrient status, and poor water-holding capacity. Populations of rhizobial species vary in their tolerance to major environmental factors; consequently, screening for tolerant strains has been pursued (**Figure 1**, isolation). Biological processes (e.g., N_2-fixation) can improve agricultural productivity; however, minimizing soil loss and ameliorating adverse edaphic conditions are essential.

2. Symbiosis under different abiotic stress

In the years 2021 and 2022, approximately the main stress reported or described was salinity. In the revision, salinity stress was 22% of total studies reported (**Figure 2**, by Google Scholar as the main trusted tool of search), followed by metals such as Fe, Al, Ni, As, Cu, Cr, Cd, water-deficient or drought, and extreme temperature. In some cases, the combination of the interrelationship between several stresses was reported. Also, exceeding levels of pesticides and fertilizers were found in low percentages due to high dependence on utilization in crop production. The remaining studies are distributed between multiples stress which mean the combination of more than three kinds of abiotic stress, and other indicating different

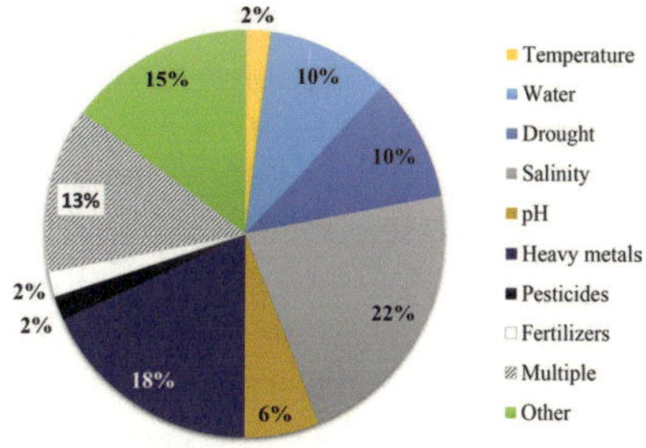

Figure 2.
Distribution of the according to the type of abiotic stress described in 2021–2022.

types of the main topic and collateral stress such as deficient of macro or micronutrients, the effect of manure of fertilization styles in combinations with symbionts under stress, individual strains studies for mechanisms, co-inoculations, and communities studies. Several of these abiotic stress mentioned will be described more deeply in the following sections.

2.1 Temperature tolerance

The temperature stress should be reported as high and low temperatures (or cold stress). First, the main report is the high temperature, which can cause different types of damage in the plant host, so reported as atmospheric or soil temperatures could be on good point from the new studies. Then, the soil temperature is a major problem for biological nitrogen fixation by legume crops [2, 3, 8, 9]. High root temperatures strongly affect rhizobia infection and N_2 fixation in several legumes, including soybean, peanut, cowpea, and beans [2, 3]. Critical temperatures reported for N_2-fixation are 30°C for *Trifolium* (clover) and *Pisum* (pea) species, and other crops such as soybean (*Glycine max*), peanut (*Arachis* sp.), *Vigna* sp. (cowpea, mungbean) are range between 35 and 40°C [10]. The optimal temperature for nodule functioning in common beans (*Phaseolus* sp.) is between 25 and 30°C and is vulnerable to soil temperatures around 30 and 33°C [11]. In addition, temperature affects root hair infection, cell differentiation, nodule structure, and the functioning of the root nodule [2]. For most rhizobia, the optimum temperature range for growth in culture is 28 to 31°C, but unable to grow at 37°C [12]. However, many rhizobial isolates have been recently described as heat-tolerant bacteria. For instance, Wei et al. [13] reported a temperature-tolerant strain of *Mesorhizobium* at 35°C. Furthermore, Bansal et al. [14] isolated rhizobia from *Vigna radiata*, and the isolates showed good growth at high temperatures.

High temperatures affect the plasmids' transmission or induce large deletions and can result in modified modulation performance [15]. The large population of nodulation plasmids in soils could be seriously affected by high temperatures and the long-term effectiveness of rhizobia as a symbiont [2, 15]. Therefore, natural genetic tolerance is necessary, and their interaction with other symbionts could lead to plasmid replications and improve rhizobia's performance in high temperatures. On the other hand, studies reported that exogenous melatonin and *Rhizobium* inoculation attenuated injuries and reduced plant oxidative damage [9]. The symbiotic relationship naturally establishes a balance between plant-symbiont under heat or cold stress; it would induce changes in the gene expressions related to cold or heat stress, affecting the oxidative stress in ROS production [9, 16, 17].

The temperature affects the response of both the roots and the shoots. However, the root temperature is more critical for symbiosis than the shoot temperature due to survival in soil by all rhizobia strains. Root hair formation is limited in some legume species, binding between symbiont and root cells, formation of infection threads, nodule initiation and growth, leghemoglobin content, and nitrogenase activity decreased [2, 8, 9]. As a consequence, low nitrogen fixation and low biomass.

For further inoculation, one possible solution is rhizobia coated with appropriate host lectin or suitable rhizobia to improve their symbiosis or interactions under extreme conditions [2, 3]. However, rhizobia as mass production and conservation pre-inoculation affect the effectiveness of the symbiosis such as low temperature 4°C conservation and inoculation in high temperatures such as >35°C leads to low effectiveness and weak competitiveness in front of native rhizobacteria [3, 16, 17].

Some studies suggest conserving after mass production inoculants at 25°C to apply in >35°C soil conditions; then, for cold or perm-frost soils, pre-cultivation in low temperatures should be better for applying in the field [3, 16, 17]. Cold stress is evidenced in soil and plants and affects the bacterial communities' richness and diversity and their enzymatic activities or interactions with their hosts [16]. However, denitrifying bacteria may have a different mechanism to adapt to low temperatures [16, 17]. Some studies reported that cold stress could increase the production of exopolysaccharide (EPS), indirectly affecting cell survival by increasing cell density and biofilm production, and maybe some quorum sense [17, 18].

2.2 Moisture deficiency and drought

In the legume-rhizobia, the symbiotic nitrogen fixation is highly sensitive to drought, which results in decreased N accumulation and yield of legume crops. Both establishment and activity of the legume-*Rhizobium* symbiosis are extremely sensitive to drought stress [19, 20]. Moisture deficiency and drought are closely related to the occurrence of rhizobial populations in desert soils, and the effective nodulation of legumes growing [1, 21] emphasizes that rhizobia can exist in soils with limiting moisture levels. However, population densities tend to be low under the most desiccated conditions and to increase as the moisture stress is relieved [2]. It is well known that some free-living rhizobia (saprophytic) can survive under drought stress or low water potential [22]. In contrast, most legume crops are sensitive to drought stress.

The symbiotic association between rhizobia and legumes is also sensitive to soil water deficiency. The responses of root nodule bacteria to water stress have been widely investigated. For example, mild water stress reduces only the number of root nodules formed on soybean, while moderate and severe water stress reduces both the number and size of nodules; it is made by culturing some species in dry conditions and after inoculation and measuring water vapor exchange and others. [1, 2, 23, 24]. *Rhizobium*-legume symbioses are currently the most important nitrogen-fixing systems, which may potentially increase N input in arid lands. The leguminous plants include species or varieties that are extremely well adapted to the harsh conditions of arid lands. For example, *Medicago sativa, Arachis hypogaea, Cyamposis tetragonoloba*, and *Melilotus* spp.; these legumes are known to be adapted to conditions prevailing in arid regions.

Moreover, naturally occurring forage legumes (annuals and perennials) are well nodulated with antibiotic-resistant rhizobia, and their root nodules are active in fixing N_2 [2, 24]. These legumes can be found in deserts or cultivated lands as wild plants. Therefore, isolating effective rhizobia from those wild legumes to inoculate other legume crops is a new strategy to improve the efficiency of the *Rhizobium*-legume symbiosis in soil with moisture deficiency [3].

The rhizobia of wild legumes may have better traits than those of cultivated legumes and can contain a marker such as antibiotic resistance and their correlation with drought tolerance [24]. When rhizobial strains are inoculated into media containing an antibiotic, a few cells may exhibit resistance because of spontaneous genetic changes or mutations. The resistance of a rhizobial strain to an antibiotic is a useful marker for keeping track of the stress-tolerance, and every antibiotic has its mechanism of action for resistance [25, 26]. In addition, resistance cells will grow on the antibiotic media, and other bacteria will be suppressed in the field. Thus, it is important that antibiotic-resistant strains, which are selected for inoculation experiments, must not lose their infectiveness (ability to form nodules) nor their effectiveness (ability to fix nitrogen) in the symbiosis with the host plant. The resistance should be stable throughout infection, nodulation, nitrogen fixation, and subsequent re-isolation [24–26].

Drought is more likely to affect the growth and more severe if the soil profile has water limitation; the pore size of soil and its distribution that is unfavorable to water retention at the potential in the available range is most likely to become a severe drought such as Entisols [3, 8]; this condition impedes the root extension, water, and gas movement. They can aggravate the restriction in the symbiosis by limiting the movement of rhizobia and the root hair extensions and lead the poor communication between symbiont and host [8]. The long dry season can only be met through a continued search for drought-tolerant genotypes of host or symbiotic microorganisms that could provide forage for a long period as the possible and maximum nutritive value in dry conditions. However, a wide range of conditions, developing somewhat comparable intimacy with other plants, such as grass/legume mixtures balances, are indeed needed [3].

Recently, many reports have described the positive effect of using symbionts or rhizobacteria in plants. From the viewpoint of the symbiont, a common example is *Bradyrhizobium japonicum,* whose cells resist dehydration by decreasing energy consumption and increasing carbon accumulation, especially by up-regulating the genes involved in trehalose synthesis. When it rehydrated, the signaling downstream restarted [27]. More deeply, it is the exopolysaccharide (EPS, extracellular polymeric substance) and the production of biofilms protecting the cells. In this way, indirectly, the survival of symbionts and their releasing compounds, as mentioned, helps the plant cell endurance through accumulation and signaling for the synthesis of osmoprotectants such as trehalose, glycine, and flavonoids, as well increasing the plant-microbe communication [27, 28].

Climate change has harsh effects on plant growth and yield. In plants, the reactive oxygen species (ROS) generation, at the same time, decreases the plant defense system and other over-activating the dependent Fe-, Ca-enzymatic system, which leads to high energy production and utilization to end in oxidative damage cells or cell necrosis or death [28]. Lectins and recognition, no nodulation, specificity in nodulation, strains selection from the host side, enzymatic process affected by temperature, and toxic seed substances [3]. There are reports describing that ROS can be positively affected through rhizobia application by producing flavonoids necessary for plant-rhizobia communication and affecting some gene expression generating a plant response such as salicylic acid (SA) production and ROS mediation in the pathway like ROS-detoxification and avoiding ROS-homeostasis [29]. In addition, fixation-N_2 and chlorophyll content increase in the plant naturally, which could allow decreasing the irrigation time or amount of water calculated to apply in the field and even decrease the inorganic fertilizers used for cultivation.

Furthermore, the most neglected system or lacks of deep research is common in southern areas worldwide due to the small-scale crop farmer support or budget deficit for researchers to continue their studies [28]. Many of these reasons and other leads to farmers adding inorganic fertilizers affecting direct or indirect rhizobia communities and their interactions or symbiosis, besides that the high application of inorganic compounds affecting human and animal health [3], and the accumulation of these inorganic compounds leads to other problems described in next sections. Thus, understanding the inorganic compound application and nutrient supply economy is needed from a basic point of view. We can comprehend the biological role of legumes and their symbiosis in the context of soil nutrient transformations, both from the host view and symbiotic microbe sight.

2.3 Salinity stress

Salinity is a serious agricultural threat in different countries, principally in arid and semiarid regions. Nearly 40% of the world's land surface has potential salinity problems [30]. Most of these areas are part of the Mediterranean and tropical regions [30]. As with most cultivated crops, the salinity response of legumes varies greatly and depends on climatic conditions, soil properties, and the growth stage. Legumes such as *Vicia faba* (broad or fava bean), *P. vulgaris* (common beans), and *G. max* (soybean) have been reported to be more salt tolerant than others, such as *P. sativum* (peas). Other legumes as *Prosopis*, *Acacia*, and *Medicago sativa* [2, 31–33], are salt tolerant, but those legumes are more susceptible to salt than their rhizobia.

The legume-*Rhizobium* symbioses and nodule formation on legumes are more sensitive to salt or osmotic stress than the rhizobia. For example, soybean showed little curling or deformation when inoculated with *Bradyrhizobium japonicum* in 170 mM NaCl, and nodulation was completely suppressed by 210 mM NaCl [2]. Salt stress had a negative effect on N_2 fixation by legumes, which is directly correlated to the decay of biomass and N content in the shoot induced by salt. The salt-induced distortions in nodule structure could also be the reason for the decline in the N_2-fixation rate by legumes subjected to salt stress. The reduction of photosynthetic activity might also affect the N_2 fixation under salt stress [2, 33]. Although the root nodule-colonizing bacteria are more salt tolerant than their legume hosts, they show marked variation in salt tolerance [2, 3, 33, 34]. Strains of *Rhizobium leguminosarum* have been reported to be tolerant to NaCl concentrations up to 350 mM NaCl in broth culture [34]. Thus, the rhizobia isolated from soybean and chickpea (*Cicer* sp.) were fast-growing tolerant strains, and the slow-growing rhizobia were sensitive at 340 mM NaCl [35]. *Rhizobium* strains from *Vigna unguiculata* were tolerant to NaCl up to 5.5%, equivalent to about 450 mM NaCl [36]. It has been found that slow-growing peanut rhizobia are more susceptible than fast-growing rhizobia [2]. Rhizobia from woody legumes also showed substantial salt tolerance: strains from *Acacia, Prosopis*, and *Leucaena* are tolerant to 500 to 850 mM NaCl [2, 3, 37, 38].

Successful *Rhizobium*-legume symbioses under salt stress require the selection of salt-tolerant rhizobia indigenous to saline soils. These salt-tolerant rhizobia usually show morphological and metabolic changes and structural modifications to cope with and adapt to salt stress. Effective salt-tolerant rhizobia were isolated from non-saline as well as saline environments. Recent reports support the finding that some rhizobia have the potential to form a successful symbiosis with legumes under salt stress, e.g., *Rhizobium tropici*, *Sinorhizobium* sp. [39, 40].

The salinity stress is closely related in soils from excessive natural accumulation or inorganic compound applications. The main salts are sodium-related compounds such as sodium chloride (NaCl), sodium nitrate ($NaNO_3$), sodium sulfate (Na_2SO_4), sodium carbonates ($NaHCO_3$, Na_2CO_3), and sulfate-related compounds such as potassium sulfate (K_2SO_4), calcium sulfate ($CaSO_4$), magnesium sulfate ($MgSO_4$) and other salts [3, 38]; the symbionts like rhizobia helps to increase the tolerance through releasing metabolites or compounds such as plant-hormones (auxins, gibberellins, cytokinins), organic acids such as succinic acid and SA [38, 41–43]. In addition, EPS accumulation helps the symbiont survival and the communication between microbe-microbe and microbe-host. These EPS indirectly helps to improve the soil structure and help the plant cell's osmotic balance and ion homeostasis. Many reports show that *B. japonicum* may involve the up-regulated expression of NH_4, NO_3, Na^{+}, and Cl^{-} transporter-related genes in rhizobia and their host [41]. For *Ensifer*, it has been reported similar to *B. japonicum* in the trehalose synthesis for increased salt tolerance. In addition, regulation of methionine and inositol metabolisms [43]. In the case of other rhizobia, it has been found that flavonoid-accumulation improves the interaction under salinity stress, and these flavonoids could be salt-depend or salt-independent induction; then, the host-interaction continuing under salt stress due to the nodulation is not flavonoid-depend in some nod genes like happen in *R. tropici* CIA899 [38]. Thus, for the mitigation of salt stress, salt-tolerance rhizobia are necessary.

Rhizobia tolerate much higher salt concentrations than agricultural plants, selection for tolerance would be useful only to eliminate the sensitive strains [8]. Studies of coastal land areas will be useful to understand the capabilities of those strains and the mechanisms for their tolerance and their symbiosis under these conditions, as commonly happened in rice cultivated in coastal areas in Asian countries.

2.4 Alkaline pH stress

The case of alkaline pH is mainly from soil conditions and is another of the major problems affecting crop yield in arid and semiarid regions, especially in the tropics and the Mediterranean regions. Alkalinity appears to be the more severe, though less frequent, stress for legumes, and it goes in hand with salinity stress.

Most leguminous plants require a neutral or slightly acidic soil for growth, especially when they depend on symbiotic N_2 fixation. Legumes and their rhizobia exhibit varying responses to soil alkalinity. Legumes like alfalfa usually show growth restriction when the pH is below 5.5–6 or more than 8. On the other hand, the rhizobial species appear more tolerant to alkaline pH than their legume hosts [2]. For instance, some isolates from *Argyrolobium uniflorum* (a legume plant commonly found in Tunisian arid regions) belonging to *Ensifer* genera showed an alkali-tolerance tendency probably related to the basic pH that characterizes most soils in arid regions of Tunisia [44].

Similarly, some other isolates from *Genista saharae* (a spontaneous legume shrub found in Sahara) showed tolerance to pH 12 [44]. The *Rhizobium*-legume symbiosis is superior to other N_2-fixing systems concerning N_2-fixing potential and adaptation to severe conditions. Several symbiotic systems involving stress-tolerant partners (rhizobia and legumes) have been characterized [45]. These associations might have sufficient traits necessary to establish successful growth and N_2 fixation under the conditions prevailing in arid regions. Most of the reports indicated that those rhizobial strains isolated from alkaline conditions or alkali soils showed the best response under alkaline stress [45–47], and their mechanism to improve the tolerance during the symbiosis process is similar, or kind of related to salinity-stress reported up to date [41–43, 45–47]. The existence of *Rhizobium*-legume symbioses, which can fix the appreciable amount of N_2 under arid conditions, is fascinating. These symbioses represent the best source of the "ideal" fertilizer in arid regions and therefore command great interest as a subject of future research.

2.5 Acid soils

Soil acidity is associated with infertility and mineral toxicities and is a major constraint to agricultural production in several parts of the world. Acid soils are distributed over extensive areas, especially in tropical and subtropical soils such as Venezuela and Brazil [2–6, 48]. Acid soil infertility can be caused by toxic levels of hydrogen, Al, and Mg, as deficiencies of phosphorous and other minerals [8, 49]. These infertility factors have been shown to affect symbiotic nitrogen fixation through their effects on any stage of the symbiosis [2, 4, 8]. Soil acidity is an important factor in restricting the occurrence of rhizobia in soils, although there is considerable species variability in the degree of acid tolerance [6, 8]. An evaluation of *Rhizobium meliloti* indicated that the competitiveness of some of the strains was affected by soil acidity [6, 8]. In the case of *B. japonicum* associated with *Glycine max* and other crops were examined and selected for growth and survived in acid soil, also selected *in vitro* for being acid, Al, and Mn-tolerant [50]. Other cases, such as *R. tropici* and *R. leguminosarum* isolated from Amazonian soils, could survive and show high efficiency in nodulating *P. vulgaris* in acid soil with Al [45, 48, 51, 52]. Other authors showed that *R. tropici* can tolerate until 4% of Al concentration *in vitro* [48, 52]. The infection of legume root hairs by rhizobia included acid-sensitive steps, with infections significantly reduced below pH [8, 51, 52], a level of H^+ that is common in acid soils. Studies about acid instability could affect root hair infection and lead to lectin binding proprieties affected.

Soil acidity is often associated with Al and Mn toxicity and Ca deficiency, as explained in the section below. All disorders are corrected by liming in traditional cropping. The symbiosis is affected by stooping rhizobia growth, nodule initiation, impairing nodule function, and slowing the plant's growth. In this section, the importance of tolerance strain and the genetics of the host are key points. The plant's sensitivity may cause symbiotic failure more often than the literature suggests; still, if there are acid tolerance rhizobia in an adequate amount, the acidity could inhibit the nodule imitation from the plant side [51, 52]. Some of the common acids responsible are carbonic, sulfuric, and nitric acids which generate by oxidation of another process.

2.6 Metals toxicity and tolerance

Many metals are commonly found in nature, though high amounts of metals and toxic elements have low mobility. Furthermore, they are a problem due to their low availability causing different kinds of problems, and sometimes it is necessary to remove them from the soil, water, and air. In this way, plants are the best way used until now; however, the beneficial organisms help the plants to grow and help the high biosorption or limiting their uptake. In this order, we can easily understand the negative effects of metal, and we can divide this chapter into heavy and light metals.

Light metals with a low density are commonly found and used; their high accumulation or interaction with other factors is highly toxic, such as Aluminum (Al), Magnesium (Mg), Titanium (Ti), and copper (Cu). In contrast, heavy metals have high density, are less common, and are used in low amounts and their interactions are highly toxic such as Chromo (Cr), Cobalt (Co), Nickel (Ni), Zn (Zinc), Lead (Pb), Cadmium (Cd), Manganese (Mn), Arsenic (As). Furthermore, using microbes-plants metal tolerant can be worth applying in fields and therefore be crucial to promote detoxification, survival, or bioremediation of contaminated soil. However, their symbiosis or interaction is poorly investigated and reported leading to a lack of understanding and underestimation.

In this order, Al is the most abundant metal, becomes more soluble as acidity increases, and is often the major toxic element in acidic soils and water. Many hypotheses have been proposed for the mechanism of Al toxicity in animals and plants [52–55]. However, Al-microorganism interactions have been receiving attention in recent years. Previous work on rhizobia has investigated direct and indirect mechanisms of Al-toxicity [53–55]. It has been shown that *Rhizobium* cells suggest that the repair mechanism used to overcome damage by Al to DNA in tolerant cells is quite different from that in sensitive cells, indicating that DNA is a possible site of action of Al in the common soil bacterium as *Rhizobium* [53]; this is an important factor for ecological applications. Many experiments demonstrated that using acid and Al-tolerant nitrogen-fixing bacteria is one alternative way to increase productivity in the field [53–55]. For example, soybean inoculated with *B. japonicum* strain BJ11 increased the production under acid-aluminum soil (pH 5.0–5.5) [56].

Al-toxicity is a major problem affecting soil fertility, microbial diversity, and nutrient uptake of plants. Rhizobia response and legume interaction under Al conditions are still unknown [54], and it is important to understand how to develop and improve legume cultivation under Al stress [55]. The rhizobia response was recorded under different Al concentrations, with decreasing rhizobial cell numbers as Al concentration increased. However, induced Al tolerance considerably depended on rhizobia types and their origins [53–55]. Accordingly, organic acids, biofilm, plant hormones, and EPS were correlated with growth rate, cell density, and improvement in plant growth [53–57]. Previous studies indicated that citric acid, GABA, and biofilm production might be a positive selective force for Al-tolerance in rhizobia tolerance [53, 57]. In addition, Al-toxicity delayed and interrupted the plant-rhizobia interaction. The effect was more pronounced under acidic conditions *Bradyrhizobium*, *Paraburkholderia,* and *Rhizobium* have been reported that significantly improve plant growth under acid-Al stress in combination with high Al concentration with soybean, alfalfa, and common bean and established an effective interaction, nodulation and N_2-fixation [54–57]. Moreover, metabolic pathways have still been little known under Al-stress.

Al-toxicity limits plant production worldwide, particularly in acidic soils. In the Americas, 70% of tropical soils experience limitations related to acidic soil, Al- and Mn-toxicity, and the deficiency of other elements such as Ca, Mg, and P [48]. Thus, many types of plants are affected by different metal concentrations in soils. Furthermore, Al and low pH disrupt plants' pathways and physiological functions, such as root elongation, root hair formation, and nutrient uptake (e.g., K, Ca, N) [58, 59]. The main symptom of Al-toxicity is the inhibition of root growth in the plant.

Moreover, Al can disrupt many other functions, including root hair elongation and nutrient uptake (especially Ca and K). Al can induce oxidative stress, disrupt cytoskeleton and apoplastic processes, and affect intracellular transport [60]. In previous studies, tolerance mechanisms were distinguished from resistance mechanisms. Exclusion mechanisms were predicted to depend on transport systems that export Al from the symplast or exudate ligands that bind Al and limit its uptake into the cytosol [60, 61]. In addition, Al and Mn toxicity are important inhibitors of the growth of rhizobia in soils; as mentioned before, this is truly affected by soil pH, and the Al toxicity has been found to stop legume growth in acid soils without inducing nodulation or by N starvation [51]. In some cases, Al toxicity happens because the Al is immobilized in humic chelates, such as in Histosols. In Mn, toxicity is more common in acid soil, where temporary or local anoxia results from wetness, high organic matter, impeding layers, or well-drained soils such as Oxisols [3, 8, 60–62].

Furthermore, heavy metals are important inorganic pollutants in recent years, such as Cu, Ni, Cd, Zn, Cr, and Pb. The studies of symbiosis with these heavy metals are more limited than light metals, mainly focused on *Rhizobium* and *Ensifer* species [60–67]. Some studies indicate that the combination *of Rhizobium* and arbuscular mycorrhizal fungi (AM) induces plant tolerance under As stress [63]. Others showed that in Ni, Cu, Cr, Cd, and Zn, the symbiont tolerance varies according to the metal type, but the strain alleviates the effect of heavy metals in soybean [64]. Some *Rhizobium* species can nodulate species of *Phaseolus*, *Trifolium,* and *Vicia* under normal conditions but also can improve the seed germination, biomass, and chlorophyll content in these species under Cd-stress, and their response level depend more on the genotype of the plant associated with these rhizobia. For example, the application of *R. leguminosarum* and *Rhizobium larrymoorei* helps to decrease the negative effect on other crops, such as wheat [65] and rice [66]. The way these *Rhizobium* species survive under Cd-stress could be due to the glutathione accumulation, proline metabolic pathways, and its plant interaction could improve because of the accumulation of several compounds such as antioxidants and putrescine protecting the plant cell besides auxins production, N_2 fixation, and others [65, 66].

What we know about natural symbiosis under heavy metals stress is the induction or improvement of plant growth, protection of plant cells received from symbiont, pigment recovery, and increasing chlorophyll content and N_2 fixation. On the other hand, from a symbiont point of view is the high production of glutathione compounds, proline, antioxidants, biofilms, and EPS that help protect cells [67]. However, the further natural interactions and the mechanism or pathways are still unknown compared to non-symbiotic rhizobacteria.

2.7 Pesticides stress

Modern agricultural practices depended on agrochemicals, such as pesticides, for crop production. Pesticides, including herbicides, insecticides, and fungicides, are heavily utilized for high quantity and quality crop yield and reduce labor and energy input in crop production [68]. However, pesticide accumulation in the soil beyond the recommended levels due to repeated application or slow degradation process poses a threat to the environment. It affects the plant host by altering the root architecture, a number of root sites for infection, and the transformation of compounds for plant uptake, such as converting ammonia into nitrates [69]. On the other hand, it also influences the growth and activity of soil microbial communities such as free-living and endophytic nitrogen-fixing bacteria [70]. In particular, rhizobia are more susceptible to fungicides, herbicides, and then insecticides [71]. Fungicides such as captan and thiram significantly reduced survival and altered the phenotypic characteristics of *Rhizobium leguminosarum bv. viceae* [72]. Also, the legume nodulation and growth were affected at the highest concentration of captan. Similarly, chickpea seeds treated with captan or arrest decreased the number of viable *Rhizobium ciceri*, nodule dry weight, and nitrogen fixation [73]. On the other hand, Moorman [74] reported the direct effect of herbicide on the rhizobial symbiont and exerted indirect effects on the host legume. In parallel, herbicides had adverse effects on the nodulation and nitrogen fixation of peas inoculated with *Rhizobium leguminosarum*, not due to its influence on the rhizobia itself but on the host plant [75]. Lastly, insecticides, including methyl parathion and dichloro-diphenyl-trichloroethane (DDT), perhaps disrupt phytochemicals from alfalfa and *nodD* receptors of *Sinorhizobium meliloti* as the mechanism of action. Consequently, it significantly delays the specific timing and initiation of signals necessary for efficient symbiotic nitrogen fixation [76, 77]. On these accounts, abnormal concentrations of pesticides negatively impacted the legume-rhizobia symbiosis and disrupted the nitrogen fixation, leading to a decrease in plant yield. Although extensive studies on the effect of pesticides on legume-rhizobia symbiosis it is still limited to in vitro and greenhouse conditions. Studies regarding it under field conditions are scanty; thus, it is important to investigate and characterize rhizobia that can be employed in the field, considering the increasing pesticide residues currently.

2.8 Others

On a global scale, there are climate changes such as acid soils and long dry seasons; in tropical countries, mainly the soil type affected by infertility are Oxisols and Ultisols [3]. Their attendant problems include deficient or unavailable nutrients such as P, K, S, Ca, Mg, Zn, Mo, Bo, and Cu. Unfortunately, this is not the main topic in this chapter, but it is important to mention that the interaction between these essential life elements, whether with the host or symbiont, is also relevant. Accordingly, the research priorities have been done according to crop consumptions and their improvement in the field, subsequently with selection and, later, breeding of tolerant genotypes of legumes, a clearer definition of mineral nutrient requirements. Several reports described that the deficiency in nutrients in soils or environmental growth conditions of hosts or symbionts negatively affected the tolerance to different stress. For example, the improvement of Bo in media through the growth conditions of *Ensifer meliloti* leads the cell to increase the tolerance to salinity [78], and other studies state similar significance with other elements [3, 9, 48, 59].

3. Conclusions and future perspectives

In conclusion, many rhizobia can be classified into relatively tolerant and sensitive isolates to different stress conditions, suggesting the application of these isolates as effective inoculants to different legumes to obtain effective inoculants in environments with abiotic stress influence. Interestingly, the distribution of resistant or tolerant rhizobia to different abiotic stress is closely related to environmental conditions of their climate soil origins, mainly for salinity, temperature, and metal stress. Therefore, many studies suggested finding suitable inoculants with multi-tolerance for legumes; those strains must be isolated in the problematic areas where they will be applied. Research into these areas is currently underway in several research groups worldwide, and it is expected that this research will provide beneficial outcomes resulting in improved sustainability and productivity in agricultural systems.

However, Tropical zones remain the most difficult zones to undertake studies of soil microorganisms. It has been suggested that tropical symbionts and wild hosts are poorly documented, although reports indicate that rhizobia diversity may be greater in tropical than in temperate regions. However, the communities' interactions are still unknown in both tropical and temperate.

Many challenges need to be addressed for future studies as the expansion of worldwide distributed legumes is needed under different countries' conditions, for example, beach bean, *Mimosa* species, and other non-common species, due to their capacity to adapt to each country's conditions and their specificity or promiscuous symbiont and their pathways of survival or plant interaction, which can lead to the key for understanding the tolerance-mechanism under stress. Furthermore, there is a large number of rhizobia reported, but 98% of the well-described symbionts fall into the alpha-proteobacteria group during these revisions. Many rhizobia have been isolated, but their response under abiotic stress are unclarified such as *Paraburkholderia, Microvirga, Phyllobacterium, Azorhizobium, Allorhizobium*, and *Cupriavidus.*

As well as further studies are needed to screen isolates resistant to other compounds such as mercury (Hg), lithium (Li), and non-renewable elements commonly used in the new decade, and it is increasing their negative effect in crops such as soybean and other legumes, as a consequence of the revision of the articles published in 2021–2022. Last but not least, there are further studies in field conditions because 92% of the recent studies are only incubator or greenhouse experiments, and less than 20% of these studies used soil from field areas or in problematic areas. Furthermore, the studies of communities' interactions can improve understanding of the changes in the enzymatic process and cycles, which can be done with new techniques such as qPCR, Next Generation Sequencing analysis (NGS), metagenomics, and metabolomics analysis. Also, the improvement of studies in another host, not only main crops such as soybean, which have been proven by the reports of symbionts applied in a non-natural host such as rice, wheat, and tomato under abiotic stress, showed improvement in the tolerance or crop production.

The symbiosis in legumes is well documented but poorly understood and known by farmers, principally in natural conditions or with wild legumes or strains. Also, many reports try to describe diversity, and many do not focus on communities or populations level. For example, a large number of studies involve soybean in which its focus from the host view as nodulation controlled by this legume. These reflect a delay in the onset of competition for energy and nutrients between the host-microbe, microbe-microbe, microbe-host-symbiont, and environment-host-microbe. Unfortunately, increasing studies on legume-rhizobia symbiosis are necessary; however, there is slow progress in detailed studies *in situ*, especially in the tropics. In addition, to develop sustainable agriculture in different ecological conditions, we need to obtain information on beneficial soil microorganisms showing nutrient accumulation, plant growth promoting, mobilizing potassium, phosphorous or iron, and nitrogen-fixing abilities associated with different legumes. New approaches are indeed needed to elucidate the mechanisms of different compounds production from symbionts, the host interactions with more than one symbiont per host or other rhizobacteria that interact in soils; these studies can be done through the research of the synergy microbe-plant, microbe-microbe, symbiont-plant-microbe. Also, the co-inoculation technology between symbionts, or symbiont and non-symbiotic bacteria, need to be investigated under different plant host.

It is necessary to strongly advocate and establish worldwide the application of natural symbionts in the field to mitigate climate change and abiotic stress as well as increase crop production globally and sustainably. In this order, the scientific work will be synergic and not need to become dead research. The global population is rapidly increasing, necessitating an increase in eco-friendly production. Climate change and its effects will be exacerbated in the coming decades, threatening plant and symbionts' survival, causing negative changes such as lower growth, low yield, and more susceptibility to damage and diseases from plants. Governments and national and international agencies need to promote the utilization of biofertilizers, bio-compounds, and other eco-friendly alternatives for crop improvement and protect soil, also showing the farmers and consumers the benefits of its natural-based compounds from natural-symbiosis relationships. It is important to display that natural symbiosis and its interactions with the soil microbiome can protect plants from abiotic stress and improve plant health and soil fertility. Soil is the base for a non-ending game of dynamic and synergic between plants and microbes (not only their symbionts). As a result of their relationship, the plant microbes establish their symbiosis or interactions, and through time under different changes, including abiotic stresses, they co-exist and co-evolve for a better future.

Acknowledgements

We are grateful to Beijing Forestry University, the University of the Ryukyus, the University of the Philippines, and their staff for their kind support in accessing many kinds of literature.

Conflict of interest

The authors declare no conflict of interest.

Notes/thanks/other declarations

We thank Dr. Zhang Haolin for the support and revision in general of this manuscript. Also, Nika Karamatic from IntechOpen for her kind support throughout the process.

Author details

Maria Daniela Artigas Ramírez[1,2]* and Jean Louise Cocson Damo[3]

1 College of Biological Science and Technology, Beijing Forestry University, Beijing, China

2 Tropical Biosphere Research Center (TBRC), University of the Ryukyus, Taketomi-cho, Yaeyama-gun, Okinawa, Japan

3 National Institute of Molecular Biology and Biotechnology, University of the Philippines Los Baños, Laguna, Philippines

*Address all correspondence to: artigasdaniela@gmail.com

References

[1] Savé R, Castell C, Terradas J. Gas exchange and water relations. In: Rodà F, Retana J, Gracia CA, Bellot J, editors. Ecology of Mediterranean Evergreen Oak Forests. Berlin: Springer-Verlag; 1999. pp. 135-147

[2] Zahran HH. *Rhizobium*-legume symbiosis and nitrogen fixation under severe conditions and in an arid climate. Microbiology and Molecular Biology Reviews. 1999;**3**(4):968-989

[3] CIAT (Centro Internacional de Agricultura Tropical). Constraints to and Opportunities for Improving Bean Production. A Planning Document 1993-98 and an Achieving Document 1987-92. Cali: CIAT; 1992

[4] Graham PH. Ecology of the root nodule bacteria of legumes. In: Dilworth MJ, James EK, Sprent JI, Newton WE, editors. Nitrogen Fixing Leguminous Symbioses. The Netherlands: Springer; 2008. pp. 23-58

[5] Allen ON, Allen EK. The Leguminosae: A Source Book of Characteristics Uses and Nodulation. Madison, London: University of Wisconsin Press and MacMillan Publishing; 1981

[6] Kosslak RM. Interactions between the Soybean Host (*Glycine max*) and its Microsymbiont (*Rhizobium japonicum*) Which Affect Competition between Strains. PhD thesis. Honolulu: University of Hawaii; 1984

[7] Gyaneshwar P, Hirsch AM, Moulin I, Chen WM, Elliott GN, Bontemps C, et al. Legume-nodulating Beta-proteobacteria: Diversity, host range and future prospects. Molecular Plant-Microbe Interactions. 2011;**24**(11):1276-1288

[8] Graham PH, Apolitano C, Ferrera R, Halliday J, Lepiz R, Menéndez D, et al. The International Bean Inoculation Trail (IBIT). Results for the 1978-1979 Trial. Biological Nitrogen Fixation Technology for Tropical Agriculture. Cali: CIAT; 1981

[9] Raza A, Charagh S, García-Caparrós P, Rahman MA, Ogwugwa VH, Saeed F, et al. Melatonin-mediated temperature stress tolerance in plants. GM Crops Food. 2022;**13**(1):196-217

[10] Michiels J, Verreth C, Vanderleyden J. Effects of temperature stress on bean-nodulating *Rhizobium* strains. Applied and Environmental Microbiology. 1994;**60**(4):1206-1212

[11] Arayankoon T, Schomberg HH, Weaver RW. Nodulation and N2 fixation of guar at high root temperature. Plant and Soil. 1990;**126**:209-213

[12] Graham PH. Stress tolerance in *Rhizobium* and *Bradyrhizobium*, and nodulation under adverse soil conditions. Canadian Journal of Microbiology. 1992;**38**:475-484

[13] Wei GH, Yang XY, Zhang ZX, Yang YZ, Lindsröm K. Strain *Mesorhizobium* sp. CCNWGX035: A stress-tolerant isolate from *Glycyrrhiza glabra* displaying a wide host range of nodulation. Pedosphere. 2008;**18**(1):102-112

[14] Bansal M, Kukreja K, Suneja S, Dudeja SS. Symbiotic effectivity of high temperature tolerant mungbean (*Vigna radiata*) rhizobia under different temperature conditions. International Journal of Current Microbiology and Applied Sciences. 2014;**3**(12):807-821

[15] Tshape H. The spread of plasmids as a function of bacterial adaptability. FEMS Microbiology Ecology. 1994;**15**:23-32

[16] Chen Y, Liu L, Kang D, Zhang D, Kou C, Mao S, et al. Large-scale evidence for microbial response and associated carbon release after permafrost thaw. Global Change Biology. 2021;**27**(14):3218-3229

[17] Wang R, Cui L, Li J, Li W, Zhu Y, Hao T, et al. Response of nir-type rhizosphere denitrifier communities to cold stress in constructed wetlands with different water levels. Journal of Cleaner Production. 2022;**362**:132377

[18] Morcillo RJ, Manzanera M. The effects of plant-associated bacterial exopolysaccharides on plant abiotic stress tolerance. Metabolites. 2021;**11**(6):337

[19] Serraj R, Sinclair TR, Purcell LC. Symbiotic N2 fixation response to drought. Journal of Experimental Botany. 1999;**50**:143-155

[20] Ali S, Tyagi A, Park S, Mir RA, Mushtaq M, Bhat B, et al. Deciphering the Plant Microbiome to Improve Drought Tolerance: Mechanisms and Perspectives. Environmental and Experimental Botany. 2022;**201**:104933

[21] Waldon HB, Jenkins MB, Virgina RA, Harding EE. Characteristics of woodland rhizobial populations from surface and deep soil environment of the Sonoran Desert. Applied and Environmental Microbiology. 1989;**55**:3068-3064

[22] Fuhrmann J, Wollum IIAG. Nodulation competition among *Bradyrhizobium* japonicum strains as influenced by rhizosphere bacteria and iron availability. Biology and Fertility of Soils. 1989;**7**:108-112

[23] Jordan WR, Morgan PW, Davenport TL. Water stress enhances ethylene-mediated leaf abscission. Plant Physiology. 1972;**50**:756-758

[24] Ramos MLG, Gordon AJ, Minchin FR, Sprent JI, Parsons R. Effect of water stress on nodule physiology and biochemistry of a drought tolerant cultivar of common bean (*Phaseolus vulgaris* L.). Annals Botany. 1999;**83**(1):57-63

[25] Yokoyama T, Ando S, Murakami T, Imai H. Genetic variability of the common nod gene in soybean bradyrhizobia isolated in Thailand and Japan. Canadian Journal of Microbiology. 1996;**42**(12):1209-1218

[26] Yokoyama T, Ando S, Tsuchiya K. Serological properties and intrinsic antibiotic resistance of soybean bradyrhizobia isolated in Thailand. Canadian Journal of Microbiology. 1999;**42**:1209-1218

[27] Zhu J, Jiang X, Guan D, Kang Y, Li L, Cao F, et al. Effects of rehydration on physiological and transcriptional responses of a water-stressed *Rhizobium*. Journal of Microbiology. 2022;**60**(1):31-46

[28] Gupta A, Mishra R, Rai S, Bano A, Pathak N, Fujita M, et al. Mechanistic insights of plant growth promoting bacteria mediated drought and salt stress tolerance in plants for sustainable agriculture. International Journal of Molecular Sciences. 2022;**23**(7):3741

[29] Veremeichik GN, Shkryl YN, Rusapetova TV, Silantieva SA, Grigorchuk VP, Velansky PV, et al. Overexpression of the A4-rolB gene from the pRiA4 of *Rhizobium rhizogenes* modulates hormones homeostasis and leads to an increase of flavonoid accumulation and drought tolerance in Arabidopsis thaliana transgenic plants. Planta. 2022;**256**(1):1-14

[30] Cordovilla MP, Ocana A, Ligero F, Lluch C. The effect of salinity on N2

fixation and assimilation in *Vicia faba*. Journal of Experimental Botany. 1994;**45**:1483-1488

[31] Fagg CW, Stewart JL. The value of *Acacia* and *Prosopis* in arid and semiarid environments. Journal of Arid Environments. 1994;**27**:3-25

[32] Zhang X, Harper R, Karsisto M, LindstroÈm K. Diversity of *Rhizobium* bacteria isolated from root nodules of leguminous trees. International Journal of Systematic Bacteriology. 1991;**41**:104-113

[33] Abdel-Wahab AM, Zahran HH. Effects of salt stress on nitrogenase activity and growth of four legumes. Biología Plantarum (Praha). 1981;**23**:16-23

[34] Breedveld MW, Dijkema C, Zevenhuizen LPTM, Zehnder AJB. Response of intracellular carbohydrates to a NaCl shock in *Rhizobium leguminosarum* biovar trifolii TA-1 and *Rhizobium meliloti* SU-47. Journal of General Microbiology. 1993;**139**:3157-3163

[35] El-Sheikh EAE, Wood M. Nodulation and N2 fixation by soybean inoculated with salt-tolerant rhizobia or salt-sensitive bradyrhizobia in saline soil. Soil Biology and Biochemistry. 1995;**27**(4):657-661

[36] Mpepereki S, Makonese F, Wollum AG. Physiological characterization of indigenous rhizobia nodulating *Vigna unguiculata* in Zimbabwean soils. Symbiosis. 1997;**22**:275-292

[37] Dobereiner J, Urquiaga S, Boddey RM. Alternatives for nitrogen nutrition of crops in tropical agriculture. In: Nitrogen Economy in Tropical Soils. Dordrecht: Springer; 1995. pp. 338-346

[38] Chakraborty S, Harris JM. At the crossroads of salinity and rhizobium-legume symbiosis. Molecular Plant-Microbe Interactions. 2022;**35**(7):540-553

[39] Marquina M, González N, Castro Y. Caracterización fenotípica y genotípica de doce rizobios aislados de diversas regiones geográficas de Venezuela. Revista de Biología Tropical. 2011;**59**(3):1017-1036

[40] Akatsu T. Evaluation of Genetic Strains of Root Nodule Bacteria Collected from Ishigakijima and Iriomote Island Islands and its Salt Tolerance Mechanism. Master Dissertation. Tokyo, Japan: Department of Biological Production Science, United Graduate School of Agriculture Science, Tokyo University of Agriculture and Technology (TUAT); 2013

[41] Wang Y, Cheng C, Du Z, Yu B. Pre-inoculation with *Bradyrhizobium japonicum* confers NaCl tolerance by improving nitrogen status and ion homeostasis in wild soybean (*Glycine soja* L.) seedlings. Acta Physiologiae Plantarum. 2022;**44**(2):1-13

[42] Kaur H, Hussain SJ, Kaur G, Poor P, Alamri S, Siddiqui MH, et al. Salicylic acid improves nitrogen fixation, growth, yield and antioxidant defence mechanisms in chickpea genotypes under salt stress. Journal of Plant Growth Regulation. 2022;**41**(5):2034-2047

[43] Belfquih M, Filali-Maltouf A, Le Quéré A. Analysis of Ensifer aridi mutants affecting regulation of methionine, Trehalose, and inositol metabolisms suggests a role in stress adaptation and Symbiosis development. Microorganisms. 2022;**10**:298. DOI: 10.3390/microorganisms10020298

[44] Mahdhi M, Nzoué A, Gueye F, Merabet C, De

Lajudie P, Mars M. Phenotypic and genotypic diversity of *Genista saharae* microsymbionts from the infra-arid region of Tunisia. Letters in Applied Microbiology. 2007;**45**(6):604-609

[45] Artigas RMD, España M, Sekimoto H, Okazaki S, Yokoyama T, Ohkama-Ohtsu N. Genetic diversity and characterization of symbiotic bacteria isolated from endemic Phaseolus cultivars located in contrasting agroecosystems in Venezuela. Microbes and Environments. 2021;**36**(2):ME20157

[46] Shah AS, Wakelin SA, Moot DJ, Blond C, Noble A, Ridgway HJ. High throughput pH bioassay demonstrates pH adaptation of *Rhizobium* strains isolated from the nodules of *Trifolium subterraneum* and T. repens. Journal of Microbiological Methods. 2022;**195**:106455

[47] Etesami H. Root nodules of legumes: A suitable ecological niche for isolating non-rhizobial bacteria with biotechnological potential in agriculture. Current Research in Biotechnology. 2022;**4**:78-86. DOI: 10.1016/j.crbiot.2022.01.003

[48] Martínez-Viera R, López M, Dibut B, Parra C, Rodríguez B. La fijación Biológica del nitrógeno atmosférico en condiciones tropicales. Caracas-Venezuela: Ministerio del Poder Popular para la Agricultura y Tierra (MPPAT); 2006. p. 172

[49] López M, Martínez-Viera R, Brossard M, Toro M. Capacity atmospheric nitrogen fixation of native strains of Venezuelan ecosystems. Agronomía Tropical. 2010;**60**(4):355-361

[50] Indrasumunar A, Menzies N, Dart P. Laboratory prescreening of *Bradyrhizobium japonicum* for low pH, Al and Mn tolerance can be used to predict their survival in acid soils. Soil Biology and Biochemistry. 2012;**48**:135-141

[51] Munns DN. Nodulation of *Medicago Sativa* in solution culture: I. acid-sensitive steps. Plant and Soil. 1968;**28**:129-146

[52] Vargas AA, Graham PH. *Phaseolus vulgaris* cultivar and *Rhizobium* strain variation in acid-pH tolerance and nodulation under acid conditions. Field Crops Research. 1988;**19**(2):91-101

[53] Johnson AC, Wood M. DNA, a possible site of action of aluminum in *Rhizobium* spp. Applied and Environmental Microbiology. 1990;**56**:3629-3633

[54] Wekesa C, Muoma JO, Reichelt M, Asudi GO, Furch ACU, Oelmüller R. The cell membrane of a novel *Rhizobium phaseoli* strain is the crucial target for aluminium toxicity and tolerance. Cell. 2022;**11**:873. DOI: 10.3390/cells11050873

[55] Shi H, Sun G, Gou L, Guo Z. Rhizobia–legume Symbiosis increases aluminum resistance in alfalfa. Plants. 2022;**11**:1275. DOI: 10.3390/plants11101275

[56] Mubarik NR, Imas T, Wahyudi AT, Widiastuti H. The use of acid-aluminum tolerant *Bradyrhizobium japonicum* formula. International Journal of Agricultural and Biosystems Engineering. 2011;**5**(5):289-292

[57] Artigas RMD, Silva JD, Ohkama-Ohtsu N, Yokoyama T. In vitro rhizobia response and symbiosis process under aluminum stress. Canadian Journal of Microbiology. 2018;**64**(8):511-526

[58] Girma M, Sundin P, Aðlsteinsson S, Jensén P. Uptake and translocation of N and P in Leucaena leucocephala grown at

different N, P and Al concentrations. In: Ando T, Fujita K, Mae T, Matsumoto H, Mori S, Sekiya J, editors. Plant Nutrition for Sustainable Food Production and Environment. Kluwer Academic Publishers: Dordrecht; 1997. pp. 135-136

[59] Nian H, Ahn SJ, Yang ZM, Matsumoto H. Effect of phosphorus deficiency on aluminium-induced citrate exudation in soybean (Glycine max). Physiologia Plantarum. 2003;**117**(2):229-236

[60] Kochian LV, Pineros MA, Hoekenga OA. The physiology, genetics and molecular biology of plant aluminum resistance and toxicity. Plant and Soil. 2005;**274**(1-2):175-195

[61] Khan MS, Zaidi A, Wani PA, Oves M. Role of plant growth promoting rhizobacteria in the remediation of metal contaminated soils. Environmental Chemistry Letters. 2009;**7**(1):1-19

[62] Moretto C, Castellane TCL, Leonel TF, Campanharo JC, De Macedo Lemos EG. Bioremediation of heavy metal–polluted environments by non-living cells from rhizobial isolates. Environmental Science and Research. 2022;**29**(36):54045-54059

[63] Ahmad I, Narayan S, Shukla J, Shirke PA, Kumar M. Endofungal *Rhizobium* species enhance arsenic tolerance in colonized host plant under arsenic stress. Archives of Microbiology. 2022;**204**(7):1-15

[64] Liu H, Cui Y, Zhou J, Penttinen P, Liu J, Zeng L, et al. Nickel mine soil is a potential source for soybean plant growth promoting and heavy metal tolerant rhizobia. Peer Journal. 2022;**10**:e13215

[65] Maslennikova D, Nasyrova K, Chubukova O, Akimova E, Baymiev A, Blagova D, et al. Effects of *Rhizobium leguminosarum* Thy2 on the growth and tolerance to cadmium stress of wheat plants. Life. 2022;**12**:1675. DOI: 10.3390/life12101675

[66] Wang Y, Wang R, Kou F, He L, Sheng X. Cadmium-tolerant facultative endophytic *Rhizobium larrymoorei* S28 reduces cadmium availability and accumulation in rice in cadmium-polluted soil. Environmental Technology and Innovation. 2022;**26**:102294

[67] Phurailatpam L, Dalal VK, Singh N, Mishra S. Heavy metal stress alleviation through omics analysis of soil and plant microbiome. Frontiers in Sustainable Food Systems. 2022;**5**:568

[68] Meena H, Meena RS, Rajput BS, Kumar S. Response of bio-regulators to morphology and yield of clusterbean [*Cyamopsis tetragonoloba* (L.) Taub.] under different sowing environments. Journal of Applied and Natural Science. 2016;**8**(2):715-718. DOI: 10.31018/jans.v8i2.863

[69] Gopalakrishnan S, Sathya A, Vijayabharathi R, Varshney RK, Gowda CLL, Krishnamurthy L. Plant growth promoting rhizobia: Challenges opportunities. Springer Verlag. Biotech. 2015;**4-5**:355-377. DOI: 10.1007/s13205-014-0241-x

[70] Mathur SC. Future of Indian pesticides industry in next millennium. Pesticides Information. 1999;**24**:9-23

[71] Drouin P, Sellami M, Prévost D, Fortin J, Antoun H. Tolerance to agricultural pesticides of strains belonging to four genera of Rhizobiaceae. Journal of Environmental Science and Health, Part B: Pesticides, Food Contaminants, and Agricultural Wastes. 2010;**45**(8):757-765. DOI: 10.1080/03601234.2010.515168

[72] Dunfield KE, Siciliano SD, Germida JJ. The fungicides thiram and captan affect the phenotypic characteristics of *Rhizobium leguminosarum* strain C1 as determined by FAME and biolog analyses. Biology and Fertility of Soils. 2000;**31**:303-309

[73] Kyei-Boahen S, Slinkard AE, Walley FL. Rhizobial survival and nodulation of chickpea as influenced by fungicide seed treatment. Canadian Journal of Microbiology. 2001;**47**(6):585-589. DOI: 10.1139/w01-038

[74] Moorman TB. A review of pesticide effects on microorganisms and microbial processes related to soil fertility. Journal of Production Agriculture. 1989;**2**(1):14-23. DOI: 10.2134/jpa1989.0014

[75] Singh G, Wright D. In vitro studies on the effects of herbicides on the growth of rhizobia. Letters in Applied Microbiology. 2002;**35**(1):12-16. DOI: 10.1046/j.1472-765X.2002.01117.x

[76] Fox JE, Starcevic M, Kow KY, Burow ME, McLachlan JA. Endocrine disrupters and flavonoid signalling. Nature. 2001;**413**(6852):128-129. DOI: 10.1038/35093163

[77] Fox JE, Gulledge J, Engelhaupt E, Burow ME, Mclachlan JA. Pesticides Reduce Symbiotic Efficiency of Nitrogen-Fixing Rhizobia and Host Plants. Proceedings of the National Academy of Sciences. 2007;**104**(24):10282-10287. DOI: 10.1073pnas.0611710104

[78] Quijada NM, Abreu I, Reguera M, Bonilla I, Bolaños L. Boron nutrition affects growth, adaptation to stressful environments, and exopolysaccharide synthesis of *Ensifer meliloti*. Rhizosphere. 2022;**22**:100534

Chapter 5

The Plant-Rhizobial Symbiotic Interactions Provide Benefits to the Host beyond Nitrogen Fixation That Promote Plant Growth and Productivity

Jemaneh Z. Habtewold and Ravinder K. Goyal

Abstract

Rhizobial symbiotic interactions are known for nitrogen fixation, providing commercial crops and other plants with self-sufficiency in nitrogen requirements. An enormous contribution from nitrogen fixation is vital to the global nitrogen cycle. The symbiotic nitrogen reduces the carbon footprint of crop cultivation, which underlines its importance in agricultural sustainability. Extensive research efforts have been made to understand the symbiotic relationship at molecular, physiological, and ecological levels. This led to the isolation and modification of symbiotic strains for enhanced nitrogen efficiency. During the evaluation of strains for nitrogen fixation in exchange for supporting the bacterium in terms of space and resources, it has been observed that the accrued benefits to the host plants extend well beyond the nitrogen fixation. The symbiotic interaction has been advantageous to the host for better growth and development, tolerating a stressful environment, and even keeping the pathogenic microbial enemies at bay. Additionally, it enabled the availability of the mineral nutrients, which otherwise were inaccessible to the host. In this chapter, we bring together the information with a focus on the role of rhizobial symbiotic interactions that promote plant growth and productivity through phytohormone synthesis, by facilitating the availability of mineral nutrients, and by improving the plant tolerance to sub-optimal growth conditions.

Keywords: rhizobia, legumes, symbiosis, growth, plant hormones, environmental adaptability, stress tolerance, heavy metals

1. Introduction

Nitrogen (N) is an essential macro-nutrient that is needed for plant growth. Plants are unable to use N in its gaseous state, which is freely and abundantly available in the atmosphere. The atmospheric nitrogen (N_2) must first be converted chemically or naturally into NO_3^- or NH_4^+ before plants can assimilate it. Nature has given many

IntechOpen

microorganisms the ability to convert N_2 into a usable form. In microorganisms, the metabolic conversion of N_2 takes place either in their free-living state or through symbiosis with the host plant. Legumes have the unique ability to form symbiotic interactions with rhizobia, which helps meet the plant's N requirement. The symbiotic interaction is very complex, beginning with a chemical conversation between the rhizobia and host plant [1]. The host plant synthesizes specific flavonoids that are perceived through *NodD* genes expressed by a symbiont. Subsequently, through an array of events, a mutualistic relationship is established in which the bacterium imports carbon metabolites from the host in exchange for a usable form of N, or fixed N. Biological N fixation has been studied extensively to understand the evolutionary logistics, chemistry, regulation, and ecological perspectives [2]. The complexity of symbiotic mechanisms indicates the coevolution of a process that led to the development of N-fixing nodules [3]. The question of whether plants selected bacteria, whether bacteria influenced evolution, or whether legumes and rhizobia coevolved is being debated. Unlike mitochondrial and chloroplast evolution, the loss of genes from the rhizobia to the host nucleus, which would have compromised bacterial survival as a free-living organism, did not occur [4]. Nonetheless, the symbiotic arrangement has evolved over time into a secure partnership.

Symbiotically fixed N plays a key role in agricultural sustainability, especially when many of the agricultural practices, including N supplementation, that were instrumental in enhancing productivity and food security may not have been in tune with safeguarding the environment and preserving our ecosystems. Legumes are high in protein content and contribute roughly one-quarter of total grain production; their N self-sufficiency and increasing the N content of soil for subsequent crops have a significant impact on reducing the need for N-fertilizers [5]. Easing the pressure on chemical-N demand is of environmental significance because the majority of the N in N-fertilizers becomes a source of aquatic system pollution [6] or a substrate for nitrous oxide (N_2O) gas in the environment through denitrification. The enormous carbon footprint of fertilizer synthesis and subsequent N-pollution endangers the sustainability of agriculture and the environment. This has generated a significant interest in improving the efficiency of both symbiotic and biological N fixation in legumes and non-legumes, respectively. Many strategies have been employed to improve the symbiotic productivity notwithstanding the challenges of competitiveness of the native strains [7, 8].

Nodulation and N fixation may have been an important factor in the diversification of legumes, enabling them to be one of the largest families of widely distributed plant species [3]. The symbionts developed the capability to synthesize plant hormones that are critical to the growth and development of plants [9]. The biosynthetic pathways of many of the hormones are long and complex, requiring dedicated genetic and cellular machinery. Their involvement in the regulation of nodule development [10] underscores the interdependence of symbiotic partners. The symbiotic interactions are known to enhance plant growth and biomass and increase the availability of micro-elements like iron (Fe) and phosphorus (P) [11]. They enable the host plants to tolerate stressful growth conditions [12], which increases the plant survivability in adverse environmental regimes. It is likely that the enhanced environmental outreach has played a role in the diversification of symbiotic hosts. Although N fixation could be a primary reason for symbiotic evolution, it is not known if the additional beneficial activities were a contributing factor at the early stage. This chapter discusses the benefits of legume-rhizobia symbiosis in plant growth and development, in addition to N fixation, through the production of hormones and increased nutrient availability. The role of symbiosis in stress tolerance and plant defense against pathogens and how this function of symbionts can be improved through molecular tools are also highlighted.

2. Plant growth and development

Plant growth and development, which is greatly influenced by the plant's capacity to carry out photosynthetic activity in favorable and adverse regimes of nutrient availability and environmental conditions, determines the plant productivity. The symbiotic relationship with rhizobia is known to promote plant growth characteristics in legume crops (ref. in [13]). Although N supply is an important factor in plant growth, numerous studies have shown that rhizobia promote plant growth by producing plant hormones. These phytohormones have a profound effect on plant cellular processes and play a critical role in plant development [14]. The symbiotic interactions may act at molecular, physiological, and cellular levels to provide such benefits [15]. Several hormones, such as jasmonates, brassinosteroids, salicylic acid, nitric oxide, strigolactones, etc., that significantly influence plant signaling or other cellular responses are not discussed in this chapter as their impact on the symbiotic process or nodulation has previously been discussed [10]. The other phytohormones, which have been traditionally recognized for their role in dictating plant growth and development, are discussed. Their representative forms are depicted in **Figure 1**.

Auxins: These are a group of phytohormones that regulate processes like growth, seed germination, cell division, elongation and differentiation, root development, and stress response [16]. Plant-associated microbes, including *Rhizobium*, *Microbacterium*, *Mycobacterium*, *Bradyrhizobium* and *Sphingomonas* are among the active producers of indole-3-acetic acid (IAA), which is the major form of auxins in plants [17, 18]. Increased production of IAA by *Bradyrhizobium japonicum* was associated with enhanced soybean yield and survival [19]. A similar stimulatory effect was observed in a different host-rhizobial symbiosis. The IAA-overproducing *Ensifer meliloti* (formerly known as *Sinorhizobium meliloti*) strain was linked with the increased plant biomass and photosynthetic activity in *Medicago sativa*. There was higher nitrogenase activity, which possibly increased the demand for carbohydrates to balance the C/N ratio for an overall positive effect on plant growth [20]. The overproduction of IAA through genetic modification of a *Rhizobium leguminosarum* strain also affected nodule development and increased N fixation [21]. The positive effects of IAA-producing rhizobia on plant growth and productivity have been observed in non-legumes. The inoculation of rice (*Oryza sativa L.)* with Rhizobium bacteria led to about a 20% increase in the dry weight of rice along with enhancements in plant growth characteristics [22, 23]. Many of the symbiotic bacteria have developed genetic infrastructure to synthesize IAA, which clearly promotes the interests of the host plant. What could be the benefit of synthesizing IAA for the symbiotic bacteria other than that the host is healthy? Studies indicate that IAA is involved in nodule development [24, 25]. Nodulated roots are found to be enriched in IAA compared to non-nodulated roots [26]. The exact role of IAA in nodule formation is not known, but the hormone may act as a signal molecule that promotes host-bacterial interaction and attachment of the latter to the host roots [27, 28]. This is supported by an earlier observation that bacteria with a higher level of IAA are found closer to the root rhizosphere than the non-producers that are distantly distributed in the soil [29]. Trehalose (α-D-glucopyranosyl-1,1-α-D-glucopyranoside) synthesis and degradation are tightly

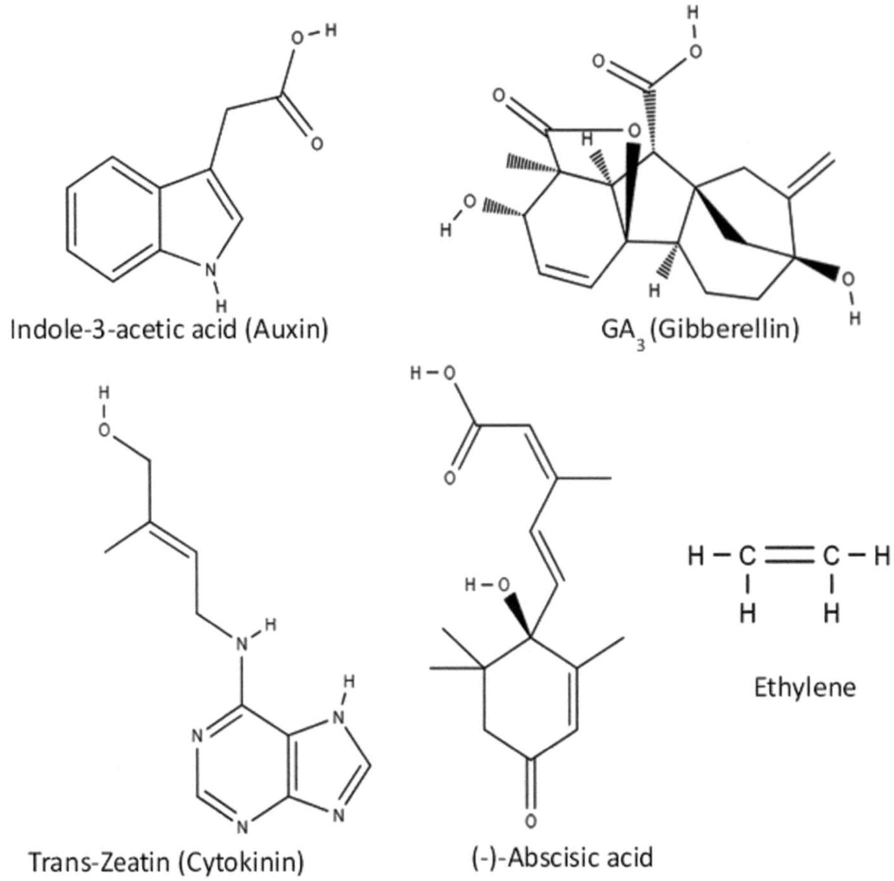

Figure 1.
The representative structure of some plant hormones. Source: PubChem.

regulated during nodule development, suggesting its involvement in nodule formation [30, 31]. The inoculation of soybean with IAA-producing Rhizobium upregulated the expression of the *otsA* gene along with an increase in nodule biomass [32]. The gene encodes trehalose-6-phosphate synthase, which synthesizes the precursor of trehalose. Similarly, the IAA-overproducing *RD64* strain of *E. meliloti* accumulated more trehalose in the nodules of *Medicago truncatula* [33]. The mutualistic benefit of IAA production by rhizobia is also observed through the regulation of N fixation genes by the hormone. There was an up-regulation of the N fixation apparatus by the IAA-overproducing endophytes [34]. A higher level of IAA in the bacteroids was associated with more abundance of transcripts of a N fixation gene, *nifH* [20].

Cytokinins (CKs): They are another group of plant hormones that regulate various aspects of plant growth and differentiation. They are implicated in cell division, seed germination, the formation of roots and shoots, the development of reproductive organs, apical dominance, senescence, stress tolerance, and more, covering almost all aspects of the plant life cycle. Earlier CKs were thought to be synthesized only in roots, but now it has been demonstrated that many plant parts synthesize the hormone. In legume roots, they are involved in nodule organogenesis. The exogenous application of CKs induced nodule-like structures and mutation in CK receptors, histidine kinase in *Lotus japonicus* hit1-1 (lhk1) and *M. truncatula* Cytokinin Response 1 (CRE1) inhibited the development of such structures (reviewed in ref. [10, 35]). The *M. truncatula* genome contains many ATP-binding cassette (ABCG) transporter proteins that enable the movement of hormones to an extracellular environment. One such transporter, MtABCG56, showed root and nodule specific expression, and the transcript level of the gene increased significantly after inoculation with *E. meliloti*, pointing to its involvement in the initial stages of legume-rhizobia symbiosis [36]. CK production was observed in *B. japonicum* [37] and secretion of their bioactive forms was demonstrated in *E. meliloti* [38]. Cytokinins are recognized as signal molecule for nodule development, but they seem to be involved in auto regulation of nodule number (AON). Plants sense the N availability in soil and control nodule formation through AON mechanism. Low N promotes nodulation, however when in abundance, the plant avoids energy intensive symbiotic N and reduce the number of nodules. A reduction in nodule number by a high concentration of CK treatment suggests a role of this hormone in AON [39], but such a possibility needs to be confirmed through further investigation. CKs are synthesized by rhizobia, but the exact role of rhizobial-derived cytokinins is unknown. It has been demonstrated that rhizobial synthesized cytokinins contribute to symbiotic interaction but were not critically required for the interaction as the *Bradyrhizobium sp. strain ORS285* mutant deficient in CK production was able to form nodules [40]. The role of CKs and other hormones in nodule development has been comprehensively reviewed [41]. Although hormones enhance plant growth in soybean and non-leguminous crops (ref. in [42]), it is not clearly established that rhizobial-derived cytokinins are transported to the plant tissue where they influence growth and differentiation.

Gibberellins (GAs): There is a large family of GAs that possess a diterpenoid structure. More than 125 GAs have been identified in plants and other organisms such as fungi and bacteria, but only a small portion of them is biologically active. GAs participates in the regulation of a wide range of plant developmental processes. The most notable are breaking the seed dormancy and enabling its germination; stem elongation; changing the vegetative phase of the plant to the reproductive stage; flower and fruit development; seed development; and senescence. The effect of GAs is mediated by their interaction with DELLA proteins. The hormone is involved in DELLA protein degradation. A certain level of these proteins is required to promote infection through an interaction with the Nod factor mediated pathway. Several mutant studies that interfered with GA biosynthesis, DELLA protein functionality, and the application of the hormone suggested that the GAs can have opposite effects on nodule development depending on the dose and mode of application [43]. In pea, a low concentration of GA application had a positive effect on nodule formation, while a higher amount inhibited the infection thread, which is one of the initial structural changes in the root to acquire the bacterium [44]. The identification of cytochromes P450 in *B. japonicum* indicated the presence of a GA biosynthetic capability in symbiotic organisms, though it was not found to be essential for a symbiotic function [45]. The protein catalyzes the steps of the pathways of many metabolites, including GAs. Subsequently, the genomic analysis of the bacterium identified diterpenoid operon which plays an important role in GA biosynthesis [46]. Later, many rhizobial species were reported to synthesize bioactive GAs (ref. in [47]). A cooperative action of both the symbiont and host led to GA biosynthesis in *Bradyrhizobium diazoefficiens*, where a precursor of a GA is converted to its bioactive form, GA9, by the host GA-3 oxidase [48]. It is speculated that rhizobial GA biosynthesis in cooperation with the host was an evolutionary event that influences nodulation in a mutually beneficial manner. Besides GAs effect on nodule formation, the exogenous application of GA3 in pea delayed the nodule senescence [49]. It is not known if the role of rhizobial-driven GAs is restricted to nodule development and senescence or if they influence other aspects of plant development.

Abscisic acid (ABA) and ethylene: Both of these hormones are central to plant defense mechanism against a variety of stresses. ABA regulates seed dormancy, senescence, stomatal opening and closing, and many other aspects of plant development. It is synthesized as a stress signal, particularly during abiotic stresses, and plays a critical role in downstream responses. Many transcription factors, including DREB2A and 2B, AREB1, RD22BP1 and MYC/ MYB, are known to regulate ABA-responsive gene expression through interactions with cis-acting elements on the DNA. The hormone alters the root architecture under water deficient conditions and modulates the ion influx to change root osmotic potential, providing drought tolerance to the plant [50]. Indirect evidence suggests that ABA might be involved in the host's drought tolerance. The inoculation of chickpea with *R. leguminosarum* led to an enhanced tolerance against drought, which was accompanied by higher synthesis of auxins and GAs with a lower level of ABA [51]. Similarly, in a symbiotic interaction of alfalfa with a IAA-overproducing Rhizobium strain, the host plant showed better adaptation to drought conditions that correlated with a higher level of IAA and downregulation of ABA biosynthesis genes [34]. ABA is known to affect nodule development. Exogenous ABA inhibited nodule formation, which appears to be mediated early in a symbiotic interaction [10]. In a report, GmbZIP1 has been shown to negatively regulate ABA-induced inhibition of nodulation by targeting GmENOD40-1 in soybean [52]. GmbZIP1 is a transcription factor that belongs to the AREB/ABF family, and GmENOD40-1 is a marker gene for nodule formation.

The role of ethylene in nodule formation has been reviewed previously [7, 53]. Ethylene is a negative regulator of nodulation, and interference in ethylene signaling led to stimulation of nodulation in alfalfa [54]. Further, several mutants and transgenic plants with altered ethylene levels and signaling showed variation in nodule size and number [53]. Ethylene is synthesized from its immediate substrate, 1-aminocyclopropane-1-carboxylic acid (ACC), through oxidation. The hormone level could be reduced by the degradation of ACC through a reaction catalyzed by acdS encoded ACC deaminase. A higher level of acdS expression in *B. japonicum* strains was linked with higher numbers of nodules, higher biomass, and delayed senescence of the nodules. The negative effect of ethylene was also demonstrated through over-expressing the ACC deaminase gene in pea and alfalfa (ref. in [7]).

3. Mineral acquisition

Most nutrients available in natural ecosystems are minimally bioavailable to plants since they are bound to inorganic molecules. Microorganisms like bacteria and fungi have the machinery necessary to depolymerize and mineralize the organic forms of these nutrients. With that, inorganic N, P, and S, including ionic species like ammonium, nitrate, phosphate, and sulfate, become available to the plant [55]. Besides the ability to fix N, soil microorganisms also provide essential nutrients by metabolizing recalcitrant forms of N, P and S. As reviewed in [56], these nutrients are needed in several metabolic processes of microorganisms including protein depolymerization and urea catabolism, phosphate and sulfate ester cleavage and phosphonate and sulfonate breakdown. Here, we discuss the role of rhizobial symbionts in P and Fe acquisition by legumes (**Figure 2**).

P availability: P is among the macronutrients essential for plant growth and development; legumes may require a greater amount of P to support N fixation in the nodules. This is likely related to the additional requirement for various metabolic processes, membrane phospholipid biosynthesis, and energy transformations in the nodules initiated by the symbiotic interaction [57]. P limitation can alter the distribution of nodule phosphatidylcholine [58], an essential component of membrane phospholipids [59]. It has also been associated with small nodules and reduced activity of Fe–S cluster enzymes [60], which are important in various regulatory systems for environmental signals [61]. The amount of plant-available P in the soil is usually low. Thus, owing to the high P requirements of legumes along with the low availability

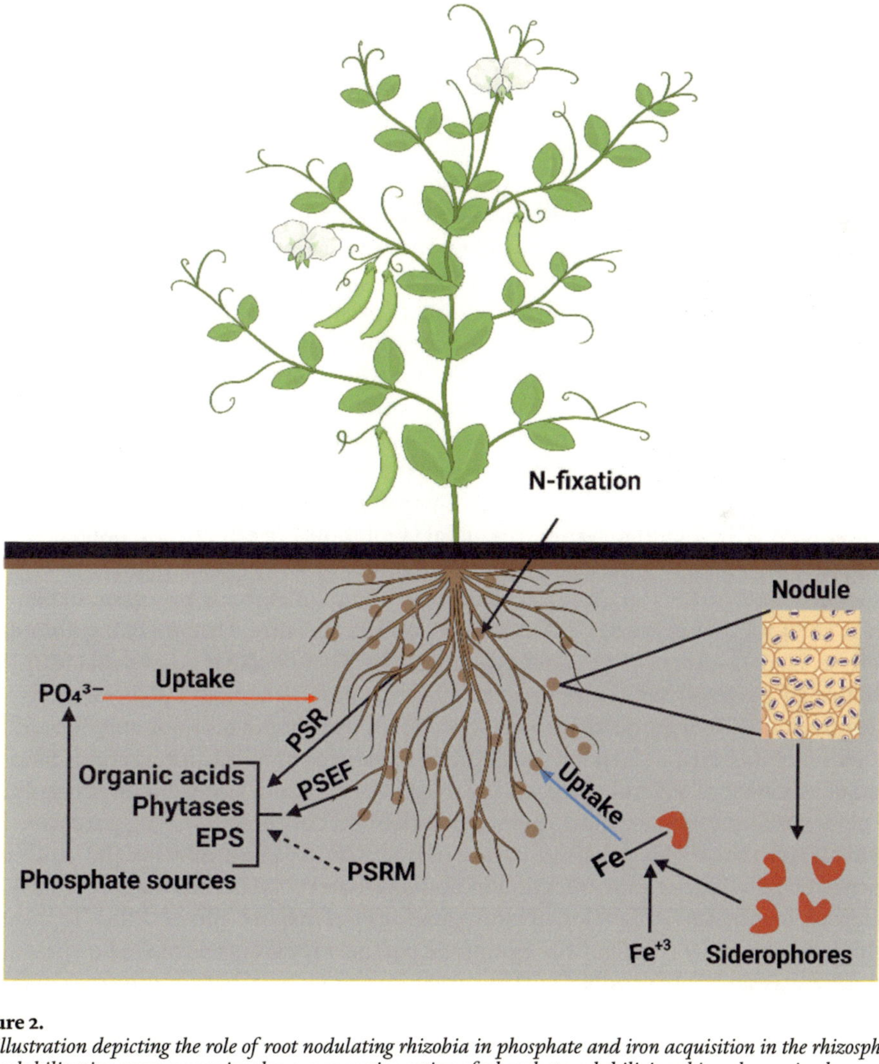

Figure 2.
An illustration depicting the role of root nodulating rhizobia in phosphate and iron acquisition in the rhizosphere. The solubilization process may involve a cooperative action of phosphate-solubilizing rhizosphere microbes (PSRM) and phosphate solubilizing endomycorrhizal fungi (PSEF) with rhizobia, using different organic acids, phytases, and extracellular polymeric substances (EPS). The red and blue arrows show phosphate and iron uptake by plant roots, respectively.

of soluble P in the soil, additional P solubilization and uptake strategies are vital for effective symbiosis. Like many rhizosphere microbes [62], rhizobial symbionts are often associated with P solubilizations [63–68]. By determining the expression level of high-affinity phosphate transporter and mutant analysis, the role of P uptake has been implicated in nitrogen fixation in soybean-*Sinorhizobium fredii* symbiosis and terminal differentiation of nodules in other rhizobial strains [69]. These studies suggest that rhizobial symbionts are important in increasing the availability and uptake of P during legume-rhizobium symbiosis.

Rhizobia-induced phosphate solubilization may involve similar mechanisms that are deployed by other rhizosphere microorganisms, such as the production of acids and phytases, proton extrusion, and extracellular oxidation [70] (**Figure 2**). Although an external application of succinic acid inhibits P solubilization in legumes (e.g., chickpea) [71], several organic acids of microbial origin have been shown to participate in P solubilization in soil [70, 72]. Gluconic acid is one of the P-solubilizing organic acids produced by root-associated rhizobia [63, 68, 73, 74]. Its production by non-rhizobial bacteria including *Azospirillum* and *Nguyenibacter* is involved in the solubilization of calcium phosphate and aluminum phosphate [73, 74]. Besides gluconic acid, the solubilization of Ca and Fe phosphates could also be mediated through other organic acids [68, 75]. Alkaline phosphatases (e.g., *Sinorhizobium meliloti* 104A14) are the other means to solubilize organic phosphates [64, 76]. It is important to note that the presence of non-rhizobial phosphate-solubilizing microorganisms in the rhizosphere can also have a significant impact on the availability of P to legumes, thereby contributing to their increased productivity [77]. Co-inoculation of *Pseudomonas putida* strains and *B. japonicum* TIIIB resulted in greater root and shoot biomass in soybean compared to the rhizobial inoculation alone [78]. Regardless of the role of non-rhizobial microbes, the phosphate solubilization ability of symbiotic rhizobia ensures P availability to support N fixation. During symbiosis, rhizobia might have developed regulation strategies for demand-based P uptake. For example, using a *PhoCDET* phosphate transport system that involves periplasmic binding proteins, *S. meliloti* may ensure P availability when environmental phosphate concentration is below 1 μM [79]. Taken together, rhizobial symbionts are effective in phosphate solubilization, which in turn ensures the availability of P to legume plants.

Iron availability: Fe is an essential element in several metabolic processes in plants such as chlorophyl synthesis, DNA synthesis, and respiration. Despite being abundant in soil, an insoluble form of Fe requires plant strategies to make it absorbable [80]. The limitation of Fe in soil has been shown to affect the growth of many plants including legumes [81]. In legume-rhizobium symbiosis, Fe is important not only for each partner's metabolism but also for maintaining effective symbiotic interaction. Because Fe is a component of several enzymes involved in N-fixation, this mineral is required in a higher amount in a symbiotic interaction than that of the free-living partners. This is evident from the highly sensitive nature of nodulation to Fe limitation [82] and the presence of Fe in higher concentration in nodules than roots, leaves, and seeds [83]. Therefore, legumes have developed strategies for Fe transport into nodules. To meet the high demand of Fe, each of the symbiotic partners is playing a role in Fe uptake. *S. meliloti* was shown to produce volatile organic compounds that function as signals for *Medicago truncatula* to increase its Fe uptake rate [84]. Rhizobial symbionts, on the other hand, released Fe reducing enzymes to acidify the rhizosphere, which facilitated Fe-uptake by the plant [85]. Recently, a study on *M. truncatula* observed an interesting phenomenon: the plant expressed protein signals, which activated the Fe import systems of *S. meliloti* by overriding the Fe-regulating system, *sufT* [86].

One of the common Fe import systems in rhizobia and many other microorganisms is the secretion of siderophores, which are low molecular-mass non-protein molecules that can scavenge Fe^{3+} (**Figure 2**). Under Fe deficiency, rhizobia can secrete siderophores, which form a soluble complex with Fe^{3+} and are imported into the cells via receptors present on the outer membrane surfaces of this Gram-negative bacteria. Rhizobia express genes encoding enzymes involved in siderophore production as well as outer membrane receptor proteins of siderophore-Fe complexes when Fe is limited. For example, enzymes involved in the biosynthesis of a catechol siderophore are encoded by *sid* genes of a rhizobial strain that can nodulate chickpea [87]. Soybean-nodulating *B. japonicum* was also shown to express the *FegA* gene, which encodes an outer membrane receptor protein for siderophore-Fe complexes [88]. From the success in root colonization, biomass, and chlorophyll content enhancement with genetically improved rhizobia, it may now be possible to improve strains having high-affinity siderophore-receptor proteins to increase the Fe uptake in legumes [89].

4. Environmental adaptability

Besides nutrient limitation, other unfavorable conditions such as, high concentrations of heavy metals, drought, soil salinity, extreme pH conditions, and diseases, negatively affect the growth and development of plants, resulting in a significant reduction in the yield of agricultural produce. With increasing anthropogenic activities, the environmental conditions are becoming unfavorable for plant growth. The chemical industries, for example, generate a tremendous amount of heavy metal waste (e.g., Ni, As, Hg, Cd, Pb, Cr, Mn, Si, Fe, and Cu). When left untreated, industrial waste becomes a source of heavy metal contamination in the environment. The negative effects of heavy-metal-contaminated soils (HMCS) on plant productivity are evident. In legumes, it resulted in poor nodulation and plant growth [90]. Nevertheless, some legumes can effectively nodulate and grow in HMCS, suggesting a protective effect of rhizobial symbionts against many of these metals. The *Rhizobium*-induced improvements in nodulation, root and shoot lengths, N-fixation, and seed yield were observed in legumes grown in soil contaminated with Ni (290 mg Ni kg^{-1} soil) [91], As [92–94], Hg [95], Cd [96], Cr [97], Cu [98], Al [99], and Pb [100].

The rhizobia display a natural adaptation to stress, as many of the isolates in HMCS showed a significant potential for heavy metal stress tolerance [101]. A number of the rhizobial strains isolated from Hg-contaminated soils showed tolerance to the metal [102]. The variation in tolerance across the strains was linked with the mercuric reductase activity that converts Hg^{2+} into its less toxic and a volatile form, Hg^0. The adaptive capacity is enhanced through cooperative action among different suites of the microflora. This has been demonstrated in faba bean plants, where a combination of Bacillus strains with those of *Rhizobium* increased the biosorption of Ni and improved plant growth [103]. Similarly, co-inoculation of rhizobia and non-rhizobial bacteria has been shown to improve tolerance against Pb [104], As [94, 105], and Cu [106, 107]. The tolerance mechanism is not well understood, but metal biosorption by rhizobia, accumulation of metals in the roots, volatilization, conversion to less toxic forms, and rhizobia-induced root uptake regulation are some of the ways the symbionts are known to alleviate the metal toxicity. Recently, a molecular investigation into metal tolerance mechanisms revealed changes in m6A RNA methylation in response to Cd stress [101, 108]. The genes that were differentially methylated and expressed were linked to Ca^{2+} homeostasis, ROS pathways, polyamine metabolism, MAPK signaling, hormones, and biotic stress responses. In another study, the primary targets of Al were genes involved in membrane biogenesis, metal ions binding and transport, carbohydrate, and amino acid metabolism and transport [99]. The *Rhizobium* strain B3, which was tolerant to Al toxicity, bound less of the ion compared to the wild-type and maintained membrane integrity. Through recombinant technology, the incorporation of mechanisms of heavy metal tolerance from unrelated species can lead to genetic improvements in the symbiotic rhizobia [109].

Drought stress can reduce nodulation, biomass, and chlorophyll contents of legumes while accumulating ROS that disrupt the structure and function of different biomolecules including DNA [110–112]. Soil salinity negatively impacts rhizobial infections and root nodulations in several legumes: bean [113], soybean [114], pea [115], and chickpea [116]. Some legumes can cope with the adverse conditions, suggesting a positive role for rhizobia in the plant's adaptation to stress. The observed protection against salinity stress by symbionts was due to the production of osmoprotectant molecules (e.g., glutamine, serine, glutamate, and proline) [117], antioxidants [118] and by changing the xylem osmotic potential and amount of aquaporins [119]. Both salinity and drought stress limit the availability of water to plants. Like salinity stress, there are many examples where the protective effect of rhizobia-host symbiosis is evident in the management of drought stress [120]. Both salinity and drought stress responses are complex and may involve non-rhizobial plant growth promoting rhizobacteria and endomycorrhiza [116–119]. On the other hand, extreme pH can have a direct impact on the establishment of symbiosis and thus on the N fixation and productivity of legumes [121, 122]. Soil pH can also influence the diversity and structure of the microbial community around the root rhizosphere, which could affect the legume-rhizobia symbiosis [123–125]. In some legumes, acidic conditions induced the expression of different rhizobial genes (e.g., *actA, typA, atvA, lpiA,* and *ubiF*), which may be playing a role in acid stress tolerance mechanism [126–130].

Symbiotic interactions are known to enhance growth and survival of the host plant against fungal and bacterial pathogens. The anti-phytopathogenic activities of different rhizobial species such as those observed in chickpea against the oomycete pathogen *Phytophthora medicaginis* [131], pigeon pea against *Fusarium* wilt diseases [132], and soybean against root rot diseases [133] have been reported.

5. Strain improvement for enhanced productivity

Advances in molecular biology have allowed genetic improvements in rhizobial strains to increase the symbiotic benefits. Rhizobial strains have been modified to increase N fixation in legumes and some cereals [7, 134] as well as to improve the growth, nutrient supply, and stress tolerance [135, 136]. The overproduction of IAA in *S. meliloti RD64* by introducing a chimeric operon containing genes from *Agrobacterium rhizogenes*, *Pseudomonas syringae pv. savastanoi*, and *Agrobacterium tumefaciens* [137] was associated with the increased production of acid phosphatase and organic acids, resulting in a higher growth of *M. truncatula* under P limitation [138]. Another *Rhizobium meliloti* strain, when modified with extra copies of the *nifA* and *dctABD* genes increased the alfalfa biomass by 13% [139]. The *nifA* gene is involved in nitrogen fixation and *dctABD* gene product is a dicarboxylate transporter. The latter is a primary carbon used by bacteria for energy and other cellular functions. An improvement in symbiotic efficiency can also lead to enhanced plant growth. This has been shown by transferring *Smed_3503*, *Smed_5985*, and *Smed_6456*, into the *S. meliloti Rm1021* strain, which led to a 60% increase in *M. truncatula* biomass [140].

Besides growth and yield, many studies have successfully demonstrated the potential of genetically improved rhizobial symbionts in alleviating adverse environmental conditions in the host legumes. The tolerance mechanisms of these microbes could be exploited for genetic engineering approaches in symbionts. By using this strategy, a *Rhizobium leguminosarum* strain transformed with an *arsenite S-adenosylmethionine methyltransferase* gene of algae, effectively methylated, and volatilized As, to a tolerance against this metal [109]. Similarly, manipulating a superoxide dismutase level reduced Cd stress and increased root and shoot length, biomass, and N content in soybean [135]. The strains have been genetically modified to overproduce cytokinin, IAA, trehalose-6-phosphate synthase, 1-aminocyclopropane-1-carboxylic acid deaminase, high-affinity cytochrome cbb3-type oxidase, etc., to increase salt or drought tolerance or to overcome the waterlogging stress in legumes [32, 34, 54, 137, 138, 141, 142]. Strain improvement followed by extensive evaluation of the strains for survival under field conditions is also critical, as gene modification may reduce the strains' competitiveness against the indigenous strains and hence their survival rate in the nodules [143].

6. Future perspectives

The rhizobial symbiotic interactions are extremely valuable for the sustainability of crop cultivation. This has gained more importance in view of climate change and its imminent threat to the ecological balance. Leguminous crops make a significant contribution to reducing the use of chemical N-fertilizers. The positive impact of commercial and wild symbiotic hosts is enhanced by their growing abilities in diverse environments. The different rhizobial species indirectly increase the N fixation productivity by promoting the growth and productivity of the host plants. A substantial number of benefits other than N fixation point to close coordination between the host and the symbiont while acquiring this capacity. The horizontal transfer of genes appears to have played an important role in improving the symbiotic functionality of the bacterium [144, 145]. Gaining insight into these evolutionary events could provide a broader base for strain improvement. The genetic engineering tools have demonstrated how specific bacterial functions can be transferred across even unrelated species. Further understanding of a symbiotic interaction can bring us closer to developing tailor-made strains.

Acknowledgements

The funding provided by Alberta Pulse Growers and Alberta Results Driven Agriculture Research is acknowledged.

Conflict of interest

The authors do not have any conflict of interests.

Author details

Jemaneh Z. Habtewold and Ravinder K. Goyal*
AAFC-Lacombe Research and Development Centre, Lacombe, AB, Canada

*Address all correspondence to: ravinder.goyal@agr.gc.ca

References

[1] Wang Q, Liu J, Zhu H. Genetic and molecular mechanisms underlying symbiotic specificity in legume-rhizobium interactions. Frontiers in Plant Science. 2018;**9**:313. DOI: 10.3389/fpls.2018.00313

[2] Bruijn F, J: Biological nitrogen fixation. Hoboken. New Jersey: Wiley-Blackwell; 2015

[3] Martínez-Romero E. Coevolution in rhizobium-legume symbiosis? DNA and Cell Biology. 2009;**28**(8):361-370. DOI: 10.1089/dna.2009.0863

[4] Coba de la Peña T, Fedorova E, Pueyo JJ, Lucas MM. The symbiosome: Legume and rhizobia co-evolution toward a nitrogen-fixing organelle? Front. Plant Science. 2017;**8**:2229. DOI: 10.3389/fpls.2017.02229

[5] Goyal R, Mattoo A, Schmidt A. Rhizobial–host interactions and symbiotic nitrogen fixation in legume crops toward agriculture sustainability. Frontiers in Microbiology. 2021;**12**:669404. DOI: 10.3389/fmicb.2021.669404

[6] Raun WR, Johnson GV. Improving nitrogen use efficiency for cereal production. Journal of Agronomy. 1999;**91**(3):357-363

[7] Goyal R, Schmidt A, Hynes M. Molecular biology in the improvement of biological nitrogen fixation by rhizobia and extending the scope to cereals. Microorganisms. 2021;**9**:125. DOI: 10.3390/microorganisms9010125

[8] Nag P, Shriti S, Das S. Microbiological strategies for enhancing biological nitrogen fixation in nonlegumes. Journal of Applied Microbiology. 2020;**129**(2):186-198. DOI: 10.1111/jam.14557

[9] Patten CL, Glick BR. Role of *pseudomonas putida* indoleacetic acid in development of the host plant root system. Applied and Environmental Microbiology. 2002;**68**(8):3795-3801. DOI: 10.1128/aem.68.8.3795-3801.2002

[10] Ferguson BJ, Mathesius U. Phytohormone regulation of legume-rhizobia interactions. Journal of Chemical Ecology. 2014;**40**(7):770-790. DOI: 10.1007/s10886-014-0472-7

[11] Rodríguez H, Fraga R. Phosphate solubilizing bacteria and their role in plant growth promotion. Biotechnology Advances. 1999;**17**(4-5):319-339. DOI: 10.1016/s0734-9750(99)00014-2

[12] Errickson W, Huang B. Roles and Mechanisms of Rhizobacteria in Regulating Plant Tolerance to Abiotic Stress. Boca Raton: CRC Press; 2020. pp. 25-39

[13] Jaiswal SK, Mohammed M, Ibny FYI, Dakora FD. Rhizobia as a source of plant growth-promoting molecules: Potential applications and possible operational mechanisms. Frontiers in Sustainable Food Systems. 2020;**4**:619676

[14] Patil A, Kale A, Ajane G, Sheikh R, Patil S. Plant growth-promoting rhizobium: Mechanisms and biotechnological prospective. In: Hansen AP, Choudhary DK, Agrawal PK, Varma A, editors. Rhizobium Biology and Biotechnology. Cham: Springer International Publishing; 2017. pp. 105-134

[15] Kasim W, Osman M, Omar N, Abd El-Daim I, Bejai S, Meijer J. Control of

drought stress in wheat using plant-growth-promoting bacteria. Journal of Plant Growth Regulation. 2012;**32**. DOI: 10.1007/s00344-012-9283-7

[16] Ullah A, Nisar M, Ali H, Hazrat A, Hayat K, Keerio AA, et al. Drought tolerance improvement in plants: An endophytic bacterial approach. Applied Microbiology and Biotechnology. 2019;**103**(18):7385-7397. DOI: 10.1007/s00253-019-10045-4

[17] Etminani F, Harighi B. Isolation and identification of endophytic bacteria with plant growth promoting activity and biocontrol potential from wild pistachio trees. Plant Pathology Journal. 2018;**34**(3):208-217. DOI: 10.5423/ppj.Oa.07.2017.0158

[18] Tsavkelova EA, Cherdyntseva TA, Klimova SY, Shestakov AI, Botina SG, Netrusov AI. Orchid-associated bacteria produce indole-3-acetic acid, promote seed germination, and increase their microbial yield in response to exogenous auxin. Archives of Microbiology. 2007;**188**(6):655-664. DOI: 10.1007/s00203-007-0286-x

[19] Masciarelli O, Llanes A, Luna V. A new PGPR co-inoculated with *Bradyrhizobium japonicum* enhances soybean nodulation. Microbiological Research. 2014;**169**(7):609-615. DOI: 10.1016/j.micres.2013.10.001

[20] Defez R, Andreozzi A, Romano S, Pocsfalvi G, Fiume I, Esposito R, et al. Bacterial IAA-delivery into Medicago root nodules triggers a balanced stimulation of c and n metabolism leading to a biomass increase. Microorganisms. 2019;7(10):403. DOI: 10.3390/microorganisms7100403

[21] Camerini S, Senatore B, Lonardo E, Imperlini E, Bianco C, Moschetti G, et al. Introduction of a novel pathway for IAA biosynthesis to rhizobia alters vetch root nodule development. Archives of Microbiology. 2008;**190**(1):67-77. DOI: 10.1007/s00203-008-0365-7

[22] Chi F, Shen SH, Cheng HP, Jing YX, Yanni YG, Dazzo FB. Ascending migration of endophytic rhizobia, from roots to leaves, inside rice plants and assessment of benefits to rice growth physiology. Applied and Environmental Microbiology. 2005;**71**(11):7271-7278. DOI: 10.1128/aem.71.11.7271-7278.2005

[23] Singh R, Mishra RPN, Jaiswal HK. Role of rhizobial endophytes as nitrogen fixer in promoting plant growth and productivity of Indian cultivated upland rice (*Oryza Sativa L.*). Plants. In: Wang YP, Lin M, Tian ZX, Elmerich C, Newton WE, editors. Biological Nitrogen Fixation, Sustainable Agriculture and the Environment. Current Plant Science and Biotechnology in Agriculture. Dordrecht: Springer; 2005;**41**. ISBN 978-1-4020-3570-8. DOI: 10.1007/1-4020-3570-5_73

[24] Pacios-Bras C, Schlaman HRM, Boot K, Admiraal P, Mateos Langerak J, Stougaard J, et al. Auxin distribution in *Lotus japonicus* during root nodule development. Plant Molecular Biology. 2003;**52**(6):1169-1180. DOI: 10.1023/b:plan.0000004308.78057.f5

[25] Subramanian S, Stacey G, Yu O. Distinct, crucial roles of flavonoids during legume nodulation. Trends in Plant Science. 2007;**12**(7):282-285. DOI: 10.1016/j.tplants.2007.06.006

[26] Ghosh S, Basu PS. Production and metabolism of indole acetic acid in roots and root nodules of *Phaseolus mungo*. Microbiological Research. 2006;**161**(4):362-366. DOI: 10.1016/j.micres.2006.01.001

[27] Spaepen S, Das F, Luyten E, Michiels J, Vanderleyden J. Indole-3-acetic acid-regulated genes in *rhizobium*

etli CNPAF512. FEMS Microbiology Letters. 2009;**291**(2):195-200. DOI: 10.1111/j.1574-6968.2008.01453.x

[28] Torres D, Benavidez I, Donadio F, Mongiardini E, Rosas S, Spaepen S, et al. New insights into auxin metabolism in *Bradyrhizobium japonicum*. Research in Microbiology. 2018;**169**(6):313-323. DOI: 10.1016/j.resmic.2018.04.002

[29] Sarwar M, Kremer RJ. Determination of bacterially derived auxins using a microplate method. Letters in Applied Microbiology. 1995;**20**(5):282-285. DOI: 10.1111/j.1472-765X.1995.tb00446.x

[30] Bargaz A, Lazali M, Amenc L, Abadie J, Ghoulam C, Farissi M, et al. Differential expression of trehalose 6-P phosphatase and ascorbate peroxidase transcripts in nodule cortex of *Phaseolus vulgaris* and regulation of nodule O2 permeability. Planta. 2013;**238**(1):107-119. DOI: 10.1007/s00425-013-1877-1

[31] Barraza A, Estrada-Navarrete G, Rodriguez-Alegria ME, Lopez-Munguia A, Merino E, Quinto C, et al. Down-regulation of PvTRE1 enhances nodule biomass and bacteroid number in the common bean. The New Phytologist. 2013;**197**(1):194-206. DOI: 10.1111/nph.12002

[32] Suárez R, Wong A, Ramírez M, Barraza A, Orozco Mdel C, Cevallos MA, et al. Improvement of drought tolerance and grain yield in common bean by overexpressing trehalose-6-phosphate synthase in rhizobia. Molecular Plant-Microbe Interactions. 2008;**21**(7):958-966. DOI: 10.1094/mpmi-21-7-0958

[33] Bianco C, Defez R. *Medicago truncatula* improves salt tolerance when nodulated by an indole-3-acetic acid-overproducing *Sinorhizobium meliloti* strain. Journal of Experimental Botany. 2009;**60**(11):3097-3107. DOI: 10.1093/jxb/erp140

[34] Defez R, Andreozzi A, Dickinson M, Charlton A, Tadini L, Pesaresi P, et al. Improved drought stress response in alfalfa plants nodulated by an IAA over-producing rhizobium strain. Frontiers in Microbiology. 2017;**8**:2466. DOI: 10.3389/fmicb.2017.02466

[35] Gamas P, Brault M, Jardinaud MF, Frugier F. Cytokinins in symbiotic nodulation: When, where, what for? Trends in Plant Science. 2017;**22**(9):792-802. DOI: 10.1016/j.tplants.2017.06.012

[36] Jarzyniak K, Banasiak J, Jamruszka T, Pawela A, Di Donato M, Novák O, et al. Early stages of legume-rhizobia symbiosis are controlled by ABCG-mediated transport of active cytokinins. Nature Plants. 2021;**7**(4):428-436. DOI: 10.1038/s41477-021-00873-6

[37] Sturtevant D, Taller B. Cytokinin Production by *Bradyrhizobium japonicum*. Plant Physiology. 1989;**89**:1247-1252. DOI: 10.1104/pp.89.4.1247

[38] Kisiala A, Laffont C, Emery RJ, Frugier F. Bioactive cytokinins are selectively secreted by *Sinorhizobium meliloti* nodulating and nonnodulating strains. Molecular Plant-Microbe Interactions. 2013;**26**(10):1225-1231. DOI: 10.1094/mpmi-02-13-0054-r

[39] Mens C, Li D, Haaima LE, Gresshoff PM, Ferguson BJ. Local and systemic effect of cytokinins on soybean nodulation and regulation of their isopentenyl transferase (IPT) biosynthesis genes following rhizobia inoculation. Frontiers in Plant Science. 2018;**9**:1150. DOI: 10.3389/fpls.2018.01150

[40] Podlešáková K, Fardoux J, Patrel D, Bonaldi K, Novák O, Strnad M, et al. Rhizobial synthesized cytokinins contribute to but are not essential for the symbiotic interaction between photosynthetic *Bradyrhizobia* and

Aeschynomene legumes. Molecular Plant-Microbe Interactions. 2013;**26**(10):1232-1238. DOI: 10.1094/mpmi-03-13-0076-r

[41] Velandia K, Reid JB, Foo E. Right time, right place: The dynamic role of hormones in rhizobial infection and nodulation of legumes. Plant Communications. 2022;**3**(5):100327. DOI: 10.1016/j.xplc.2022.100327

[42] Swarnalakshmi K, Yadav V, Tyagi D, Dhar DW, Kannepalli A, Kumar S. Significance of plant growth promoting rhizobacteria in grain legumes: Growth promotion and crop production. Plants (Basel). 2020;**9**(11):1596. DOI: 10.3390/plants9111596

[43] McGuiness PN, Reid JB, Foo E. The role of gibberellins and brassinosteroids in nodulation and arbuscular mycorrhizal associations. Frontiers in Plant Science. 2019;**10**:269. DOI: 10.3389/fpls.2019.00269

[44] Ferguson BJ, Ross JJ, Reid JB. Nodulation phenotypes of gibberellin and brassinosteroid mutants of pea. Plant Physiology. 2005;**138**(4):2396-2405. DOI: 10.1104/pp.105.062414

[45] Tully RE, Keister DL. Cloning and mutagenesis of a cytochrome P-450 locus from *Bradyrhizobium japonicum* that is expressed anaerobically and symbiotically. Applied and Environmental Microbiology. 1993;**59**(12):4136-4142. DOI: 10.1128/aem.59.12.4136-4142.1993

[46] Kaneko T, Nakamura Y, Sato S, Minamisawa K, Uchiumi T, Sasamoto S, et al. Complete genomic sequence of nitrogen-fixing symbiotic bacterium *Bradyrhizobium japonicum* USDA110. DNA Research. 2002;**9**(6):189-197. DOI: 10.1093/dnares/9.6.189

[47] Keswani C, Singh SP, García-Estrada C, Mezaache-Aichour S, Glare TR, Borriss R, et al. Biosynthesis and beneficial effects of microbial gibberellins on crops for sustainable agriculture. Journal of Applied Microbiology. 2022;**132**(3):1597-1615. DOI: 10.1111/jam.15348

[48] Nett RS, Bender KS, Peters RJ. Production of the plant hormone gibberellin by rhizobia increases host legume nodule size. The ISME Journal. 2022;**16**(7):1809-1817. DOI: 10.1038/s41396-022-01236-5

[49] Serova TA, Tsyganova AV, Tikhonovich IA, Tsyganov VE. Gibberellins inhibit nodule senescence and stimulate nodule meristem bifurcation in pea (*Pisum sativum L.*). front. Plant Science. 2019;**10**:285. DOI: 10.3389/fpls.2019.00285

[50] Sah SK, Reddy KR, Li J. Abscisic acid and abiotic stress tolerance in crop plants. Frontiers in Plant Science. 2016;7:571. DOI: 10.3389/fpls.2016.00571

[51] Bano A, Batool R, Dazzo FB. Adaptation of chickpea to desiccation stress is enhanced by symbiotic rhizobia. Symbiosis. 2010;**50**:129-133

[52] Xu S, Song S, Dong X, Wang X, Wu J, Ren Z, et al. GmbZIP1 negatively regulates ABA-induced inhibition of nodulation by targeting GmENOD40-1 in soybean. BMC Plant Biology. 2021;**21**(1):35. DOI: 10.1186/s12870-020-02810-9

[53] Guinel FC. Ethylene, a hormone at the center-stage of nodulation. Frontiers in Plant Science. 2015;**6**:1121. DOI: 10.3389/fpls.2015.01121

[54] Peters NK, Crist-Estes DK. Nodule formation is stimulated by the ethylene inhibitor aminoethoxyvinylglycine. Plant Physiology. 1989;**91**(2):690-693. DOI: 10.1104/pp.91.2.690

[55] Van Der Heijden MGA, Bardgett RD, Van Straalen NM. The unseen

majority: Soil microbes as drivers of plant diversity and productivity in terrestrial ecosystems. Ecology Letters. 2008;**11**(3):296-310. DOI: 10.1111/j.1461-0248.2007.01139.x

[56] Jacoby R, Peukert M, Succurro A, Koprivova A, Kopriva S. The role of soil microorganisms in plant mineral nutrition-current knowledge and future directions. Frontiers in Plant Science. 2017;**8**:1617. DOI: 10.3389/fpls.2017.01617

[57] Mitran T, Meena RS, Lal R, Layek J, Kumar S, Datta R. In: Meena A, Das GSY, Lal R, editors. Role of Soil Phosphorus on Legume Production, in Legumes for Soil Health and Sustainable Management, R.S. Springer Singapore: Singapore; 2018. pp. 487-510

[58] Dokwal D, Romsdahl TB, Kunz DA, Alonso AP, Dickstein R. Phosphorus deprivation affects composition and spatial distribution of membrane lipids in legume nodules. Plant Physiology. 2021;**185**(4):1847-1859. DOI: 10.1093/plphys/kiaa115

[59] Botella C, Jouhet J, Block MA. Importance of phosphatidylcholine on the chloroplast surface. Progress in Lipid Research. 2017;**65**:12-23. DOI: 10.1016/j.plipres.2016.11.001

[60] Qin L, Wang M, Chen L, Liang X, Wu Z, Lin Z, et al. Soybean Fe-S cluster biosynthesis regulated by external iron or phosphate fluctuation. Plant Cell Reports. 2015;**34**(3):411-424. DOI: 10.1007/s00299-014-1718-0

[61] Mettert EL, Kiley PJ. Fe-S proteins that regulate gene expression. Biochimica et Biophysica Acta. 2015;**1853**(6):1284-1293. DOI: 10.1016/j.bbamcr.2014.11.018

[62] Kucey RMN, Janzen HH, Leggett ME. Microbially mediated increases in plant-available phosphorus. Advances in Agronomy. 1989;**42**:198-228

[63] Alikhani HA, Saleh-Rastin N, Antoun H. Phosphate solubilization activity of rhizobia native to Iranian soils. Plant and Soil. 2006;**287**(1):35. DOI: 10.1007/s11104-006-9059-6

[64] Deng S, Elkins JG, Da LH, Botero LM, McDermott TR. Cloning and characterization of a second acid phosphatase from *Sinorhizobium meliloti* strain 104A14. Archives of Microbiology. 2001;**176**(4):255-263. DOI: 10.1007/s002030100311

[65] Kang X, Yu X, Zhang Y, Cui Y, Tu W, Wang Q, et al. Inoculation of *Sinorhizobium saheli YH1* leads to reduced metal uptake for *Leucaena leucocephala* grown in mine tailings and metal-polluted soils. Frontiers in Microbiology. 2018;**9**:1853. DOI: 10.3389/fmicb.2018.01853

[66] Kumar G, Ram R. Phosphate solubilizing rhizobia isolated from *Vigna trilobata*. American Journal of Microbiological Research. 2014;**2**:105-109. DOI: 10.12691/ajmr-2-3-4

[67] Sridevi M, Mallaiah KV. Phosphate solubilization by rhizobium strains. Indian Journal of Microbiology. 2009;**49**(1):98-102. DOI: 10.1007/s12088-009-0005-1

[68] Yadav A, Singh RP, Singh AL, Singh M. Identification of genes involved in phosphate solubilization and drought stress tolerance in chickpea symbiont *Mesorhizobium ciceri Ca181*. Archives of Microbiology. 2021;**203**(3):1167-1174. DOI: 10.1007/s00203-020-02109-1

[69] Hu Y, Jiao J, Liu LX, Sun YW, Chen WF, Sui XH, et al. Evidence for phosphate starvation of rhizobia without terminal differentiation in legume nodules. Molecular Plant-Microbe Interactions. 2018;**31**(10):1060-1068. DOI: 10.1094/MPMI-02-18-0031-R

[70] Rawat P, Das S, Shankhdhar D, Shankhdhar SC. Phosphate-solubilizing microorganisms: Mechanism and their role in phosphate solubilization and uptake. Journal of Soil Science and Plant Nutrition. 2020;**21**(1):49-68. DOI: 10.1007/s42729-020-00342-7

[71] Iyer B, Rajput MS, Rajkumar S. Effect of succinate on phosphate solubilization in nitrogen fixing bacteria harbouring chickpea and their effect on plant growth. Microbiological Research. 2017;**202**:43-50. DOI: 10.1016/j.micres.2017.05.005

[72] Alori ET, Glick BR, Babalola OO. Microbial phosphorus solubilization and its potential for use in sustainable agriculture. Frontiers in Microbiology. 2017;**8**:971. DOI: 10.3389/fmicb.2017.00971

[73] Li XL, Zhao XQ, Dong XY, Ma JF, Shen RF. Secretion of gluconic acid from *Nguyenibacter sp. l1* is responsible for solubilization of aluminum phosphate. Frontiers in Microbiology. 2021;**12**:784025. DOI: 10.3389/fmicb.2021.784025

[74] Rodriguez H, Gonzalez T, Goire I, Bashan Y. Gluconic acid production and phosphate solubilization by the plant growth-promoting bacterium *Azospirillum spp*. Die Naturwissenschaften. 2004;**91**(11):552-555. DOI: 10.1007/s00114-004-0566-0

[75] Marra LM. Fonsêca Sousa Soares C R, de Oliveira S M, Avelar Ferreira P a, Lima Soares B, de Fráguas Carvalho R, de Lima J M, and de Souza Moreira F M: Biological nitrogen fixation and phosphate solubilization by bacteria isolated from tropical soils. Plant and Soil. 2012;**357**(1-2):289-307. DOI: 10.1007/s11104-012-1157-z

[76] Cerecetto V, Beyhaut E, Amenc L, Trives C, Altier N, Drevon J-J. Contrasting expression of rhizobial phytase in nodules of two soybean cultivars grown under low phosphorus availability. Frontiers in Sustainable Food Systems. 2021;**4**. DOI: 10.3389/fsufs.2020.607678

[77] Prabhu N, Borkar S, Garg S. Phosphate solubilization by microorganisms: Overview, mechanisms, applications and advances (Chapter 11). Advances in Biological Science Research: A Practical Approach. 2019;161-176. DOI: 10.1016/B978-0-12-817497-5.00011-2

[78] Rosas SB, Andrés JA, Rovera M, Correa NS. Phosphate-solubilizing *pseudomonas putida* can influence the rhizobia–legume symbiosis. Soil Biology and Biochemistry. 2006;**38**(12):3502-3505. DOI: 10.1016/j.soilbio.2006.05.008

[79] Bardin S, Dan S, Osteras M, Finan TM. A phosphate transport system is required for symbiotic nitrogen fixation by *rhizobium meliloti*. Journal of Bacteriology. 1996;**178**(15):4540-4547. DOI: 10.1128/jb.178.15.4540-4547.1996

[80] Colombo C, Palumbo G, He J-Z, Pinton R, Cesco S. Review on iron availability in soil: Interaction of Fe minerals, plants, and microbes. JSS. 2013;**14**(3):538-548. DOI: 10.1007/s11368-013-0814-z

[81] Lopez-Millan AF, Grusak MA, Abadia A, Abadia J. Iron deficiency in plants: An insight from proteomic approaches. Frontiers in Plant Science. 2013;**4**:254. DOI: 10.3389/fpls.2013.00254

[82] Tang CK, Chin J, Harford JB, Klausner RD, Rouault TA. Iron regulates the activity of the iron-responsive element binding protein without changing its rate of synthesis or degradation. JBC. 1992;**267**(34):24466-24470. DOI: 10.1016/S0021-9258(18)35789-2

[83] Burton JW, Harlow C, Theil EC. Evidence for reutilization of nodule iron in soybean seed development. Journal of Plant Nutrition. 2008;**21**(5):913-927. DOI: 10.1080/01904169809365453

[84] Orozco-Mosqueda Mdel C, Macías-Rodríguez LI, Santoyo G, Farías-Rodríguez R, Valencia-Cantero E. *Medicago truncatula* increases its iron-uptake mechanisms in response to volatile organic compounds produced by *Sinorhizobium meliloti*. Folia Microbiologia (Praha). 2013;**58**(6):579-585. DOI: 10.1007/s12223-013-0243-9

[85] Orozco-Mosqueda M. d C, Rocha-Granados M d C, Glick B R, and Santoyo G: Microbiome engineering to improve biocontrol and plant growth-promoting mechanisms. Microbiological Research. 2018;**208**:25-31

[86] Sankari S, Babu VMP, Bian K, Alhhazmi A, Andorfer MC, Avalos DM, et al. A haem-sequestering plant peptide promotes iron uptake in symbiotic bacteria. Nature Microbiology. 2022;**7**(9):1453-1465. DOI: 10.1038/s41564-022-01192-y

[87] Datta B, Chakrabartty PK. Siderophore biosynthesis genes of *rhizobium sp.* isolated from *Cicer arietinum L. 3*. Biotech. 2014;**4**(4):391-401. DOI: 10.1007/s13205-013-0164-y

[88] LeVier K, Guerinot ML. The Bradyrhizobium japonicum fegA gene encodes an iron-regulated outer membrane protein with similarity to hydroxamate-type siderophore receptors. Journal of Bacteriology. 1996;**178**(24):7265-7275. DOI: 10.1128/jb.178.24.7265-7275.1996

[89] Khan A, Archana G, Desai A. Engineering heterologous iron siderophore complex utilization in rhizobia: Effect on growth of peanut and pigeon pea plants. Applied Soil Ecology. 2012;**53**. DOI: 10.1016/j.apsoil.2011.10.015

[90] Jach ME, Sajnaga E, Ziaja M. Utilization of legume-nodule bacterial symbiosis in phytoremediation of heavy metal-contaminated soils. Biology (Basel). 2022;**11**(5):676. DOI: 10.3390/biology11050676

[91] Wani PA, Khan MS, Zaidi A. Effect of metal-tolerant plant growth-promoting rhizobium on the performance of pea grown in metal-amended soil. Archives of Environmental Contamination and Toxicology. 2008;**55**(1):33-42. DOI: 10.1007/s00244-007-9097-y

[92] Mandal SM, Gouri SS, De D, Das BK, Mondal KC, Pati BR. Effect of arsenic on nodulation and nitrogen fixation of blackgram (*Vigna mungo*). Indian Journal of Microbiology. 2011;**51**(1):44-47. DOI: 10.1007/s12088-011-0080-y

[93] Reichman SM. The potential use of the legume–rhizobium symbiosis for the remediation of arsenic contaminated sites. Soil Biology and Biochemistry. 2007;**39**(10):2587-2593. DOI: 10.1016/j.soilbio.2007.04.030

[94] Ye L, Yang P, Zeng Y, Li C, Jian N, Wang R, et al. Rhizobium symbiosis modulates the accumulation of arsenic in *Medicago truncatula* via nitrogen and NRT3.1-like genes regulated by ABA and linalool. Journal of Hazardous Materials. 2021;**415**:125611. DOI: 10.1016/j.jhazmat.2021.125611

[95] Quiñones MA, Ruiz-Díez B, Fajardo S, López-Berdonces MA, Higueras PL, Fernández-Pascual M. *Lupinus albus* plants acquire mercury tolerance when inoculated with an Hg-resistant *Bradyrhizobium strain*. Plant Physiology and Biochemistry. 2013;**73**:168-175. DOI: 10.1016/j.plaphy.2013.09.015

[96] Zribi K, Djébali N, Mrabet M, Khayat N, Smaoui A, Mlayah A, et al. Physiological responses to cadmium, copper, lead, and zinc of *Sinorhizobium sp.* strains nodulating *Medicago sativa* grown in Tunisian mining soils. Annales de Microbiologie. 2012;**62**(3):1181-1188. DOI: 10.1007/s13213-011-0358-7

[97] Wani PA, Khan MS, Zaidi A. Chromium-reducing and plant growth-promoting *Mesorhizobium* improves chickpea growth in chromium-amended soil. Biotechnology Letters. 2008;**30**(1):159-163. DOI: 10.1007/s10529-007-9515-2

[98] Duan C, Mei Y, Wang Q, Wang Y, Li Q, Hong M, et al. Rhizobium inoculation enhances the resistance of alfalfa and microbial characteristics in copper-contaminated soil. Frontiers in Microbiology. 2021;**12**:781831. DOI: 10.3389/fmicb.2021.781831

[99] Wekesa C, Muoma JO, Reichelt M, Asudi GO, Furch ACU, Oelmüller R. The cell membrane of a novel *rhizobium phaseoli* strain is the crucial target for aluminum toxicity and tolerance. Cell. 2022;**11**(5):873. DOI: 10.3390/cells11050873

[100] Rother JA, Millbank JW, Thornton I. Nitrogen fixation by white clover (Trifolium repens) in grasslands on soils contaminated with cadmium, lead and zinc. Journal of Soil Science. 1983;**34**(1):127-136. DOI: 10.1111/j.1365-2389.1983.tb00819.x

[101] Fan M, Liu Z, Nan L, Wang E, Chen W, Lin Y, et al. Isolation, characterization, and selection of heavy metal-resistant and plant growth-promoting endophytic bacteria from root nodules of *Robinia pseudoacacia* in a Pb/Zn mining area. Microbiological Research. 2018;**217**:51-59. DOI: 10.1016/j.micres.2018.09.002

[102] Arregui G, Hipólito P, Pallol B, Lara-Dampier V, García-Rodríguez D, Varela HP, et al. Coba de la Peña T, Lucas M M, and Pueyo J J: Mercury-tolerant *Ensifer medicae* strains display high mercuric reductase activity and a protective effect on nitrogen fixation in *Medicago truncatula* nodules under mercury stress. Frontiers in Plant Science. 2020;**11**:560768. DOI: 10.3389/fpls.2020.560768

[103] Elbagory M, El-Nahrawy S, Omara AE. Synergistic interaction between symbiotic N_2 fixing bacteria and *bacillus* strains to improve growth, physiological parameters, antioxidant enzymes and ni accumulation in faba bean plants (*Vicia faba*) under nickel stress. Plants (Basel). 2022;**11**(14):1812. DOI: 10.3390/plants11141812

[104] Jebara SH, Abdelkerim S, Fatnassi IC, Chiboub M, Saadani O, Jebara M. Identification of effective Pb resistant bacteria isolated from Lens culinaris growing in lead contaminated soils. Journal of Basic Microbiology. 2015;**55**(3):346-353. DOI: 10.1002/jobm.201300874

[105] Armendariz AL, Talano MA, Olmos Nicotra MF, Escudero L, Breser ML, Porporatto C, et al. Impact of double inoculation with *Bradyrhizobium japonicum E109* and *Azospirillum brasilense Az39* on soybean plants grown under arsenic stress. Plant Physiology and Biochemistry. 2019;**138**:26-35. DOI: 10.1016/j.plaphy.2019.02.018

[106] Fatnassi IC, Chiboub M, Saadani O, Jebara M, Jebara SH. Impact of dual inoculation with rhizobium and PGPR on growth and antioxidant status of Vicia faba L. under copper stress. Comptes Rendus Biologies. 2015;**338**(4):241-254. DOI: 10.1016/j.crvi.2015.02.001

[107] Ju W, Liu L, Fang L, Cui Y, Duan C, Wu H. Impact of co-inoculation with

plant-growth-promoting rhizobacteria and rhizobium on the biochemical responses of alfalfa-soil system in copper contaminated soil. Ecotoxicology and Environmental Safety. 2019;**167**:218-226. DOI: 10.1016/j.ecoenv.2018.10.016

[108] Han X, Wang J, Zhang Y, Kong Y, Dong H, Feng X, et al. Changes in the m6A RNA methylome accompany the promotion of soybean root growth by rhizobia under cadmium stress. Journal of Hazardous Materials. 2022;**441**:129843. DOI: 10.1016/j.jhazmat.2022.129843

[109] Zhang J, Xu Y, Cao T, Chen J, Rosen BP, Zhao FJ. Arsenic methylation by a genetically engineered rhizobium-legume symbiont. Plant and Soil. 2017;**416**(1-2):259-269. DOI: 10.1007/s11104-017-3207-z

[110] Cruz de Carvalho MH. Drought stress and reactive oxygen species: Production, scavenging and signaling. Plant Signaling & Behavior. 2008;**3**(3):156-165. DOI: 10.4161/psb.3.3.5536

[111] Sarker U, Oba S. Drought stress effects on growth, ros markers, compatible solutes, phenolics, flavonoids, and antioxidant activity in *Amaranthus tricolor*. Applied Biochemistry and Biotechnology. 2018;**186**(4):999-1016. DOI: 10.1007/s12010-018-2784-5

[112] You J, Chan Z. ROS regulation during abiotic stress responses in crop plants. Frontiers in Plant Science. 2015;**6**:1092. DOI: 10.3389/fpls.2015.01092

[113] Zahran HH, Sprent JI. Effects of sodium chloride and polyethylene glycol on root-hair infection and nodulation of *Vicia faba L.* plants by *rhizobium leguminosarum*. Planta. 1986;**167**(3):303-309. DOI: 10.1007/bf00391332

[114] Tu JC. Effect of salinity on rhizobium-root-hair interaction, nodulation and growth of soybean. Canadian Journal of Plant Science. 1981;**61**(2):231-239. DOI: 10.4141/cjps81-035

[115] Borucki W, Sujkowska M. The effects of sodium chloride-salinity upon growth, nodulation, and root nodule structure of pea (*Pisum sativum L.*) plants. Acta Physiologiae Plantarum. 2007;**30**(3):293-301. DOI: 10.1007/s11738-007-0120-8

[116] L'Taief B, Sifi B, Zaman-Allah M, Drevon JJ, Lachaâl M. Effect of salinity on root-nodule conductance to the oxygen diffusion in the *Cicer arietinum-Mesorhizobium ciceri* symbiosis. Journal of Plant Physiology. 2007;**164**(8):1028-1036. DOI: 10.1016/j.jplph.2006.05.016

[117] Chien CT, Maundu J, Cavaness J, Dandurand LM, Orser CS. Characterization of salt-tolerant and salt-sensitive mutants of *rhizobium leguminosarum biovar viciae strain C1204b*. FEMS Microbiology Letters. 1992;**69**(2):135-140. DOI: 10.1016/0378-1097(92)90617-w

[118] Dong R, Zhang J, Huan H, Bai C, Chen Z, Liu G. High salt tolerance of a *Bradyrhizobium* strain and its promotion of the growth of *Stylosanthes guianensis*. International Journal of Molecular Sciences. 2017;**18**(8):1625. DOI: 10.3390/ijms18081625

[119] Franzini VI, Azcón R, Ruiz-Lozano JM, Aroca R. Rhizobial symbiosis modifies root hydraulic properties in bean plants under non-stressed and salinity-stressed conditions. Planta. 2019;**249**(4):1207-1215. DOI: 10.1007/s00425-018-03076-0

[120] Chakraborty S, Harris JM. At the crossroads of salinity and rhizobium-legume symbiosis.

Molecular Plant-Microbe Interactions. 2022;**35**(7):540-553. DOI: 10.1094/mpmi-09-21-0231-fi

[121] Miransari M, Balakrishnan P, Smith D, Mackenzie AF, Bahrami HA, Malakouti MJ, et al. Overcoming the stressful effect of low pH on soybean root hair curling using lipochitooligosacharides. Communications in Soil Science and Plant Analysis. 2006;**37**(7-8):1103-1110. DOI: 10.1080/00103620600586391

[122] Richardson AE, Simpson RJ, Djordjevic MA, Rolfe BG. Expression of nodulation genes in *rhizobium leguminosarum biovar trifolii* is affected by low pH and by Ca and Al ions. Applied and Environmental Microbiology. 1988;**54**(10):2541-2548. DOI: 10.1128/aem.54.10.2541-2548.1988

[123] Checcucci A, Marchetti M. The rhizosphere talk show: The rhizobia on stage. Frontiers in Agronomy. 2020;**2**. DOI: 10.3389/fagro.2020.591494

[124] Han Q, Ma Q, Chen Y, Tian B, Xu L, Bai Y, et al. Variation in rhizosphere microbial communities and its association with the symbiotic efficiency of rhizobia in soybean. The ISME Journal. 2020;**14**(8):1915-1928. DOI: 10.1038/s41396-020-0648-9

[125] Mehboob I, Naveed M, Zahir ZA, Sessitsch A. Potential of rhizosphere bacteria for improving rhizobium-legume symbiosis. In: Arora N, editor. Plant Microbe Symbiosis: Fundamentals and Advances. New Delhi: Springer; 2013:305-349. DOI: 10.1007/978-81-322-1287-4_12. ISBN: 978-81-322-1286-7

[126] Kiss E, Huguet T, Poinsot V, Batut J. The *typA* gene is required for stress adaptation as well as for symbiosis of *Sinorhizobium meliloti 1021* with certain *Medicago truncatula* lines. Molecular Plant-Microbe Interactions. 2004;**17**(3):235-244. DOI: 10.1094/mpmi.2004.17.3.235

[127] Martini MC, Vacca C, Torres Tejerizo GA, Draghi WO, Pistorio M, Lozano MJ, et al. *ubiF* is involved in acid stress tolerance and symbiotic competitiveness in *rhizobium favelukesii LPU83*. Brazilian Journal of Microbiology. 2022;**53**(3):1633-1643. DOI: 10.1007/s42770-022-00780-8

[128] Reeve WG, Bräu L, Castelli J, Garau G, Sohlenkamp C, Geiger O, et al. The *Sinorhizobium medicae WSM419 lpiA* gene is transcriptionally activated by *FsrR* and required to enhance survival in lethal acid conditions. Microbiology (Reading). 2006;**152**(Pt 10):3049-3059. DOI: 10.1099/mic.0.28764-0

[129] Tiwari S, Sarangi BK, Thul ST. Identification of arsenic resistant endophytic bacteria from *Pteris vittata* roots and characterization for arsenic remediation application. Journal of Environmental Management. 2016;**180**:359-365. DOI: 10.1016/j.jenvman.2016.05.029

[130] Vinuesa P, Neumann-Silkow F, Pacios-Bras C, Spaink HP, Martínez-Romero E, Werner D. Genetic analysis of a pH-regulated operon from *rhizobium tropici CIAT899* involved in acid tolerance and nodulation competitiveness. Molecular Plant-Microbe Interactions. 2003;**16**(2):159-168. DOI: 10.1094/mpmi.2003.16.2.159

[131] Plett JM, Solomon J, Snijders F, Marlow-Conway J, Plett KL, Bithell SL. Order of microbial succession affects rhizobia-mediated biocontrol efforts against Phytophthora root rot. Microbiological Research. 2021;**242**:126628. DOI: 10.1016/j.micres.2020.126628

[132] Tewari S, Sharma S. Rhizobial-metabolite based biocontrol of fusarium wilt in pigeon pea. Microbial

Pathogenesis. 2020;**147**:104278. DOI: 10.1016/j.micpath.2020.104278

[133] Parveen G, Noreen R, Shafique HA, Sultana V, Ehteshamul-Haque S, Athar M. Role of rhizobia in suppressing the root diseases of soybean under soil amendment. Planta Daninha. 2019;**37**. DOI: 10.1590/s0100-83582019370100038

[134] Paau AS. Improvement of rhizobium inoculants by mutation, genetic engineering and formulation. Biotechnology Advances. 1991;**9**(2):173-184. DOI: 10.1016/0734-9750(91)90002-D

[135] Geetha SJ, Joshi SJ. Engineering rhizobial bioinoculants: A strategy to improve iron nutrition. Scientific World Journal. 2013;**2013**:315890. DOI: 10.1155/2013/315890

[136] Kibido T, Kunert K, Makgopa M, Greve M, Vorster J. Improvement of rhizobium-soybean symbiosis and nitrogen fixation under drought. Food and Energy Security. 2020;**9**(1):e177. DOI: 10.1002/fes3.177

[137] Pii Y, Crimi M, Cremonese G, Spena A, Pandolfini T. Auxin and nitric oxide control indeterminate nodule formation. BMC Plant Biology. 2007;**7**:21. DOI: 10.1186/1471-2229-7-21

[138] Ali S, Kim WC. Plant growth promotion under water: Decrease of waterlogging-induced acc and ethylene levels by acc deaminase-producing bacteria. Frontiers in Microbiology. 2018;**9**:1096. DOI: 10.3389/fmicb.2018.01096

[139] Bosworth AH, Williams MK, Albrecht KA, Kwiatkowski R, Beynon J, Hankinson TR, et al. Alfalfa yield response to inoculation with recombinant strains of *rhizobium meliloti* with an extra copy of *dctABD* and/or modified *nifA* expression. Applied and Environmental Microbiology. 1994;**60**(10):3815-3832. DOI: 10.1128/aem.60.10.3815-3832.1994

[140] Ghosh P, Adolphsen KN, Yurgel SN, Kahn ML. *Sinorhizobium medicae WSM419* genes that improve Symbiosis between *Sinorhizobium meliloti Rm1021* and *Medicago truncatula Jemalong A17* and in other symbiosis systems. Applied and Environmental Microbiology. 2021;**87**(15):e0300420. DOI: 10.1128/aem.03004-20

[141] Nascimento F, Brígido C, Alho L, Glick BR, Oliveira S. Enhanced chickpea growth-promotion ability of a *Mesorhizobium* strain expressing an exogenous ACC deaminase gene. Plant and Soil. 2011;**353**(1-2):221-230. DOI: 10.1007/s11104-011-1025-2

[142] Saad MM, Michalet S, Fossou R, Putnik-Delić M, Crèvecoeur M, Meyer J, et al. Loss of nifq leads to accumulation of porphyrins and altered metal-homeostasis in nitrogen-fixing symbioses. Molecular Plant-Microbe Interactions. 2019;**32**(2):208-216. DOI: 10.1094/mpmi-07-18-0188-r

[143] Hirsch PR, Spokes JD. Survival and dispersion of genetically modified rhizobia in the field and genetic interactions with native strains. FEMS Microbiology Ecology. 1994;**15**(1-2):147-159. DOI: 10.1111/j.1574-6941.1994.tb00239.x

[144] Andrews M, De Meyer S, James EK, Stępkowski T, Hodge S, Simon MF, et al. Horizontal transfer of Symbiosis genes within and between Rhizobial genera: Occurrence and importance. Genes (Basel). 2018;**9**(7):321. DOI: 10.3390/genes9070321

[145] Epstein B, Tiffin P. Comparative genomics reveals high rates of horizontal transfer and strong purifying selection on rhizobial symbiosis genes. Proceedings of the Biological Sciences. 1942;**2021**(288):20201804. DOI: 10.1098/rspb.2020.1804

Chapter 6

Exploring the Role of Microbial Live Factories in Post-Harvest Management of Potatoes-Possible Solution to the Optimization of Supply Chain

Pallavi Mansotra

Abstract

Potato (*Solanum tuberosum* L.) is the fourth most important food crop in the world with annual production of nearly 300 million tonnes. However, significant amount of the product (20–25%) is compromised to postharvest losses. Significant amount of the product (20–25%) is compromised to postharvest losses, therefore, alleviation of food security problems can be achieved through reduction in postharvest losses. Role of plant growth-promoting (PGP) microbes for the enhancement of potato production has been subject of extensive research. However, their impact on postharvest quality of horticultural crops has largely been unexplored, with limited research conducted on plant–microbe interactions in postharvest crops and their impact on storage stability. Although, microbial control has emerged as one of the most promising alternatives to chemical fungicides in several studies, however, significant research and development are required in development of sustainable microbial bio formulations for effective management of the crops under storage, in keeping with the quality of the produce. Therefore, manipulation of the bacterial microbiome, specially during crop storage, might provide microbial solutions as cleaner and sustainable alternatives to chemicals for plant production along the whole food chain. This chapter would elucidate functional analysis of the dynamics and potential of microbial live formulations for reducing the crop losses due to various diseases and status of the crop

Keywords: potatoes, postharvest challenges, cold storage, microbial formulation, biocontrol

1. Introduction

Food loss worldwide is estimated to be around 13% of the food production. Food loss is distinguished from food waste as the former occurring between harvest and supply to retail and the latter as losses in retail and consumer consumption [1]. Overall, postharvest food loss (PHL) is defined as measurable qualitative and quantitative food loss along the supply chain, starting at the time of harvest till its consumption or other end uses that varies largely between regions and also between food chains [2]. Potatoes are the world's third most important food crop and are regularly consumed by billions of people. Potatoes are cultivated on more than 20 million hectares in 150 countries with a total global output of 359 million tonnes in 2020 [3]. The total potato production in Asia, Africa, and Latin America increased from 32 million tons in 1961 to 226 million tons in 2020, displaying six times increase, with an average annual growth rate of 3.37% [2]. According to FAO, about 30% of cereals, 20% of dairy products, 35% of fish and seafood, 45% of fruits and vegetables, 20% of meat, 20% of oilseed and pulses, and 45% of roots and tubers are lost or wasted [1]. Postharvest loss (20–25%) is one of the major problems in the potato production incurring significant constraints in food security and income generation. Therefore, application of good agricultural practices *viz.* good quality and disease resistance seeds, organic and inorganic soil amendments, and advanced irrigation farming techniques to meet the market requirements is only worth the investment once postharvest management is effective. In order to realize additional income by various stakeholders across the supply chain, complete food supply system comprising the postharvest chain and the production system needs to be optimized for uniform product quality. Another considerable aspect is that postharvest interventions often culminate into overemphasis on technical aspects at the expense of social, cultural, environmental, and political issues, and economic feasibility. Interventions are often not adapted to the level of development of food systems, leading to project investments that are not economically feasible for the market that is part of the food system and thus are not being used after the project ends [4].

Several factors affect food losses during storage *viz.* pathogens (insects, bacteria, and molds), environmental conditions (e.g. rain, humidity, heat, and frost), sprouting and quality loss (rancidity, water loss, and saccharification), or animals (rodents and birds) [5]. In potatoes, about 53–55% of the initial fresh potato production and 41–46% of the initial processing potato production are lost mainly due to pathogen infection, saccharification, water loss of tubers, and early sprouting during storage, leading to poor quality and appearance leading to rejection from consumers [6]. However, diseases due to pathogens account for the leading cause of losses during storage apart from cultivation, production, handling, transportation, and storage [7]. Some of the most important diseases worldwide are late blight (*Phytophthora infestans*), early blight (*Alternaria solani*), stem canker (*Rhizoctonia solani*), potato wart (*Synchytrium endobioticum*), powdery scab (*Spongospora subterranea*), bacterial wilt (*Ralstonia solanacearum*), black leg (*Pectobacterium* spp.), potato virus Y (PVY), potato leaf roll virus (PLRV), and yellow potato cyst nematode (*Globodera rostochiensis*) [8, 9]. To control these diseases, agricultural crops are effectively treated with synthetic pesticides, which is the most common control management strategy due to dominance of varieties with low or moderate resistance to late blight as well as low marketability and acceptance of resistant cultivars [10]. As a direct consequence of late blight, potato is one of the most fungicide-dependent crops [11]. These chemical alternatives incur significant economic and environmental costs as they leave harmful residues in the soil, water, and atmosphere, and also induce resistance in phytopathogenic strains [12, 13]. Thus, in view of the long-term health and environmental impact, there is need for development of efficient and environmental friendly technologies eventually leading to elimination or reduction of the application of synthetic pesticides in agriculture [14, 15]. Also from the perspective of increased demand for organic produce over the global markets, new technological alternatives to the use of chemicals might be an effective area of research for enhanced acceptability among the consumers. Toward this goal, microbial agents can emerge as promising alternatives.

Plant microbiomes are composed of several different types of organisms, including bacteria, fungi, protozoa, archaea, and viruses, which can play a beneficial role by protecting the plant from potential pathogens, simultaneously, improving growth, health, and production, along with conferring an adaptive advantage to the plants [16, 17]. Extensive research has been conducted on the plant-beneficial microbiome for sustainable agriculture. Species, such as *Pseudomonas fluorescens*, *Pseudomonas putida*, *Pseudomonas aeruginosa*, *Bacillus subtilis*, and other *Bacillus* sp., *Azotobacter* spp., *Rhizobium* spp. Arbuscular mychorrhizal (AM) fungi, etc. are widely used for the commercial production of PGPR [18]. However, limited research has been carried out on plant–microbe interactions in postharvest crops and their impact on storage stability. Therefore, manipulation of the bacterial microbiome, which might be especially important during crop storage, might provide microbial solutions as cleaner and sustainable alternatives to chemicals for plant production along the whole food chain from farm to fork.

2. Preharvest factors and storage conditions affecting quality of tuber

Abiotic factors in the field are very crucial for establishing final tuber quality, which can be only preserved post-storage by sustainable agricultural practices such as balanced fertilizer regimes that improved tuber yield and size [19]. One of the major factors is the variability of cultivar and season as it has reported improvement in the processing quality (fry color) of younger tubers (late planting) than that of the tubers planted earlier [20]. For good tuber quality, soil demands high nutrient content, which is ultimately related to the high organic matter and nitrogen input [21, 22]. Nitrogen (N) is an important nutrient for increased potato tuber yield. However, excess nitrogen fertilization levels remain one of the adverse environmental factors for tuber quality, as they enhance the proneness to loss of nutritional value during storage of ready-to-fry potato sticks [23]. Zhang et al. observed negative effect of excessive N treatment on tuber quality, with increased invertase activity along with accumulation of reducing sugars after cold storage [24]. Also, plant growth conditions affect basal metabolism in fresh potatoes, thereby inducing differences in cold adaptation and tolerance. It has been observed that heat stress during plant growth results in physiological disorders of potato tubers such as shortening of dormancy period and higher sugar accumulation after storage [25]. Vine desiccation/killing by physical or chemical means is another factor that strongly impacts quality. This aide in controlling the spread of viral or bacterial infections from vines to the tubers and hence, limit further occurrence during storage and in seed potato production, it can also control tuber size. Also, it triggers both maturation of the tuber periderm and stolon release, thereby reducing bruising that also reduces the danger of diseases in storage as many pathogens enter tubers through breaks in the skin [26].

3. Factors affecting postharvest storage

Free amino acids and reducing sugars are the important components of tubers. Irrational nutrient management during cultivation and cold stress during postharvest storage are two abiotic stresses that potatoes may often encounter [24, 27]. After harvest, mature potato tubers enter a stage of deep dormancy for a certain period of time, which is defined as a developmental stage in which bud growth will not occur even under favorable growth conditions. The length of dormancy varies among cultivars and is affected by pre- and postharvest factors, but mainly by the temperature conditions during growth and storage [28]. Cultivated potatoes are mainly consumed by the fresh market, however, an increased demand by the food processing industry has prompted the need for long-term storage of tubers for maintaining a constant flow of demand and supply across the food chain. Postharvest cold storage is a key solution to extend the year-round supply of potatoes, and it can effectively prevent the loss caused by decay and sprouting. Temperature management depends on the intended market: tubers for the fresh market can be stored at temperatures below 7°C while tubers destined for the processing market need higher temperature (8–13°C) to preserve frying quality [29]. Control of sprouting is critical to potato storage because it leads to alterations in weight, increases in respiration, changes in texture, and nutritional value, softening, shrinkage, and the formation of toxic glycoalkaloids [30]. Also, during sprouting, there is a fast buildup of soluble sugars and increased activity of oxidative enzymes [31]. These processes result in lower quality and intense browning of French fries and potato chips. A major quality requirement for these tubers is the ability to produce light-colored, flavorful fried products desired for consumption, which depends mainly on the amount of reducing sugars, primarily glucose, and fructose, in raw tubers [32, 33]. During frying, these reducing sugars react with free amino acids in a nonenzymatic Maillard reaction that results in dark-colored, bitter-tasting products [34]. Alternately, there is production of acrylamide, a neurotoxin, and suspected carcinogen [35]. It has been suggested that decreasing the reducing sugar content in tubers is a highly effective way to minimize acrylamide production in potato products [36].

3.1 Storage at low temperature

Postharvest storage of crops at low temperature (4°C) results in “cold-induced sweetening” (CIS) due to saccharification of starch, when starch is converted to the reducing sugars glucose and fructose compromising the quality of stored potatoes. As averse to senescent sweetening (natural irreversible process occurring as a result of cellular breakdown due to tuber aging); cold-induced sweetening can be partially reversed by temperature reconditioning [37]. Since cold is also an abiotic stress; as an adaptive behavior of tubers to cold stress, it can induce enhanced starch–sugar metabolism and cause sugar accumulation [38]. The contents and basal metabolism of small–molecule metabolites such as sugars and free amino acids directly affect tuber adaptability to cold. This is attributed to their crucial roles in maintaining cell osmotic pressure, preventing cell damage, and restoring homeostasis in plant cells [39]. The problem arises during processing of starchy crops, such as frying, roasting, or baking, when the reducing sugars react with free amino acids (asparagine) to acrylamide, a potential carcinogen [40]. One of the effective interventions to reduce the accumulation of reducing sugars in tubers would be preventing sucrose accumulation as well as sucrose conversion into reducing sugars during cold storage through manipulation of the expression of genes encoding the key regulatory enzymes, transcriptional factors, and other regulatory molecules that are directly or indirectly involved in CIS [41, 42]. In a recent study, cold storage significantly affected certain metabolites like sugars, sugar alcohols, amino acids, and organic acids [43]. Based on this, it has been indicated that metabolites such as glucose, fructose, sucrose, asparagine, glutamine, citrate, malate, proline, and 4-aminobutyrate can be potentially utilized for the selection and development of potato cultivars for long-term storage, nutritional, and processing attributes [37–40, 43].

3.2 Sprout control

Among major challenges, premature sprouting accounts for the leading cause of loss during postharvest storage of potatoes. It not only reduces the number of marketable tubers but also fresh weight due to water loss from sprout surfaces, and the remobilization of starch [44]. Dormancy break in potato tubers is a physiological phenomenon that is regulated by both exogenous (environmental factors) and endogenous signals [45]. The relative concentration of several biochemical compounds such as plant growth regulators (*viz.* abscisic acid (ABA), auxins, cytokinins (CKs), gibberellins (GAs), ethylene, and strigolactones (SLs)) and other compounds (*viz.* carbohydrates and organic acids) are believed to coordinate the onset and further development of dormancy break [46]. Endogenous ethylene is required at the earliest stage of dormancy initiation (endodormancy induction) [47]; however, its role during dormancy and sprouting is still unclear. Exogenous ethylene has been reported to break endodormancy when applied for short intervals and inhibit sprout growth and promote ecodormancy when supplied continuously—either starting immediately after harvest or at first instance of sprouting [48]. However, work carried out on mini tubers showed that ethylene was not involved in hormone-induced dormancy break [47]. Apparently, the effect of ethylene depends partly on the physiological state of potato tubers [44]. Moreover, there is cross-talk between ABA and other phytohormones, as well as with sugar metabolic pathways, which facilitates the onset of dormancy break and further sprouting [49–51]. Auxins are essential for their role in vascular development as they favor the symplastic reconnection of the apical bud region—a discrete cell domain that remains symplastically isolated throughout tuberisation that is essential for sucrose to reach the meristematic apical bud [52, 53]. High sucrose levels promote trehalose-6-phosphate accumulation (T6P), which supports sprouting probably decreasing sensitivity to ABA [54]. Nevertheless, the increase in ABA as a result of exogenous ethylene application has been postulated to delay dormancy break [55, 56]. Concomitant to the ABA decline, there is an increase in sucrose contents, which is considered a prerequisite for bud outgrowth [48]. It has also been demonstrated that CKs and GAs are required for the reactivation of meristematic activity and sprout growth [57]. Just prior to dormancy break, an increase in both cytokinin concentration and sensitivity has been reported as key factors for meristematic reactivation [58]. Furthermore, CKs coordinated with auxins stimulate sprout elongation [59]. Sensitivity to GAs, which is negatively affected by SLs, increases throughout postharvest storage and is possibly responsible for sprout vigor [60]. SLs act as key regulators of lateral bud development and are related to paradormancy establishment instead of eco- and endodormancy [46]. To address this issue, the cold storage industry has relied on the application of commercial sprout suppressant isopropyl-N-(3-chlorophenyl) carbamate (chlorpropham or CIPC), as thermal hotfog (single or multiple treatments) during prolonged potato storage [61].

However, there has been an alternative trend in use of chemical sprout suppressants *viz.* hydrogen peroxide plus (HPP) [62], 1,4-dimethylnaphthalene (1,4-DMN) [63], UV-C [64], gamma radiation [65], ethylene [66], and essential oils such as mint and monoterpene (carvone) derived from caraway seed [67]. These alternative chemical sprout suppressants during postharvest aim at damaging the meristematic tissue to cease or disrupt cell proliferation; for example, local necrosis of the bud meristem was found after the application of mint essential oils [67]. However, such alternative solutions present a number of other challenges in their applications for effective sprout control like the number of applications required that render them ineffective as compared to traditional applicants like CIPC [61].

3.3 Role of microbes in postharvest stored potatoes

Worldwide, roughly 70 diseases and physiological disorders are known to severely damage potato crops, particularly the tubers [68]. Some of the main symptoms of the diseases that affect tubers during storage like spots, and rotting caused by fungi and bacteria have been listed in **Table 1**. Surface spots only affect the epidermis of the tubers, and do not alter their taste or nutritional properties [69]. However, some pathogenic species such as *Colletotrichum coccodes, Helminthosporium solani, Rhizoctonia solani, Spongospora* f. sp. subterranea, and *Streptomyces* spp. may affect the peridermis of the tubers providing means of entry for opportunist microorganisms that lead to the rotting of tubers [70]. Fungi species such as *Fusarium* sp. frequently

	Potato storage disease	Pathogenic microorganism	Symptom in tuber
Bacterial diseases			
1.	Soft rot	*Erwinia carotovora*	Tan- to brown-colored water-soaked areas of granular, mushy tissue often outlined by brown to black margins. Soft-rotted tubers tend to spread bacteria to surrounding tubers invading mainly through the lenticels, resulting in localized pockets of rot or "hot spots." The process is favored by high temperatures and anaerobic conditions. Under these conditions.
2.	Ring rot	*Cornybaeterium sepedonieum*	Yellowish color of the vascular ring just beneath the skin Secondary infections It is primarily detected by cutting open the tuber in advanced stage.
Fungal diseases			
3.	Dry rot	*Fusarium solani, F. roseum, F.* sambucinum, F. *coeruleum*	Characteristic symptoms are sunken areas of brown, firm rot often involving a large portion of the tuber. The surface of the affected area is sunken, wrinkled, and frequently with blue or white protuberances. Infected tuber tissues initially have firm, wet texture with dark brown-black appearance, eventually shrivel and mummify. The papery rot stage is dry, crumbly, and brown in color with collapsed tissue often laced with secondary white- or other-colored fungal growth.

	Potato storage disease	Pathogenic microorganism	Symptom in tuber
4.	Pink rot (water rot) and leak	*Phytophthora erythroseptica* and *Pythium* species	Normal tuber shape, outer skin turns dark, and internal flesh when cut has a rubbery, "boiled potato" consistency that turns pink after about 15 to 20 minutes of exposure to warm air, distinctive ammonia odor.
5.	Leak	*Pythium* species	Outer cortex, or "shell" of the tuber is intact, with gray/brown/black rot in the interior of the tuber, with internal tissues wet and rotted, clear liquid to stream out on squeezing, typically needs a wound to infect.
6.	Vascular discoloration	*Verticillium albo-atrum* and *Fusarium oxysporum*	Vascular ring discoloration, particularly at the stem end. The disease does not spread from one tuber to another in storage, but secondary infection may occur in affected tubers.
7.	Late blight	*Phytophthora infestans*	Reddish or tanbrown, dry, and granular rot that extends from the skin of the tuber inward an inch or more, internal tissues are firm to the touch, unlike wet or mushy, and peeling the skin over the affected areas presents characteristic reddish color.
8.	Silver scurf	*Helminthosporium solani*	Gray to silvery blotches on the surface of the tuber with unaffected internal tuber tissues. Diseased skin has a silvery sheen, appears thicker, and can be unappealing to consumers. Pathogens disrupt the periderm and lose water at a higher rate of water loss than healthy tubers.
9.	Black dot	*Colletotrichum coccodes*	Lesions are usually more prominent and thicker than those associated with silver scurf. Within the lesions, small black sclerotia may be visible with the aid of a magnifying lens.
10.	Early blight	*Alternaria solani, A. alternata*	Characterized by shallow, gray to black, dry lesions, often associated with wounding often when tubers with immature skins are harvested from sandy soils. Lesion margins may appear watersoaked and have a yellow color.

Table 1.
Common potato storage diseases and symptoms.

damage relevant damages to tubers on fields and in storage worldwide, which may affect up to 60% of the production [71]. Other species of fungi such as *Fusarium, Verticillium,* and bacteria such as *Pectobacterium carotovorum subsp. carotovorum, P. carotovorum subsp. atrosepticum, and R. solanacearum* cause rotting or more severe damage to the peridermis of tubers as a result of wide range of enzymes such as pectinases, cellulases, xylanases, and proteases, responsible for the maceration of tissues and cell death [72, 73]. On the contrary, several bacteria and fungi possess enzymes such as a-amylases, which hydrolyze starch molecules into polymers composed of glucose units [74] and so the activity of plant-associated microorganisms could play a role in the accumulation of reducing sugars in plants.

Phytohormones play a key role in the regulation of sprouting—cytokinins and indole-3-acetic acid signaling induce the onset of sprouts, and gibberellin stimulates sprout growth [75]. Many plant-associated bacteria are able to produce plant hormone-like metabolites, and the role of auxins, cytokinins, and gibberellins produced by bacteria in plant growth regulation is well documented [76, 77]. In contrast with this, Slininger et al. described an inhibitory effect of six bacterial strains on potato sprouting [78]. Further trials on bacterial strains of *P. fluorescens* and *Enterobacter* cloacae for disease suppressiveness in tubers stored for a week at 15°C and 90% relative humidity had been shown to control late blight in addition to dry rot and sprouting [79]. Another aspect was isolation and identification of polysachharide called marginalan, which could be effective in biocontrol formulations for stored potatoes [78].

4. Role of microbial bioinoculants in mitigating postharvest challenges

4.1 Nutrient management and water use efficiency

It is generally reported that the adequate application of fertilizers enhances water use efficiency by increasing the transpiration efficiency of the crop plants [80, 81]. This in turn helps in optimizing crop yields with higher drought tolerance [82]. In several studies, soil nitrogen (N) level was positively related to water use efficiency [83–85]. Potato cultivation is quite demanding in inorganic nutrients and adequate fertilization is a key factor for maximizing yield and producing tubers of high quality [86]. Most farmers apply a nitrogen-phosphorus-potassium 15–15-15 at the time of planting. Potassium and nitrogen fertilization is required for maximum potato production. As a rule of thumb, potato plants have greater needs in nitrogen (N-P-K 34–0-0) during the first two months (when the foliar part of the plant develops rapidly). From the second month until two weeks before harvest, the plants need more potassium (12–12-17 or 14–7-21) in order to create well-shaped potatoes. Similarly, potash (K) fertilizers are directly involved in the water management of the plant since it reduces water loss through transpiration. In sandy soils, water use efficiency for total dry matter production is increased by K application [87]. However, there has been indiscriminate use of fertilizers in the past. In an attempt to maximize the yield, farmers increase the doses of N fertilizer, leading to high application rates [88]. Moreover, negative effects of excessive N application on potato growth and tuber yield have been demonstrated [81, 88]. Low phosphorus (P) availability is a major limiting factor for potatoes while the residual and unavailable P in some soils is estimated to sustain crop yield for the next century [89]. In a balanced soil fertility program, P essentially increases the water use efficiency of the crop and helps in attaining optimal performance under limited moisture conditions [90]. However, it may become toxic when plants accumulate at high concentrations. It has already been demonstrated that scientific application of chemical fertilizer may contribute to the obvious, although not full, recovery of the root growth by reversing the central metabolite changes [91]. Hence, there is widespread interest in finding alternate ways to utilize accumulated soil nutrient resources for sustainable crop production. Biofertilizer application improves the resilience of the cropping system by maintaining proper nutrient balance and with increased use efficiencies [92]. Diazotropic bacteria are able to fix atmospheric nitrogen (N) to plant available forms and include the endophytes *Azoarcus* sp., *Burkholderia* sp., *Gluconacetobacter diazotrophicus* and *Herbaspirillum* sp., and the rhizospheric bacteria *Azotobacter* sp. and *Paenibacillus* (*Bacillus*) *polymyxa*, *Klebsiella spp.*, and *Enterobacter spp.* [93, 94]. Phosphorus solubilizing bacteria (*Pseudomonas spp*. and Bacillus spp.) and fungi (*Aspergillus* spp. and *Penicillium* spp.), etc. help in faster release of soluble P from fixed P through the release of organic acids, phosphatase enzymes, process of chelation and exchange reaction., etc. for improving [95]. Further, use of *Bacillus altitudinis* and *Tridens flavus var. flavus* has been shown to improve plant growth and enrich zinc in tubers [96]. Plethora of microorganisms has been reported in plants for improved mineral solubilization and uptake of nutrients with reduced dependence on chemical fertilizers [97]. In potatoes, several reports have confirmed that application of PGPR significantly enhances the available nutrients such as nitrogen, transferable phosphorus, magnesium, and dissolved K [98–101]. Alternately, it has also been reported by Zhang et al. that sustainable agricultural practices such as balanced fertilizer regimes improved not only tuber yield but also marketing tuber size and quality of potatoes [24]. Additionally, these PGPR release certain chemical substances like indole acetic acid, which increase the plant development and enhance the acquisition of other nutrients and water for plants [102].

4.2 Alleviating the effect of drought stress

Water availability, water use, and nutrient supply to plants are closely interacting factors influencing plant growth and yield production [103]. Drought stress during plant growth results in transpiration reduction and this continues postharvest [104]. As compared to other crops, potato is more sensitive to drought stress; especially during early tuberization, yield is affected by decreasing both number and weight of tubers [105]. Water deficiency, especially during the tuber bulking period, decreases yield to larger extents than water limitation during other growth stages of potatoes [106]. Similarly during storage moisture loss and consequently weight loss through transpiration is one of the main reasons for optical quality constraints with potatoes. A symbiotic relationship exists between plants and beneficial soil microorganisms wherein the microbes help the plants in nitrogen acquisition, water uptake, and survival during stress [81]. In view of this, different strains of PGPR have been reported to be involved in drought stress resistance through various physiological and biochemical changes [107, 108]. The major mechanisms adopted by PGPR to overcome drought stress include alteration in root morphology and production of osmolytes like sorbitol, polyols, mannitol, sucrose, fructan, proline, and ectoine, antioxidants, phytohormones, extracellular polymeric substance (EPS), and volatile organic compounds (VOCs), siderophores, and 1-aminocyclopropane-1-carboxylate (ACC) deaminase [109]. It is well known that inoculation with certain plant-beneficial bacteria can minimize symptoms of drought stress in plants and one of the effects is that the water content in plant tissue remains higher during drought stress periods than in non-inoculated control plants [110]. The application of microbial inoculants that reduce drought stress on the field could, therefore, also have positive effects on water content of crops during storage. Several PGPR (*Bacillus, Rhizobium, Azospirillum, Pseudomonas, Flavobacterium,* and *Arthrobacter*) are reported to display improved osmoregulation, oligotrophic, endogenous metabolism, resistance to starvation, and thus showed their efficient metabolic processes to adapt dry and saline environments [111]. Recent studies demonstrated that inoculation with ACC deaminase-producing rhizobacteria increased root–shoot length, root–shoot mass, and the lateral number of roots of wheat plants compared with that of the control resulting in improved growth and yield under drought stress water and nutrient uptake [112]. Similar positive effects of ACC deaminase-producing bacteria against drought-induced oxidative damage have been reported in tomato plants with *B. Subtilis* [113] and in maize with *E. cloacae* and *A. xylosoxidans* [114]. In potatoes, PGPR strain of *B. subtilis* was observed to enhance the growth and yield under drought stress by maintaining physiological functions and antioxidant enzyme activities of the plant [115] with increase in stress-related enzymatic activity of catalase (CAT), peroxidase (POD) and superoxide dismutase (SOD), contents of total soluble sugars, soluble proteins, and proline.

4.3 Abiotic and biotic stress tolerance

The abiotic stress tolerance in a plant is a network of complex signaling pathways where numerous cell molecules, enzymes, transcription factors, hormones, and metabolites are involved. Under different abiotic and abiotic stress conditions, PGPR like *Bacillus* spp. has been shown to synthesize small molecules such as carotenoids, ascorbic acid, tocopherols, anthocyanins, and antioxidant enzymes that protect potato plants from oxidative injury through elimination of stress-induced reactive oxygen species (ROS) [116]. PGPRs are also involved in the initiation of defense mechanisms like phenylpropanoid pathways and lignin biosynthesis by inducing the production of certain plant molecules such phytoalexins, salicylic acid, jasmonic acid, methyl salicylate, and methyl jasmonate that are formed under stress [115, 117]. Such molecules act as signaling molecules that trigger a cascade of the stress signaling pathways.

It has been discussed that rhizobacteria confer stress tolerance through biocontrol of phytopathogens in the rhizosphere and by the production of phytohormones and ACC deaminase, favoring osmolyte accumulation in plants [118]. Plants produce ethylene during ripening from 1-aminocyclopropane-1-carboxylate (ACC) *via* ACC oxidase activity [119, 120]. Many bacteria in soil and plants possess ACC deaminase activity, which cleaves ACC to 2-oxobutyrate and ammonia and can contribute to ethylene balancing in plants [111, 118]. An alternative role of ethylene in microbial-driven modulation of ripening was introduced by a recent study exploring the potential of a *Rhodococcus rhodochrous* strain to delay ripening in different species of climacteric fruits [121]. In this case, the proposed mechanism is ethylene-induced nitrile hydratase and/or nitrilase activity in the bacterium, which resulted in maintenance of fruit firmness and reduced spotting and spoilage [121].

Stress-tolerant and competitive species are physically adapted to limited environmental resources or carrying-capacity environments and are more stable and permanent members of the community. Pathogens that tolerate environmental stress often have few competitors, since few species can exist under such conditions [122]. Some postharvest rots result from preharvest latent infections, especially in tropical and subtropical regions where environmental conditions in the field are particularly conducive to infection. Latent contamination involves fungal spores on the surface, which fail to germinate until the host reaches maturity or senescence [123]. Nevertheless, successful control of latent infections by postharvest applications has been reported. For example, Grossi et al. has recently reported *Methylobacterium* as potential PGPR to alleviate salt stress and restrict *P.infestans* infection in potato plants [124]. Therefore, selection of stress-tolerant microbial strains that have competitive advantage over the quiescent pathogens becomes very useful for preparation of different microbial formulations for various purposes including biocontrol against the spoilage microorganism.

4.4 Biofilm production in disease management

Biofilms consist of cells and matrices where complex exopolysaccharides and proteins are major components, and they can provide an important bacterial survival strategy in natural systems [125]. It is generally recognized that surface-associated bacteria colonize as biofilms, which are microniches entirely different from their surroundings. This allows the bacteria to work as a functional unit, accomplishing tasks not possible in their planktonic state. The biocontrol ability of *Paenibacillus polymyxa* has been attributed to the large bacterial pool of bioactive compounds such as nonribosomal peptides/polyketides (NRPs/PKs) [126]. It has been indicated that along with the lipopeptides, the bacterial biofilm EPS compounds *viz.* uronates are capable of antagonizing *Fusarium graminearum* causing Fusarium Head Blight (FHB) in cereals [126]. Bacterial EPS is an important compound that is dependent upon the perception of numerous environmental signals from the host and the ecosystem. Such categories of sustainable biopolymers of microbial origin could be further explored in addressing various industrial and agricultural necessities.

4.5 Other compounds

Among various research efforts, application of elicitors or resistance inducers (RIs) has been of much interest as alternative control methods. Many of these compounds are nontoxic to environment and with no reported negative effects on fruit quality [71]. Resistance inducers can be divided into two groups: natural and synthetic compounds such as benzothiadiazole, β-aminobutyric acid, salicylic acid, chitosan, and chitin [127]. Among the most widely used natural compounds, chitosan is explored as an elicitor and an antifungal agent [128]. Chitosan is a natural carbohydrate polymer, nontoxic, and biodegradable, which can be applied in solution and as powder [127]. This β-(1,4) N-acetyl-glucosamine polymer is collected from the chitin that has been deacetylated to provide 70% free amino groups [129]. In the past, the antifungal activity of chitosan has been successfully used for the control of several plant diseases such as fusarium wilt of tomato [130], gray mold of table grapes [131], and orange-green molds [132]. Also, chitosan was first used at 400 μg/ml of acetic acid-distilled water solution for the control of FDR, where a significant reduction in disease severity was noted on potato tubers inoculated with *F. sambucinum* [133]). On the basis of inhibitory activity of chitosan against *Fusarium* spp. involved in potato tuber dry rot and wilt, and the capacity to activate defense-related enzymes and to increase the total phenolic contents and oxidative enzymes in potato tubers, suggestions have been made for potential use of plant RIs, as alternatives to synthetic fungicides. This eco-friendly approach could be easily applied and generally recognized safe and cheap method for controlling some soil-borne plant pathogens.

4.6 Biocontrol agents in storage disease management

Fresh potatoes are exposed to an intensive postharvest microbial colonization leading to storage rots. This is attributed to the latent contamination involving fungal spores on the surface, which fail to germinate until the host reaches maturity or senescence. However, there is intense interplay of individual members of the diverse microbial community of the crop for effective colonization of pathogens consequently variation in the degree of spoilage [16]. As evidence of this, Liebe et al. observed an unspecific resistance

Microbial antagonists	Pathogens/Diseases controlled
Pseudomonas putida	*Erwinia carotovora* [135]
Muscodor albus	*B. cinerea*, *Colletotrichum coccodes*, *Cerastium acutatum*, *Fusarium sambucinum*, *G. candidum*, *Helminthosporium solani*, *Monilinia fructicola*, *Pectobacterium atrosepticum*, *P. expansum*, *Rhizopus* spp. [135]
Bacillus subtilis and *Trichoderma harzianum*	Common scab disease caused by *Streptomyces* spp. [136]
Pseudomonas aeruginosa, *Alcaligenes feacalis* and *Serratia marcescens*	Fusarium wilt caused by *Fusarium* sp. [137]
Enterobacter amnigenus, *Serratia plymuthica*, *Serratia rubidaea* and *Rahnella aquatilis*	Soft rot caused by *E. carotovora* [138]
commercial arbuscular mycorrhiza fungi (AMF)	Bacterial wilt [139]
Bacillus cereus, B.subtilis, *B.weihenstephanensis*, *Pseudomonas* sp.	*Pectobacterium* spp. and *Dickeya* spp. [140–143]
Bacillus thuringiensis, *Rhodococcus erythropolis*, *S.plymuthica*, *Delftia acidovorans*, *Ochrobactrum* sp.	*Piliocolobus parmentieri* and *Pectobacterium carotovorum* subsp. *carotovorum* [140–143]

Table 2.
Biocontrol agents with antagonistic properties against postharvest pathogens of potatoes.

mechanism that has been suggested for slowing down the spread of pathogens in more resistant genotypes [134]. Several workers have studied antagonistic properties of different biocontrol agents against postharvest pathogens of potatoes some of which have been listed in **Table 2** [137–143]. Studies on evolutionary genomics have shown that species are characterized by the paleome (core genes, which allow basic metabolism), whereas strains are defined by their cenome (genes, which allow cells to live in and explore a particular niche); that is responsible for strain-specific properties [144]. In this context, each strain of one species has specific antagonistic properties that can show different biocontrol effects in the field [145]. Therefore, a selection strategy considering these strain-specific effects becomes necessary for an effective outcome. So far several workers have deployed different approaches in manipulation of microbial agents for devising successful biocontrol strategies. In yet, another attempt to develop efficient biocontrol strategy, Furkranz et al. studied the diversity of endophytic microbial communities isolated from diverse microhabitats *viz.* from seeds, roots, flowers, and fruits of three different pumpkin cultivars [146]. On the basis of broad-spectrum antagonism, *Pseudomonas chlororaphis* treatment, and combined treatment of strains (*Lysobacter gummosus*, *P. chlororaphis*, *P. polymyxa, and Serratia plymuthica*) significantly reduced the disease severity on pumpkins. In concordance with this, it has been observed that the seed microbiome of oilseed rape is cultivar-specific and cultivars hosting a higher indigenous microbial diversity showed better resistance toward colonization by pathogenic microorganisms [147]. In a recent study, it has been concluded that PGPR strain of *P. chlororaphis* can be used as probiotics in order to increase plant tolerance/resistance to abiotic/biotic stresses [145]. Additionally, many species of endophytic *Bacillus* are known to exhibit antagonistic properties against the tested plant pathogens in melons, which have been hinted at protecting plants not only in the field but also for postharvest [148, 149]. Further, the efficacy of biocontrol agents can be enhanced by isolation of the antagonists and can be improved by using undisturbed habitats, where natural populations have not been disturbed by conventional management with agrochemicals, with rich pool of potential antagonists. A variety of enrichment procedures have been used that favor isolation of microorganisms; one such rapid and cost-effective method of antagonist isolation was adopted by Wilson and colleagues in 1993 with application of rinsing waters from tomatoes and apples directly on wounds inoculated with the pathogen [7]; subsequently isolating antagonists from asymptomatic wounds. One of the drawbacks of this strategy is that it favors the selection of antagonists that are generally fast growers with the ability to colonize a specific niche rich in nutrients that mainly exhibit protective rather than curative activity, with seemingly poor effect on latent infections [135].

5. Conclusions

Postharvest loss is one of the major problems in the potato production incurring significant constraints in food security and income generation. Despite the fact, integrated soil and water management can potentially optimize yields and marketable quality of potatoes; additional income can be realized by various stakeholders through uniform product quality during storage and transfer across the supply chain. In the previous studies, wide range of plant growth-promoting (PGP) microbes with potential biocontrol abilities have been identified for improvement of plant growth and yield in potatoes. Also, much progress has been made in the development of various biocontrol agents (BCAs) for control of diseases in potatoes, especially during storage. However, additional research is required to validate their stability for their wide use at the commercial level. Moreover, limited research has been carried out on plant–microbe interactions in postharvest crops and their impact on storage stability. Manipulation of the bacterial microbiome, which might be especially important during crop storage, might provide microbial solutions as cleaner and sustainable alternatives to chemicals for plant production along the whole food chain from farm to fork. Therefore, concerted research efforts are required to validate their stability for wide use at the commercial level.

Conflict of interest

The authors declare no conflict of interest.

Author details

Pallavi Mansotra
Sangha Innovation Centre, Jalandhar, India

*Address all correspondence to: p.mansotra@gmail.com

References

[1] FAO. Crop Prospects and Food Situation. Rome: Food and Agriculture Organization, Global Information and Early Warning System (Trade and Markets Division (EST); 2015

[2] Kuyu CG, Tola YB, Abdi GG. Study on post-harvest quantitative and qualitative losses of potato tubers from two different road access districts of Jimma zone, south West Ethiopia. Heliyon. 2019;**5**(8):e02272. DOI: 10.1016/j.heliyon.2019.e02272

[3] Dongyu Q. Role and potential of potato in global food security challenges of global food security contribution of potato to the world potential of global potato production strategies for promoting potato development. 2022.

[4] Verschoor J, Oostewechel R, NJJP Koenderink dr, Pereira da Silva F, Hetterscheid S. Postharvest Interventions, Key for Improvement of Food Systems Exploring the Impact of Postharvest Interventions on Increasing Food Availability, Stability and Income Generation in Developing Economies. Available from: www.wur.eu/wfbr

[5] Aksenova NP, Sergeeva LI, Konstantinova TN, Golyanovskaya SA, Kolachevskaya OO, Romanov GA. Regulation of potato tuber dormancy and sprouting. Russian Journal of Plant Physiology. 2013;**60**(3):301-312. DOI: 10.1134/S1021443713030023

[6] Charkowski A, Sharma K, Parker ML, Secor GA, Elphinstone J. Bacterial diseases of potato BT. In: Campos H, Ortiz O, editors. The Potato Crop: Its Agricultural, Nutritional and Social Contribution to Humankind. New York: Springer International Publishing; 2020. pp. 351-388. DOI: 10.1007/978-3-030-28683-5_10

[7] Vilvert E, Stridh L, Andersson B, Olson A, Alden L, Berlin A. Evidence based disease control methods in potato production: A systematic map protocol. Environment Evidence. 2022;**11**(1):6. DOI: 10.1186/s13750-022-00259-x

[8] Kreuze JF, Souza-Dias JAC, Jeevalatha A, Figueira AR, Valkonen JPT, Jones RAC. Viral diseases in potato BT. In: Campos H, Ortiz O, editors. The Potato Crop: Its Agricultural, Nutritional and Social Contribution to Humankind. New York: Springer International Publishing; 2020. pp. 389-430. DOI: 10.1007/978-3-030-28683-5_11

[9] Adolf B, Andrade-Piedra J, Bittara Molina F, et al. Fungal, oomycete, and Plasmodiophorid diseases of potato BT. In: Campos H, Ortiz O, editors. The Potato Crop: Its Agricultural, Nutritional and Social Contribution to Humankind. New York: Springer International Publishing; 2020. pp. 307-350. DOI: 10.1007/978-3-030-28683-5_9

[10] Liljeroth E, Lankinen A, Wiik L, Burra DD, Alexandersson E, Andreasson E. Potassium phosphite combined with reduced doses of fungicides provides efficient protection against potato late blight in large-scale field trials. Crop Protection. 2016;**86**:42-55. DOI: 10.1016/j.cropro.2016.04.003

[11] Yuen J. Pathogens which threaten food security: Phytophthora infestans, the potato late blight pathogen. Food Security. 2021;**13**(2):247-253. DOI: 10.1007/s12571-021-01141-3

[12] Brauer VS, Rezende CP, Pessoni AM, et al. Antifungal agents in agriculture: Friends and foes of public health. Biomolecules. 2019;**9**(10):521. DOI: 10.1016/j.micres.2018.01.005

[13] Gikas GD, Parlakidis P, Mavropoulos T, Vryzas Z. Particularities of fungicides and factors affecting their fate and removal efficacy: A review. Sustainability. 2022;**14**(7):4056. DOI: 10.3390/su14074056

[14] Palmieri D, Ianiri G, Conte T, Castoria R, Lima G, De Curtis F. Influence of biocontrol and integrated strategies and treatment timing on plum Brown rot incidence and fungicide residues in fruits. Agriculture. 2022;**12**(10):1656. DOI: 10.3390/agriculture12101656

[15] Tudi M, Ruan HD, Wang L, et al. Agriculture development, pesticide application and its impact on the environment. International Journal of Environmental Research and Public Health. 2021;**18**(3):1-24. DOI: 10.3390/ijerph18031112

[16] Berg G, Koberl M, Rybakova D, Muller H, Grosch R, Smalla K. Plant microbial diversity is suggested as the key to future biocontrol and health trends. FEMS Microbiology Ecology. 2017;**93**(5). DOI: 10.1093/femsec/fix050

[17] del Orozco-Mosqueda MC, del Rocha-Granados MC, Glick BR, Santoyo G. Microbiome engineering to improve biocontrol and plant growth-promoting mechanisms. Microbiological Research. 2018;**208**:25-31. DOI: 10.1016/j.micres.2018.01.005

[18] Goswami M, Deka S. Plant growth-promoting rhizobacteria-alleviators of abiotic stresses in soil: A review. Pedosphere. 2020;**30**(1):40-61. DOI: 10.1016/S1002-0160(19)60839-8

[19] Tan B, Huang Z, Yin Z, et al. Preparation and thermal properties of shape-stabilized composite phase change materials based on polyethylene glycol and porous carbon prepared from potato. RSC Advances. 2016;**6**(19):15821-15830. DOI: 10.1039/C5RA25685B

[20] Driskill EP, Knowles LO, Knowles NR. Temperature-induced changes in potato processing quality during storage are modulated by tuber maturity. American Journal of Potato Research. 2007;**84**:367-383. DOI: 10.1007/BF02987183

[21] Shakeel SN, Gao Z, Amir M, Chen YF, Rai MI, Haq NU, et al. Ethylene regulates levels of ethylene receptor/CTR1 signaling complexes in Arabidopsis thaliana. The Journal of Biological Chemistry. 2015;**290**:12415-12424. DOI: 10.1074/jbc.M115.652503

[22] Nesbitt JE, Adl SM. Differences in soil quality indicators between organic and sustainably managed potato fields in eastern Canada. Ecological Indicators. 2014;**37**:119-130. DOI: 10.1016/j.ecolind.2013.10.002

[23] Licciardello F, Lombardo S, Rizzo V, Pitino I, Strano MG, Muratore G, et al. Integrated agronomical and technological approach for the quality maintenance of ready-to-fry potato sticks during refrigerated storage. Postharvest Biology and Technology. 2018;**136**:23-30. DOI: 10.1016/j.postharvbio.2017.10.001

[24] Zhang H, Liu X, Song B, Nie B, Zhang W, Zhao Z. Effect of excessive nitrogen on levels of amino acids and sugars, and differential response to post-harvest cold storage in potato (*Solanum tuberosum* L.) tubers. Plant Physiology and Biochemistry. 2020;**157**:38-46. DOI: 10.1016/j.plaphy.2020.09.040

[25] Zommick DH, Knowles LO, Pavek MJ, Knowles NR. In-season heat stress compromises postharvest quality and low-temperature sweetening resistance in potato (*Solanum tuberosum*

L.). Planta. 2014;**239**(6):1243-1263. DOI: 10.1007/s00425-014-2048-8

[26] Naerstad R, Dees MW, Le VH, Holgado R, Hermansen A. Occurrence of skin blemish diseases (scab and scurf) in Norwegian potato production. Potato Research. 2012;**55**:225-239. DOI: 10.1007/s11540-012-9221-x

[27] Dourado C, Pinto FJ, Barba JM, Lorenzo I, Delgadillo J, Saraiva A. Innovative non-thermal technologies affecting potato tuber and fried potato quality. Trends in Food Science and Technology. 2019;**88**:274-289. DOI: 10.1016/j.tifs.2019.03.015

[28] Alamar MC, Tosetti R, Landahl S, Bermejo A, Terry LA. Assuring potato tuber quality during storage: A future perspective. Frontiers in Plant Science. 2017;8:2034. DOI: 10.3389/fpls.2017.02034

[29] Elmore JS, Briddon A, Dodson AT, Muttucumaru N, Halford NG, Mottram DS. Acrylamide in potato crisps prepared from 20 UK-grown varieties: Effects of variety and tuber storage time. Food Chemistry. 2015;**182**:1-8

[30] Suttle JC, Abrams SR, De Stefano-Beltran L, Huckle LL. Chemical inhibition of potato ABA-80 -hydroxylase activity alters in vitro and in vivo ABA metabolism and endogenous ABA levels but does not affect potato microtuber dormancy duration. Journal of Experimental Botany. 2012;**63**:5717-5725. DOI: 10.1093/jxb/ers146

[31] Nefissi Ouertani R, Abid G, Karmous C, et al. Evaluating the contribution of osmotic and oxidative stress components on barley growth under salt stress. AoB Plants. 2021;**13**(4):plab034. DOI: 10.1093/aobpla/plab034

[32] Rosen C, Sun N, Olsen N, Thornton M, Pavek M, Knowles L, et al. Impact of agronomic and storage practices on acrylamide in processed potatoes. American Journal of Potato Research. 2018;**95**(4):319-327. DOI: 10.1007/s12230-018-9659-8

[33] Muttucumaru N, Powers SJ, Elmore JS, Dodson A, Briddon A, Mottram DS, et al. Acrylamide-forming potential of potatoes grown at different locations, and the ratio of free asparagine to reducing sugars at which free asparagine becomes a limiting factor for acrylamide formation. Food Chemistry. 2017;**220**:76-86. DOI: 10.1016/j.foodchem.2016.09.199

[34] Negoita M, Mihai AL, Horneț GA. Influence of water, NaCl and citric acid soaking pre-treatments on acrylamide content in french fries prepared in domestic conditions. Food. 2022;**11**(9):1204. DOI: 10.3390/foods11091204

[35] Bertuzzi T, Mulazzi A, Rastelli S, Sala L, Pietri A. Mitigation measures for acrylamide reduction in dough-based potato snacks during their expansion by frying. Food Additives & Contaminants. Part A, Chemistry, Analysis, Control, Exposure & Risk Assessment. 2018;**35**(10):1940-1947. DOI: 10.1080/19440049.2018.1512757

[36] Liyanage DWK, Yevtushenko DP, Konschuh M, Bizimungu B, Lu ZX. Processing strategies to decrease acrylamide formation, reducing sugars and free asparagine content in potato chips from three commercial cultivars. Food Control. 2021;**119**:107452. DOI: 10.1016/j.foodcont.2020.107452

[37] Jaiswal S, Paul K, Raman KV, et al. Amelioration of cold-induced sweetening in potato by RNAi mediated silencing

of StUGPase encoding UDP-glucose pyrophosphorylase. Frontiers in Plant Science. 2023:14 https://www.frontiersin.org/articles/10.3389/fpls.2023.1133029

[38] Kamillah WO, Kusumaningrum HP, Ferniah RS, Herida AP, Nasansia GL. Physiological response and detection of Inh2 gene in Dieng red potato (Solanum tuberosum L.) affected by frost. Jordan Journal of Biological Sciences. 2022;**15**(1):101-106. DOI: 10.54319/jjbs/150113

[39] Zivanovic B, Komic SM, Tosti T, Vidovic M, Prokic L, Jovanovic SV. Leaf soluble sugars and free amino acids as important components of abscisic acid—Mediated drought response in tomato. Plants. 2020;**9**(9):1-17. DOI: 10.3390/plants9091147

[40] Sarion C, Codina GG, Dabija A. Acrylamide in bakery products: A review on health risks, legal regulations and strategies to reduce its formation. International Journal of Environmental Research and Public Health. 2021;**18**(8):4332. DOI: 10.3390/ijerph18084332

[41] Shi W, Ma Q, Yin W, et al. The transcription factor StTINY3 enhances cold-induced sweetening resistance by coordinating starch resynthesis and sucrose hydrolysis in potato. Journal of Experimental Botany. 2022;**73**(14):4968-4980. DOI: 10.1093/jxb/erac171

[42] Yu J, Tseng Y, Pham K, Liu M, Beckles DM. Starch and sugars as determinants of postharvest shelf life and quality: Some new and surprising roles. Current Opinion in Biotechnology. 2022;**78**:102844. DOI: 10.1016/j.copbio.2022.102844

[43] Datir SS, Yousf S, Sharma S, Kochle M, Ravikumar A, Chugh J. Cold storage reveals distinct metabolic perturbations in processing and non-processing cultivars of potato (*Solanum tuberosum* L.). Scientific Reports. 2020;**10**(1):6268. DOI: 10.1038/s41598-020-63329-5

[44] Sonnewald S, Sonnewald U. Regulation of potato tuber sprouting. Planta. 2014;**239**:27-38. DOI: 10.1007/s00425-013-1968-z

[45] El-Awady Aml A, Moghazy AM, Gouda AEA, Elshatoury RSA. Inhibition of sprout growth and increase storability of processing potato by antisprouting agent. Trends Horticulture Researh. 2014;**4**:31-40. DOI: 10.3923/thr.2014.31.40

[46] Pasare SA, Ducreux LJ, Morris WL, Campbell R, Sharma SK, Roumeliotis E, et al. The role of the potato (*Solanum tuberosum*) CCD8 gene in stolon and tuber development. The New Phytologist. 2013;**198**:1108-1120. DOI: 10.1111/nph.12217

[47] Suttle JC. Ethylene is not involved in hormone-and bromoethane-induced dormancy break in russet Burbank minitubers. American Journal of Potato Research. 2009;**86**:278-285. DOI: 10.1007/s12230-009-9081-3

[48] Foukaraki SG, Cools K, Chope GA, Terry LA. Effect of the transition between ethylene and air storage on the post-harvest quality in six UK-grown potato cultivars. The Journal of Horticultural Science and Biotechnology. 2014;**89**:599-606. DOI: 10.1080/14620316.2014.11513126

[49] Chang KN, Zhong S, Weirauch MT, Hon G, Pelizzola M, Li H, et al. Temporal transcriptional response to ethylene gas drives growth hormone cross-regulation in Arabidopsis. eLife. 2013;**2**:e00675. DOI: 10.7554/eLife.00675

[50] Brady SM. When the time is ripe. eLife. 2013;**2**:e00958. DOI: 10.7554/eLife.00958

[51] Kolachevskaya OO, Sergeeva LI, Flokova K, Getman IA, Lomin SN, Alekseeva VV, et al. Auxin synthesis gene tms1 driven by tuber-specific promoter alters hormonal status of transgenic potato plants and their responses to exogenous phytohormones. Plant Cell Reports. 2017;**36**(3):419-435. DOI: 10.1007/s00299-016-2091-y

[52] Kolachevskaya OO, Lomin SN, Arkhipov DV, Romanov GA. Auxins in potato: Molecular aspects and emerging roles in tuber formation and stress resistance. Plant Cell Reports. 2019;**38**(6):681-698. DOI: 10.1007/s00299-019-02395-0

[53] Ordaz-Ortiz JJ, Foukaraki S, Terry LA. Assessing temporal flux of plant hormones in stored processing potatoes using high definition accurate mass spectrometry. Horticulture Research. 2015;**2**:15002. DOI: 10.1038/hortres.2015.2

[54] Debast S, Nunes-Nesi A, Hajirezaei MR, et al. Altering trehalose-6-phosphate content in transgenic potato tubers affects tuber growth and alters responsiveness to hormones during sprouting. Plant Physiology. 2011;**156**(4):1754-1771. DOI: 10.1104/pp.111.179903

[55] Esztergalyos A, Polgar Z. The effect of chemical treatments on the tuber dormancy of hungarian potato cultivars. Potato Research. 2021;**64**(3):327-337. DOI: 10.1007/s11540-020-09479-5

[56] Foukaraki SG, Cools KC, GA, Terry LA. Impact of ethylene and 1-MCP on sprouting and sugar accumulation in stored potatoes. Postharvest Biology Technology. 2016;**114**:95-103. DOI: 10.1016/j.postharvbio.2015.11.013

[57] Teo CJ, Takahashi K, Shimizu K, Shimamoto K, Taoka KI. Potato tuber induction is regulated by interactions between components of a Tuberigen complex. Plant & Cell Physiology. 2017;**58**(2):365-374. DOI: 10.1093/pcp/pcw197

[58] Gazzarrini S, Tsai AY-L. Trehalose-6-phosphate and SnRK1 kinases in plant development and signaling: The emerging picture. Frontiers in Plant Science. 2014;**5**:119 https://www.frontiersin.org/articles/10.3389/fpls.2014.00119

[59] Raspor M, Motyka V, Ninkovic S, et al. Endogenous levels of cytokinins, indole-3-acetic acid and abscisic acid in vitro grown potato: A contribution to potato hormonomics. Scientific Reports. 2020;**10**(1):3437. Published 2020 Feb 26. DOI: 10.1038/s41598-020-60412-9

[60] Saidi A, Hajibarat Z. Phytohormones: Plant switchers in developmental and growth stages in potato. Journal, Genetic Engineering & Biotechnology. 2021;**19**(1):89. DOI: 10.1186/s43141-021-00192-5

[61] Blenkinsop RW, Copp LJ, Yada RY, Marangoni AG. Changes in compositional parameters of tubers of potato (*Solanum tuberosum*) during low-temperature storage and their relationship to chip processing quality. Journal of Agricultural and Food Chemistry. 2002;**50**(16):4545-4553. DOI: 10.1021/jf0255984

[62] Al-Mughrabi K. Post harvest treatment with hydrogen peroxide supresses silver scurf (*Helminthosporium solani*), dry rot (*Fusarium sambucinum*), and soft rot (*Erwinia carotovora* subsp. *carotovora*) of stored potatoes. American

Journal of Plant Science Biotechnology. 2010;**4**:74-81

[63] Gumbo N, Magwaza LS, Ngobese NZ. Evaluating ecologically acceptable sprout suppressants for enhancing dormancy and potato storability: A review. Plants (Basel). 2021;**10**(11):2307. DOI: 10.3390/plants10112307

[64] Cools K, Alamar MC, Terry LA. Controlling sprouting in potato tubers using ultraviolet-C irradiance. Postharvest Biology Tecnology. 2014;**98**:106-114. DOI: 10.1016/j.postharvbio.2014.07.005

[65] Sharma S, Sanyal SK, Sushmita K, Chauhan M, Sharma A, Anirudhan G, et al. Modulation of phototropin signalosome with artificial illumination holds great potential in the development of climate-smart crops. Current Genomics. 2021;**22**(3):181-213. DOI: 10.2174/1389202922666210412104817

[66] Tosetti R, Waters A, Chope GA, Cools K, Alamar MC, McWilliam S, et al. New insights into the effects of ethylene on ABA catabolism, sweetening and dormancy in stored potato tubers. Postharvest Biology and Technology. 2021;**173**:111420. DOI: 10.1016/j.postharvbio.2020.111420

[67] Teper-Bamnolker P, Dudai N, Fischer R, et al. Mint essential oil can induce or inhibit potato sprouting by differential alteration of apical meristem. Planta. 2010;**232**(1):179-186. DOI: 10.1007/s00425-010-1154-5

[68] García-Ávila Clemente de Jesus et al. Organisms associated with damage to post-harvest potato tubers. Mexican Journal of Phytopathology 2018;36(2) ISSN 2007-8080. Available at: https://www.smf.org.mx/rmf/ojs/index.php/RMF/article/view/109

[69] Jemison JM Jr, Sexton P, Camire ME. Factors influencing consumer preference of fresh potato varieties in Maine. American Journal of Potato Research. 2008;**85**:140-149. DOI: 10.1007/s12230-008-9017-3

[70] Fiers M, Edel-Hermann V, Chatot C, Le Hingrat Y, Alabouvette C, Steinberg C. Potato soil-borne diseases. A review. Agronomy for Sustainable Development. 2012;**32**:93-132. DOI: 10.1007/s13593-011-0035-z

[71] Mejdoub-Trabelsi B, Touihri S, Ammar N, Riahi A, Daami-Remadi M. Effect of chitosan for the control of potato diseases caused by *Fusarium* species. Journal of Phytopathology. 2020;**168**(1):18-27. DOI: 10.1111/jph.12847

[72] Czajkowski R, Perombelon MC, van Veen JA, van der Wolf JM. Control of blackleg and tuber soft rot of potato caused by *Pectobacterium* and *Dickeya* species: A review. Plant Pathology. 2011;**60**:999-1013. DOI: 10.1111/j.1365-3059.2011.02470.x

[73] Buchholz F, Kostić T, Sessitsch A, Mitter B. The potential of plant microbiota in reducing postharvest food loss. Microbial Biotechnology. 2018;**11**(6):971-975. DOI: 10.1111/1751-7915.13252

[74] Buchholz F, Junker R, Samad A, et al. 16S rRNA gene-based microbiome analysis identifies candidate bacterial strains that increase the storage time of potato tubers. Scientific Reports. 2021;**11**(1):3146. DOI: 10.1038/s41598-021-82181-9

[75] Naveed M, Mitter B, Reichenauer TG, Wieczorek K, Sessitsch A. Increased

drought stress resilience of maize through endophytic colonization by *Burkholderia phytofirmans* PsJN and *Enterobacter* sp. FD17. Environmental and Experimental Botany. 2014;**97**:30-39. DOI: 10.1016/j.envexpbot.2013.09.014

[76] Matilla AJ. Auxin: Hormonal signal required for seed development and dormancy. Plants. 2020;**9**(6):705. DOI: 10.3390/plants9060705

[77] Mir AR, Siddiqui H, Alam P, Hayat S. Foliar spray of auxin/IAA modulates photosynthesis, elemental composition, ROS localization and antioxidant machinery to promote growth of *Brassica juncea*. Physiology and Molecular Biology of Plants. 2020;**26**(12):2503-2520. DOI: 10.1007/s12298-020-00914-y

[78] Slininger PJ, Dunlap CA, Schisler DA. Polysaccharide production benefits dry storage survival of the biocontrol agent *Pseudomonas fluorescens* S11:P:12 effective against several maladies of stored potatoes. Biocontrol Science and Technology. 2010;**20**(3):227-244. DOI: 10.1080/09583150903469525

[79] Slininger PJ, Schisler DA. High-throughput assay for optimising microbial biological control agent production and delivery. Biocontrol Science and Technology. 2013;**23**(8):920-943. DOI: 10.1080/09583157.2013.808739

[80] Heffer P, Magen H, Mikkelsen R, Wichelns D. International Water Management Institute (IWMI), International Plant Nutrition Institute (IPNI), and International Potash Institute IPI. 2015. Available from: www.ipipotash.org

[81] Wang X, Sale P, Franks A, et al. An insight into the effect of organic amendments on the transpiration efficiency of wheat plant in a sodic duplex soil. Frontiers in Plant Science. 2021;**12**:722000. DOI: 10.3389/fpls.2021.722000

[82] Hatfield JL, Dold C. Water-use efficiency: Advances and challenges in a changing climate. Frontiers in Plant Science. 2019;**10**:103. DOI: 10.3389/fpls.2019.00103

[83] Ierna A, Mauromicale G. Sustainable and profitable nitrogen fertilization management of potato. Agronomy. 2019;**9**(10):582. DOI: 10.3390/agronomy9100582

[84] Kaur A, Singh KB, Gupta RK, Alataway A, Dewidar AZ, Mattar MA. Interactive effects of nitrogen application and irrigation on water use, growth and tuber yield of potato under subsurface drip irrigation. Agronomy. 2023;**13**(1):11. DOI: 10.3390/agronomy13010011

[85] Ospina Nieto CA, Lammerts van Bueren ET, Allefs S, Vos PG, van der Linden G, Maliepaard CA, et al. Association mapping of physiological and morphological traits related to crop development under contrasting nitrogen inputs in a diverse set of potato cultivars. Plants. 2021;**10**(8):1727. DOI: 10.3390/plants10081727

[86] Blecharczyk A, Kowalczewski PŁ, Sawinska Z, Rybacki P, Radzikowska-Kujawska D. Impact of crop sequence and fertilization on potato yield in a long-term study. Plants (Basel). 2023;**12**(3):495. DOI: 10.3390/plants12030495

[87] Neshev N, Manolov I. Content and uptake of nutrients with plant biomass of potatoes depending on potassium fertilization. Agriculture and Agricultural Science Procedia. 2015;**6**:63-66. DOI: 10.1016/j.aaspro.2015.08.039

[88] Xu Y, He P, Xu X, et al. Estimating nutrient uptake requirements for

potatoes based on quefts analysis in China. Agronomy Journal. 2019;**111**(5):2387-2394. DOI: 10.2134/agronj2018.09.0572

[89] Chea L, Pfeiffer B, Schneider D, Daniel R, Pawelzik E, Naumann M. Morphological and metabolite responses of potatoes under various phosphorus levels and their amelioration by plant growth-promoting rhizobacteria. International Journal of Molecular Sciences. 2021;**22**(10):5162. DOI: 10.3390/ijms22105162

[90] Pereira SIA, Abreu D, Moreira H, Vega A, Castro PML. Plant growth-promoting rhizobacteria (PGPR) improve the growth and nutrient use efficiency in maize (*Zea mays* L.) under water deficit conditions. Heliyon. 2020;**6**(10):e05106. DOI: 10.1016/j.heliyon.2020.e05106

[91] Wu XW, Liu GD, Riaz M, Yan L, Jiang CC. Metabolic changes in roots of trifoliate orange [*Poncirus trifoliate* (L.) *Raf*.] as induced by different treatments of boron deficiency and resupply. Plant and Soil. 2018;**434**:217-229. DOI: 10.1007/s11104-018-3684-8

[92] Munda S, Shivakumar BG, Rana DS, et al. Inorganic phosphorus along with biofertilizers improves profitability and sustainability in soybean (*Glycine max*)–potato (*Solanum tuberosum*) cropping system. Journal of the Saudi Society of Agricultural Sciences. 2018;**17**(2):107-113. DOI: 10.1016/j.jssas.2016.01.008

[93] Gouda S, Kerry RG, Das G, Paramithiotis S, Shin HS, Patra JK. Revitalization of plant growth promoting rhizobacteria for sustainable development in agriculture. Microbiological Research. 2018;**206**:131-140. DOI: 10.1016/j.micres.2017.08.016

[94] Guo DJ, Li DP, Singh RK, et al. Comparative transcriptome analysis of two sugarcane varieties in response to diazotrophic plant growth promoting endophyte Enterobacter roggenkampii ED5. Journal of Plant Interactions. 2022;**17**(1):75-84. DOI: 10.1080/17429145.2021.2012608

[95] Gulati A, Sharma N, Vyas P, et al. Organic acid production and plant growth promotion as a function of phosphate solubilization by Acinetobacter rhizosphaerae strain BIHB 723 isolated from the cold deserts of the trans-Himalayas. Archives of Microbiology. 2010;**192**(11):975-983. DOI: 10.1007/s00203-010-0615-3

[96] Gupta S, Gupta R, Sharma S. Impact of pesticides on plant growth promotion of Vigna radiata and non-target microbes: Comparison between chemical- and bio-pesticides. Ecotoxicology. 2014;**23**(6):1015-1021. DOI: 10.1007/s10646-014-1245-3

[97] Vessey JK. Plant growth promoting rhizobacteria as biofertilizers. Plant and Soil. 2003;**255**:571-558

[98] Emad A. Abdeldaym, El-Sawy, M.B.I, M.A.El-Helaly. Combined application of different sources of nitrogen fertilizers for improvement of potato yield and quality. Plant Archives. 2019

[99] Khan A, Singh J, Upadhayay VK, Singh AV, Shah S. Microbial Biofortification: A green technology through plant growth promoting microorganisms. In: Shah S, Venkatramanan V, Prasad R, editors. Sustainable Green Technologies for Environmental Management. Singapore: Springer; 2019. p.255-269. DOI: 10.1007/978-981-13-2772-8_13

[100] Mukhongo RW, Tumuhairwe JB, Ebanyat P, AbdelGadir AH, Thuita M,

Masso C. Combined application of biofertilizers and inorganic nutrients improves sweet potato yields. Frontiers in Plant Science. 2017;**8**:219 https://www.frontiersin.org/articles/10.3389/fpls.2017.00219

[101] Pantigoso HA, He Y, Manter DK, Fonte SJ, Vivanco JM. Phosphorus-solubilizing bacteria isolated from the rhizosphere of wild potato *Solanum bulbocastanum* enhance growth of modern potato varieties. Bulletin of the Natinal Research Centre. 2022;**46**(1):224. DOI: 10.1186/s42269-022-00913-x

[102] Tirry N, Kouchou A, El Omari B, Ferioun M, El Ghachtouli N. Improved chromium tolerance of *Medicago sativa* by plant growth-promoting rhizobacteria (PGPR). Journal, Genetic Engineering & Biotechnology. 2021;**19**(1):149. DOI: 10.1186/s43141-021-00254-8

[103] Drechsel P, Heffer P, Magen H, Mikkelsen R, Wichelns D, editiors. Managing Water and Fertilizer for Sustainable Agricultural Intensification. International Fertilizer Industry Association (IFA), International Water Management Institute (IWMI), International Plant Nutrition Institute (IPNI), and International Potash Institute (IPI). 2015.1st ed. Paris, France.

[104] Saed-Moucheshi A, Sohrabi F, Fasihfar E, Baniasadi F, Riasat M, Mozafari AA. Superoxide dismutase (SOD) as a selection criterion for triticale grain yield under drought stress: A comprehensive study on genomics and expression profiling, bioinformatics, heritability, and phenotypic variability. BMC Plant Biology. 2021;**21**(1):148. DOI: 10.1186/s12870-021-02919-5

[105] Gangadhar BH, Sajeesh K, Venkatesh J, Baskar V, Abhinandan K, Yu JW, et al. Enhanced tolerance of transgenic potato plants over-expressing non-specific lipid transfer protein-1 (StnsLTP1) against multiple abiotic stresses. Frontiers in Plant Science. 2016;7:1228. DOI: 10.3389/fpls.2016.01228

[106] Abid M, Ali S, Qi LK, et al. Physiological and biochemical changes during drought and recovery periods at tillering and jointing stages in wheat (*Triticum aestivum* L.). Scientific Reports. 2018;**8**(1):4615. DOI: 10.1038/s41598-018-21441-7

[107] Msimbira LA, Smith DL. The roles of plant growth promoting microbes in enhancing plant tolerance to acidity and alkalinity stresses. Frontiers in Sustainability Food System. 2020;4:106. Available from: https://www.frontiersin.org/articles/10.3389/fsufs.2020.00106

[108] Xiong Q, Hu J, Wei H, Zhang H, Zhu J. Relationship between plant roots, rhizosphere microorganisms, and nitrogen and its special focus on rice. Agriculture. 2021;**11**(3):234. DOI: 10.3390/agriculture11030234

[109] Sprenger H, Erban A, Seddig S, et al. Metabolite and transcript markers for the prediction of potato drought tolerance. Plant Biotechnology Journal. 2018;**16**(4):939-950. DOI: 10.1111/pbi.12840

[110] Gupta S, Pandey S. ACC deaminase producing bacteria with multifarious plant growth promoting traits alleviates salinity stress in french bean (*Phaseolus vulgaris*) plants. Frontiers in Microbiology. 2019;**10**:1506. Available from: https://www.frontiersin.org/articles/10.3389/fmicb.2019.01506

[111] Niu X, Song L, Xiao Y, Ge W. Drought-tolerant plant growth-promoting rhizobacteria associated with foxtail millet in a semi-arid

agroecosystem and their potential in alleviating drought stress. Frontiers in Microbiology. 2018;**8**:2580. DOI: 10.3389/fmicb.2017.02580

[112] Shen H, He X, Liu Y, Chen Y, Tang J, Guo T. A complex inoculant of N2-fixing, P- and K-solubilizing bacteria from a purple soil improves the growth of kiwifruit (*Actinidia chinensis*) plantlets. Frontiers in Microbiology. 2016;7:841. Available from: https://www.frontiersin.org/articles/10.3389/fmicb.2016.00841

[113] Gowtham HG, Singh B, Murali M, et al. Induction of drought tolerance in tomato upon the application of ACC deaminase producing plant growth promoting rhizobacterium *Bacillus subtilis* Rhizo SF 48. Microbiological Research. 2020;**234**:126422. DOI: 10.1016/j.micres.2020.126422

[114] dos Santos RM, Diaz PAE, Lobo LLB, Rigobelo EC. Use of plant growth-promoting rhizobacteria in maize and sugarcane: Characteristics and applications. Frontiers in Sustainability Food System. 2020;**4**:136. Available from: https://www.frontiersin.org/articles/10.3389/fsufs.2020.00136

[115] Batool T, Ali S, Seleiman MF, et al. Plant growth promoting rhizobacteria alleviates drought stress in potato in response to suppressive oxidative stress and antioxidant enzymes activities. Scientific Reports. 2020;**10**(1):16975. DOI: 10.1038/s41598-020-73489-z

[116] Ahmad HM, Fiaz S, Hafeez S, et al. Plant growth-promoting rhizobacteria eliminate the effect of drought stress in plants: A review. Frontiers Plant Science. 2022;**13**:875774. DOI: 10.3389/fpls.2022.875774

[117] Dahal K, Li XQ, Tai H, Creelman A, Bizimungu B. Improving potato stress tolerance and tuber yield under a climate change scenario - a current overview. Frontiers in Plant Science. 2019;**10**:563. DOI: 10.3389/fpls.2019.00563

[118] Pastore C, Pii Y, Liu X, et al. Aminocyclopropane-1-carboxylate deaminase-producing plant growth-promoting rhizobacteria improve drought stress tolerance in grapevine (*Vitis vinifera* L.). Frontiers in Plant Science www.frontiersin.org. 2021;**12**:706990. DOI: 10.3389/fpls.2021.706990

[119] Tiwari G, Duraivadivel P, Sharma SPH. 1-Aminocyclopropane-1-carboxylic acid deaminase producing beneficial rhizobacteria ameliorate the biomass characters of *Panicum maximum Jacq*. By mitigating drought and salt stress. Scientific Reports. 2018;**8**(1):17513. DOI: 10.1038/s41598-018-35565-3

[120] Mozumder AB, Chanda K, Chorei R, Prasad HK. An evaluation of aluminum tolerant *Pseudomonas aeruginosa* A7 for *In vivo* suppression of *Fusarium* wilt of chickpea caused by *Fusarium oxysporum* f. sp. *ciceris* and growth promotion of chickpea. Microorganisms. 2022;**10**(3):568. DOI: 10.3390/microorganisms10030568

[121] Pierce GE, Tucker TA, Wang C, Swensen K, Crow SA. Delayed ripening of climacteric fruit by catalysts prepared from induced cells of *Rhodococcus rhodochrous* DAP 96253: A case for the biological modulation of yang-cycle driven processes by a prokaryote. Indian Journal of Biotechnology. 2014;**10**(5):354-362. DOI: 10.1089/ind.2014.0016

[122] Gashgari RM, Gherbawy YA. Pathogenicity of some *Fusarium* species associated with superficial blemishes of potato tubers. Polish Journal of Microbiology. 2013;**62**(1):59-66

[123] Kowalski KP, Bacon C, Bickford W, et al. Advancing the science of microbial symbiosis to support invasive species management: A case study on phragmites in the great lakes. Frontiers in Microbiology. 2015;**6**:95. DOI: 10.3389/fmicb.2015.00095

[124] Grossi CEM, Fantino E, Serral F, Zawoznik MS, Fernandez Do Porto DA, Ulloa RM. *Methylobacterium* sp. 2A is a plant growth-promoting rhizobacteria that has the potential to improve potato crop yield under adverse conditions. Frontiers in Plant Science. 2020;11:71. DOI: 10.3389/fpls.2020.00071

[125] Costa OYA, Raaijmakers JM, Kuramae EE. Microbial extracellular polymeric substances: Ecological function and impact on soil aggregation. Frontiers in Microbiology. 2018;**23**(9):1636. DOI: 10.3389/fmicb.2018.01636

[126] Timmusk S, Copolovici D, Copolovici L, Teder T, Nevo E, Behers L. *Paenibacillus polymyxa* biofilm polysaccharides antagonise *Fusarium graminearum*. Scientific Reports. 2019;**9**(1):662. DOI: 10.1038/s41598-018-37718-w

[127] Hosseini MS, Zahedi SM, Abadía J, Karimi M. Effects of postharvest treatments with chitosan and putrescine to maintain quality and extend shelf-life of two banana cultivars. Food Science & Nutrition. 2018;29;**6**(5):1328-1337. DOI: 10.1002/fsn3.662

[128] Feliziani E, Smilanick JL, Margosan DA, et al. Preharvest fungicide, potassium sorbate, or chitosan use on quality and storage decay of table grapes. Plant Disease. 2013;**97**(3):307-314. DOI: 10.1094/PDIS-12-11-1043-RE

[129] Stasinska-Jakubas M, Hawrylak-Nowak B. Protective, biostimulating, and eliciting effects of chitosan and its derivatives on crop plants. Molecules. 2022;28;**27**(9):2801. DOI: 10.3390/molecules27092801

[130] Constantin ME, de Lamo FJ, Vlieger BV, Rep M, Takken FLW. Endophyte-mediated resistance in tomato to *Fusarium oxysporum* is independent of ET, JA, and SA. Frontiers in Plant Science. 2019;**31**(10):979. DOI: 10.3389/fpls.2019.00979

[131] Lopez-Moya F, Suarez-Fernandez M, Lopez-Llorca LV. Molecular mechanisms of chitosan interactions with fungi and plants. International Journal of Molecular Sciences. 2019;**20**(2):E332. DOI: 10.3390/ijms20020332

[132] Long LT, Tan LV, Boi VN, Trung TS. Antifungal activity of water-soluble chitosan against *Colletotrichum capsici* in postharvest chili pepper. Journal of Food Processing and Preservation. 2018;**42**:e13339

[133] Mejdoub-Trabelsi B, AbdallahR AB, AmmarN DR. Antifungal potential of extracellular metabolites from *Penicillium* spp. and *aspergillus* spp. naturally associated to potato against Fusarium species causing tuber dry rot. Journal of microbial and biochemical. Technology. 2017;**9**:181-190

[134] Liebe S, Wibberg D, Winkler A, Pühler A, Schlüter A, Varrelmann M. Taxonomic analysis of the microbial community in stored sugar beets using high-throughput sequencing of different marker genes. FEMS Microbiology Ecology. 2016;**92**(2):1-12. DOI: 10.1093/femsec/fiw004

[135] Carmona-Hernandez S, Reyes-Perez JJ, Chiquito-Contreras RG,

Rincon-Enriquez G, Cerdan-Cabrera CR, Hernandez-Montiel LG. Biocontrol of postharvest fruit fungal diseases by bacterial antagonists: A review. Agronomy. 2019;**9**(3):121. DOI: 10.3390/agronomy9030121

[136] Wang Z, Li Y, Zhuang L, et al. A rhizosphere-derived consortium of *Bacillus subtilis* and *Trichoderma harzianum* suppresses common scab of potato and increases yield. Computational and Structural Biotechnology Journal. 2019;**17**:645-653. DOI: 10.1016/j.csbj.2019.05.003

[137] Devi AR, Sharma GD, Majumdar PB, Pandey P. A multispecies consortium of bacteria having plant growth promotion and antifungal activities, for the management of *Fusarium wilt* complex disease in potato (*Solanum tuberosum* L.). Biocatalysis and Agricultural Biotechnology. 2018;**16**:614-624

[138] Maciag T, Krzyzanowska DM, Jafra S, Siwinska J, Czajkowski R. The great five-an artificial bacterial consortium with antagonistic activity towards *Pectobacterium* spp. and *Dickeya* spp.: Formulation, shelf life, and the ability to prevent soft rot of potato in storage. Applied Microbiology and Biotechnology. 2020;**104**(10):4547-4561. DOI: 10.1007/s00253-020-10550-x

[139] Aguk JA, Karanja N, Schulte-Geldermann E, Bruns C, Kinyua Z, Parker M. Control of bacterial wilt (*Ralstonia solanacearum*) in potato (*Solanum tuberosum*) using rhizobacteria and arbuscular mycorrhiza fungi. African Journal of Food, Agriculture, Nutrition and Development. 2018;**18**(2):13371-13387. DOI: 10.18697/ajfand.82.16905

[140] Jafra S, Przysowa J, Czajkowski R, Michta A, Garbeva P, van der Wolf JM. Detection and characterization of bacteria from the potatorhizosphere degrading N-acyl-homoserine lactone. Canadian Journal of Microbiology. 2006;**52**:1006-1015

[141] Jafra S, Przysowa J, Gwizdek-Wi Sniewska A, van der Wolf JM. Potential of bulb-associated bacteria for biocontrol of hyacinth soft rot caused by *Dickeya zeae*. Journal of Applied Microbiology. 2009;**106**:268-277

[142] Osei R, Yang C, Cui L, Wei L, Jin M. Antagonistic bioagent mechanisms of controlling potato soft rot. Plant Protection Science. 2022;**58**(1):18-30. DOI: 10.17221/166/2020-PPS

[143] Krzyzanowska DM, Maciag T, Siwinska J, Krychowiak M, Jafra S, Czajkowski R. Compatible mixture of bacterial antagonists developed to protect potato tubers from soft rot caused by *Pectobacterium* spp. and *Dickeya* spp. Plant Disease. 2019;**103**(6):1374-1382. DOI: 10.1094/PDIS-10-18-1866-RE

[144] Danchin A, Sekowska A. The role of information in evolutionary genomics of bacteria. In: Caetano-Anolles G, editor. Evolutionary Genomics and Systems Biology. New Jersey: Wiley; 2010. pp. 81-95

[145] Bertani I, Zampieri E, Bez C, Volante A, Venturi V, Monaco S. Isolation and characterization of pseudomonas chlororaphis strain ST9; rhizomicrobiota and in planta studies. Plants. 2021;**10**(7):1466. DOI: 10.3390/plants10071466

[146] Furnkranz M, Lukesch B, Müller H, Huss H, Grube M, Berg G. Microbial diversity inside pumpkins: Microhabitat-specific communities display a high antagonistic potential against phytopathogens. Microbial Ecology.

2012;**63**(2):418-428. DOI: 10.1007/s00248-011-9942-4

[147] Rybakova D, Mancinelli R, Wikström M, et al. The structure of the *Brassica napus* seed microbiome is cultivar-dependent and affects the interactions of symbionts and pathogens. Microbiome. 2017;**5**(1):104. DOI: 10.1186/s40168-017-0310-6

[148] Glassner H, Zchori-Fein E, Compant S, et al. Characterization of endophytic bacteria from cucurbit fruits with potential benefits to agriculture in melons (*Cucumis melo* L.). FEMS Microbiology Ecology. 2015;**91**(7):fiv074. DOI: 10.1093/femsec/fiv074

[149] Tahir HA, Gu Q, Wu H, Niu Y, Huo R, Gao X. *Bacillus volatiles* adversely affect the physiology and ultra-structure of *Ralstonia solanacearum* and induce systemic resistance in tobacco against bacterial wilt. Scientific Reports. 2017;**7**:40481. DOI: 10.1038/srep40481

Section 2

Plant Growth Fungi

Chapter 7

Development and Resource Exchange Processes in Root Symbioses of Legumes

Debatosh Das, Salina Tripathi, Prema Mutyala, Dhruv Aditya Srivastava and Heike Bücking

Abstract

Plants are associated with complex microbiomes, and many of the microorganisms that reside on plant surfaces (epiphytes) or within plant tissues (endophytes) are beneficial for the host plant and improve plant growth or stress resistance by a variety of plant growth-promoting capabilities. The plant microbiome could serve as a tool box to design synthetic microbiomes to enhance plant growth and crop resiliency under stress or to integrate benefits of plant microbiomes as important traits into plant breeding programs. For legumes, the most important members of the plant microbiome are nitrogen (N)-fixing rhizobia and arbuscular mycorrhizal (AM) fungi. Legumes harbor rhizobia in specialized root nodules, in which the bacteria fix gaseous N from the atmosphere and transfer plant available forms of N to host. AM fungi play a key role for the uptake of nutrients such as phosphate and nitrogen and improve the resistance of plants against abiotic (e.g. drought, salinity, and heavy metals) and biotic (herbivores and pathogens) stresses. Both partners compete with these benefits for photosynthetically fixed carbon from the host. In this review, we will summarize our current understanding of these interactions and will also focus on cooperative or competitive interactions between these two root symbionts in tripartite interactions.

Keywords: arbuscular mycorrhizal symbiosis, biological nitrogen fixation, mutualism, nutrient uptake, plant biotic interaction, rhizobia

1. Introduction

Plants are sessile organisms, and to compensate for their lack in mobility, plants evolved specialized mechanisms that allowed them to adapt to changing environments and to a variety of abiotic and biotic stresses. Arguably, the most important adaptation to stress was the development of beneficial plant microbe interactions. Due to recent developments in sequencing technologies, we have a better understanding of the concept that plants are meta-organisms, whose phenotype, particularly under stress, is not only shaped by plant traits but also by their associated microbiomes. The plant microbiome represents "the second plant genome" and consists of 10 times more genes than typical plant genomes and is an unexplored resource for a wide range of potentially plant growth-promoting capabilities [1]. A better understanding of beneficial plant microbe interactions could be a key to the development of microbial fertilizers or microbial pesticides, as well as new biotechnological tools that increase the nutrient efficiency and stress tolerance of crops in environments that are increasingly affected by climate change and other stresses.

The most important root symbioses for legumes are interactions with nitrogen-fixing rhizobia and arbuscular mycorrhizal (AM) fungi. Legumes harbor rhizobia in specialized root organs or nodules, and their biological nitrogen (N) fixation can contribute with up to 77% to the total N nutrition of crop legumes [2]. AM fungi colonize the majority of land plants and transfer nutrients such as phosphate (P), N, and potassium (K) to their host and improve the host plant's resistance against abiotic (drought, salinity, and heavy metals) and biotic (herbivores and root pathogens) stresses [3]. Arbuscule-like structures found in the cells of early land plant fossils suggests that the AM symbiosis played a key role during the evolution of land plants, and the AM symbiosis is therefore also called "the mother of all root endosymbioses" [4, 5]. We will focus here on the symbiosis between legumes, rhizobia, and AM fungi and will summarize our current knowledge about the development and nutritional benefits of these interactions and discuss knowledge gaps that still limit their application potential.

2. Beneficial root symbioses of legumes

2.1 Arbuscular mycorrhizal symbiosis

The arbuscular mycorrhizal (AM) symbiosis is a mutualistic interaction between approximately 70% of all known land plant species and fungi of the phylum *Glomeromycota* [6]. AM fungi are ubiquitous in soils, and the extraradical mycelium (ERM) of the fungus can account for up to 50% of the microbial biomass in soils [7]. The ERM of the AM fungus can explore a larger soil volume, mine for nutrients beyond any root depletion zone, and transfer these nutrients to the intraradical mycelium (IRM) in the host root, where these nutrients are exchanged against carbohydrates and lipids from the host. The AM fungus provides the host plant with soil nutrients, such as P, N, K, and sulfur (S), but also trace elements, such as copper and zinc. In addition, AM fungi provide nonnutritional benefits to their host plant and improve the resistance of plants against various abiotic (drought, salinity, and heavy metals) and biotic (pathogens and herbivores) stresses [3]. Due to the key role that AM fungi play for the survival and fitness of plants, they have also been described as "ecosystem engineers" of plant communities [8].

Roots that are colonized with AM fungi have two pathways for nutrient uptake: the plant uptake pathway (PP) and the mycorrhizal uptake pathway (MP). The PP involves the uptake of nutrients from the soil via high- or low-affinity uptake transporters in the epidermis or root hairs. However, nutrients such as P are relatively immobile in the soil, and the efficiency of the PP is often limited by the development of depletion zones around the roots. The MP, on the other hand, is characterized by the uptake of nutrients from the soil via high-affinity nutrient transporters in the ERM, followed by the translocation of nutrients from the ERM to the IRM in the root cortex, and the uptake of nutrients from the mycorrhizal interface through AM-inducible plant uptake transporters. In AM roots, the MP represents often the dominant pathway for plant nutrient uptake [9, 10].

AM fungi and their plant partners form a complex network of many-to-many interactions; each plant host is colonized by communities of AM fungi, and AM fungi colonize multiple host plants and connect plants via common mycorrhizal networks (CMNs). CMNs are involved in the long-distance transport of nutrients, water, stress chemicals, and allelochemicals and allow plants to "communicate" with other plants of the same or of different species that share the same CMN [11–13]. Many-to-many interactions allow both partners in the AM symbiosis to choose among multiple trading partners and force both partners to compete with other partners for nutrient or carbon resources.

The colonization of the root by AM fungi is initiated through a bidirectional exchange of signals. Before the AM symbiosis is established, plants release the carotenoid-derived plant hormone strigolactone, which activates fungal metabolism and stimulates hyphal branching during the pre-symbiotic growth phase (**Figure 1**) [14–16]. In addition to their role in plant partner recognition in parasitic and beneficial plant microbe interactions, strigolactones play a key role for the adaptation of plants against a variety of abiotic stresses and actively participate within regulatory networks of plant stress adaptation regulated by phytohormones. Abiotic stresses that have been shown to affect strigolactone production and/or the expression of genes involved in strigolactone biosynthesis include P and N deficiency, heat, UV, wounding, salinity, and drought [17]. For example, as part of the P starvation signal activated by PHR proteins in plants, strigolactone biosynthesis increases [18]. In response to strigolactones, AM fungi secrete Myc factors that are composed of lipo-chitooligosaccharides (LCOs) and chitin oligomers and are likely recognized by LysM-receptor-like kinases of the plant (RLKs) and prepare the root for colonization.

The developmental events in the establishment of the AM symbiosis include epidermal penetration, restructuring of the underlying epidermal and cortical cells, the assembly of a pre-penetration apparatus (PPA), intraradical colonization by hyphal elongation, and the development of arbuscules in inner cortical cells (**Figure 1**). After fungal hyphae have established contact with the host root, the AM fungus forms an appressorium or hyphopodium, a specialized cell or adhesion structure on the root surface. After the recognition of AM fungal appressoria, root epidermal cells develop

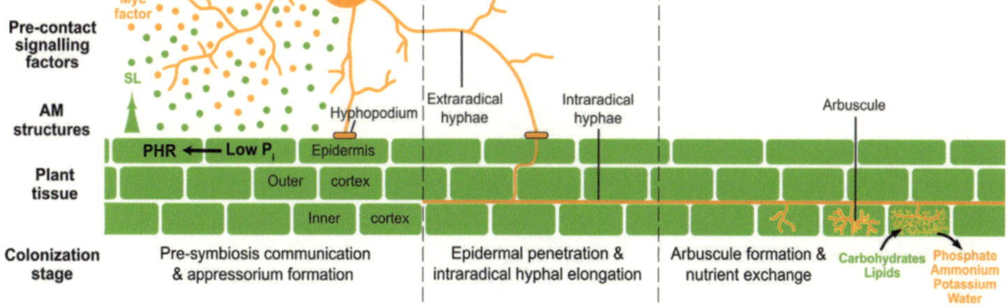

Figure 1.
Pre-symbiotic signaling and intraradical development in the AM symbiosis. Under phosphate starvation, PHR-induced strigolactones (SL) are released by roots and activate the AM fungus, which in response releases Myc factors that are perceived by LysM-receptor-like kinases of the plant. The fungus forms an appressorium on the root surface after successful plant recognition. When AM fungal appressoria are detected, epidermal cells produce a pre-penetration apparatus (PPA) that guides AM fungal hyphae through the epidermis into the root cortex. Characteristic for the AM colonization is the development of arbuscules, highly branched structures in cortical cells, which play a critical role for the bidirectional exchange of nutrients across the mycorrhizal interface.

a PPA, which directs AM hyphae through the epidermis into the root cortex and controls the intracellular path of hyphal penetration [19]. Arbuscules are highly branched structures that act as sites of nutrient exchange between the fungal and plant partner. Concurrent with these morphological events, a signaling cascade involving receptor kinases, nucleoproteins, ion channels, and a transcriptional complex takes place to accommodate the fungus inside roots (see 2.3 below).

2.2 The rhizobium-legume symbiosis

Legumes are characterized by their ability to establish symbiotic interactions with diazotrophic soil bacteria (known as rhizobia). Rhizobia can fix gaseous N_2 and convert it into plant-available forms of N in specialized root structures or nodules. Critical for the establishment of the symbiosis between legumes and rhizobia is a two-way recognition process between both partners. A key component of this recognition process is the secretion of flavonoid compounds by plant roots, which are recognized through bacterial NodD receptors and induce the biosynthesis of bacterial lipo-chitooligosaccharides or Nod-factors. The production of flavonoids and the expression of chalcone synthase (the first committed enzyme of flavonoid biosynthesis) and isoflavone reductase (conversion of flavanones to isoflavones) is triggered by N deficiency of the plant [20]. NodD belongs to the family of LysR-type transcriptional regulators that mediate the expression of nodulation (*Nod*) genes in rhizobia. Common in all bacteria that nodulate legumes are the *Nod* genes *nodABC,* which are all responsible for Nod factor biosynthesis [21]. Nod factors released by rhizobia are composed of lipo-chitooligosaccharides and are perceived by LysM-domain receptor-like kinases, such as NFR1 and NFR5 in *Lotus japonicus* [22, 23]. The perception and recognition of Nod factors by the host plant triggers a series of plant responses involved in nodule development and infection.

Unlike the AM symbiosis, the rhizobium-legume (RL) symbiosis is highly specific, and each rhizobial strain establishes a symbiosis with only a limited number of host plants and vice versa [24]. The ability of rhizobial species to recognize and interpret specific flavonoid signals produced by compatible host plants through NodD is in part responsible for the host specificity in RL interactions [25]. There is also a high level of specificity at the N-fixing stage in the RL symbiosis. Bacterial strains can form N-fixing root nodules on one plant genotype (Nod^+/Fix^+) and nodulate other plant genotypes, but the formed nodules of this plant are unable to fix N (Nod^+/Fix^-). This specificity is caused by the nodule-specific cysteine-rich (NCR) peptides NFS1 and NFS2 of the plant, which induce bacterial cell death, and early nodule senescence dependent on the rhizobial strain and the genetic background of the host. It has been suggested that NCR peptides possess prosymbiotic and antisymbiotic properties and that they play a role in fine-tuning the activity of rhizobia for optimum symbiotic performance [26].

Parallel to the change in root morphology, which includes cortical cell division and differentiation into nodule primordia, rhizobia are directed into the cortex. Rhizobia can enter roots through nod-dependent infection pockets formed during root curling via tubular infection threads (ITs) or cracks in the root surface. In addition, rhizobia can also gain entry via intercellular spaces through a nod-factor-independent process [27]. Necessary for most infections in legumes, however, is root hair curling, which involves root hairs that are tightly bending and entrap bacteria in a "shepherds crook" structure [28]. This deformation of root hairs is typically only induced by the release of Nod factors from bacteria closely attached to root hair tips. In some legumes, the Nod factor (NodRm-1 in *Medicago sativa*) alone can elicit cortical cell division and root hair deformation [29]. In other cases, Nod factors induce oscillations in intracellular calcium levels, which later activate calcium- and calmodulin-dependent kinase (CCaMK), a symbiotic signaling pathway that is important in the activation of transcription factors critical for nodule organogenesis [30].

In legumes, two types of root nodules can be distinguished. Indeterminate nodules develop after Nod factor perception as a result of periclinal cell divisions in the pericycle followed by inner cortical cell proliferation. Indeterminate nodules are characterized by a persistent meristem and the development of distinct zones within the root nodule: meristem, infection zone, interzone, fixation zone, and senescence zone. By contrast, determinate nodules have a defined lifespan and lose their central meristem. Determinate nodules develop by cell divisions in the outer root cortex, but the cells lose their meristematic activity when the nodule matures [31]. The bacteria are taken up by a process similar to endocytosis of the plasma membrane where bacteria proliferate into N-fixing bacteroids and form a symbiosome, an organelle-like structure consisting of 1–10 bacteroids enclosed by a plant-derived peribacteroid membrane or symbiosome membrane (SYM) and the symbiosome space that is located between the bacteroids and the SYM [32].

2.3 The common symbiosis signaling pathway

AM or RL symbiosis play a critical role for the nutrient supply of their host, and low P_i or N supply conditions have been shown to stimulate the release of strigolactones or flavonoids by host plants to recruit AM fungi or rhizobia. AM fungi and rhizobia respond to these signals with the release of Myc or Nod factors, respectively, which are composed of sulfated or non-sulfated lipo-chitooligosaccharides [33]. The perception and interpretation of Myc or Nod factors by the host plant plays a critical role in the establishment of both root symbioses. Compared to the AM symbiosis, which evolved around 450 to 480 million years ago, the mutualistic RL symbiosis is much younger and evolved approximately 58 million years ago [34]. The similarities in the signaling pathways of the AM and RL symbiosis has led to the assumption that the RL symbiosis evolved via the adoption of a signaling pathway that had previously been established for the successful colonization of plants by AM fungi. Similarly, the same signaling pathway can also be found in ectomycorrhizal associations that evolved also much later than the AM symbiosis [35, 36]. The common symbiotic signaling pathway (CSSP) is a conserved molecular signaling pathway that plays a key role in the establishment of the AM and RL symbiosis and acts downstream of Myc and Nod factor perception and upstream of the activation of processes required for the root colonization by specific root symbionts. Mutations in the CSSP prevent both fungal and bacterial entry into the host root [37].

The first component of the CSSP are heteromers of plasma membrane-localized lysin motif-type receptor-like kinases, such as the Nod factor receptors NFR1/NFR5 in *Lotus*, or LYK3/NFP in *Medicago* [38, 39], or the Myc factor receptor MYR1/LYK2 and CERK1 in rice [40]. NFR1 and NFR5 can both bind to Nod factors, but only NFR1 has kinase activity and can phosphorylate NFR5 and initiate the signal transduction cascade [38]. Kinase activities are essential for their function in rhizobial symbiosis. Some receptor kinases also seem to play a role in the recognition of chitin oligosaccharides in the cell walls of pathogenic fungi. For example, the fact that *CERK1* knockout mutants in rice are not only impaired in AM development but also in chitin-triggered defense responses indicates that CERK1 plays not only a role in mutualistic but also in parasitic interaction recognition [41].

Downstream of Myc or Nod factor perception, the symbiosis receptor-like kinase SYMRK is indispensable for the development of the AM or rhizobial symbiosis. In *Lotus japonicus*, SYMRK forms a receptor complex with NFR5 [42], and it was shown that SYMRK is able to suppress BAK1 kinase activity (BRASSINOSTEROID INSENSITIVE 1-Associated receptor Kinase 1). BAK1 is a positive regulator of plant innate immunity, and its inhibition by SYMRK could indicate that rhizobia actively suppress the host's immune response [43]. Only minutes after Nod factor perception, an activated oscillation in cytoplasmic Ca^{2+} concentrations in the perinuclear region can be detected. In *Lotus japonicus*, two ion channels (CASTOR and POLLUX) [44, 45] and three nuclear pore proteins (NUP85, NUP133, and SEH1) [46, 47] act in a signaling cascade to generate this epidermal calcium spiking. This calcium spiking leads to a derepression of a nuclear-localized calcium calmodulin-dependent kinase (CCaMK) [48]. Subsequently, CCaMK (DMI3) phosphorylates the transcription factor CYCLOPS (IPD3) [49]. The result is a symbiont-specific transcriptional reprogramming, which is necessary to support the colonization of the root by AM fungi or start nodule organogenesis (**Figure 2**).

After the CSSP, specific signaling pathways regulate the development of the AM or RL symbiosis. In the AM symbiosis, molecular signaling coincides with the initiation of a PPA in plant cells to accommodate the invading fungus [50]. Subsequent fungal infection involves hyphal elongation and cortical root colonization and arbuscule formation. Arbuscule formation is regulated by a complex consisting

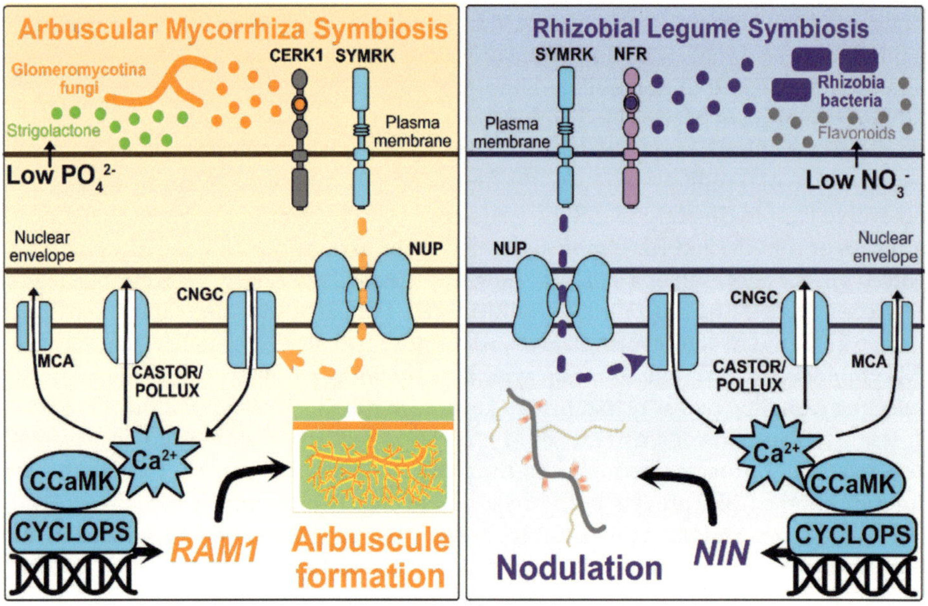

Figure 2.
Common Symbiosis Signaling pathway (CSSP). In pre-symbiotic stages, SL and flavonoids are released by plants to attract AM fungi and rhizobia. In return, both symbionts release Myc or Nod factors that are perceived by CERK1 or NFR in the plasma membrane. Signals generated at SYMRK pass through the nucleopore protein complex (NUP) and lead to the activation of ion channels and ATPases (CNGC, CASTOR/POLLUX, and MCA) and calcium spikes in the nucleus. Calcium spiking activates the kinase CCaMK, which phosphorylates and activates the transcription factor CYCLOPS. The CCaMK-CYCLOPS complex regulates AM or RL symbiosis-specific transcription through the RAM1 or NIN transcription factor, respectively. This leads to intraradical colonization and the development of arbuscules in the AM symbiosis and infection thread formation and primordia development in the RL symbiosis.

of CCaMK, CYCLOPS, and DELLA proteins, which induce the essential GRAS gene *RAM1* and are accompanied by an induction of AM transporter genes such as *PT4* and *AMT2;3* [51]. In rice, DELLA interacts with another GRAS protein, DIP1 (DELLA Interacting Protein 1), which in turn interacts with RAM1, and affects AM-induced gene expression [52].

By contrast, in the RL symbiosis, several nuclear-associated transcriptional regulators are essential for the expression of Nod-factor-induced genes and the initiation of nodulation, including nodule inception (NIN) [53], an ERF family protein (ERN) [54], and two GRAS family proteins, nodulation signaling pathway 1 (NSP1) and NSP2 [55, 56]. DELLA proteins promote nodule development and infection thread formation during root nodule symbiosis by promoting CCaMK-IPD3/CYCLOPS complex formation and increasing IPD3/CYCLOPS phosphorylation. DELLAs can also form a protein complex with NSP2 and NSP1 and bridge a protein complex containing IPD3/CYCLOPS and NSP2. It has been suggested that the combination of transcription factors such as NSP2 and NSP1 and CCaMK-IDP3 may act in tandem to control the expression of early nodulin genes [57]. First, the phosphorylated form of CYCLOPS (IPD3 in *Medicago truncatula*) binds to the NIN promoter and induces nodulation even in the absence of rhizobia [49], followed by the NSP1-NSP2 hetero complex, which binds to the promoters of the Nod-factor-inducible genes *ENOD11*, *ERN1*, and NIN [58]. Interestingly, NSPs are not required for CYCLOPS-induced NIN expression but for CYCLOPS-induced nodule organogenesis [49].

3. Nutrient uptake and transport across the mycorrhizal interface

3.1 Phosphate uptake and transport

The total phosphate (P) contents in soils can be high, but a large percentage of this P is not plant available. In addition to organic forms of P, inorganic phosphate (P_i) readily binds with iron, aluminum, and manganese when the pH in the soil is acidic or binds with calcium or calcium carbonate when the soil is alkaline and then becomes plant unavailable [59]. Due to the low soil concentrations, P is often a growth-limiting nutrient for plants (accounts for 0.2% of dry weight). In addition, due to the low P mobility in soils, plant P uptake leads rapidly to the development of depletion zones around the roots that further limit plant P uptake to the slow rate of diffusion. The ability of AM fungi to grow with their ERM beyond these depletion zones combined with their efficient P uptake systems is the main basis for their positive impact on P uptake and plant growth. Plants and fungi absorb P as negatively charged P ions ($H_2PO_4^-$). To take up P against the concentration gradient, high-affinity transporter proteins from the Pht1 family are required, which transport P into cells via a proton gradient generated by a plasma membrane H^+-ATPase [60].

During the development of arbuscules, the colonized cell undergoes a reorganization and starts to enclose newly formed hyphal branches of the arbuscule by a novel symbiosis-specific membrane, the periarbuscular membrane (PAM), which is an extension of the plant plasma membrane. The PAM plays a vital role in the nutrient exchange between the symbiotic partners and is composed of a variety of transport proteins specifically designed to facilitate the nutrient uptake from the symbiotic interface (**Figure 3**). The most extensively studied membrane proteins in the PAM are inorganic phosphate (P_i) transporters of the PHT1 gene family, such as MtPT4, LjPT4, and OsPT11 that were characterized in *Medicago truncatula*, *Lotus japonicus*,

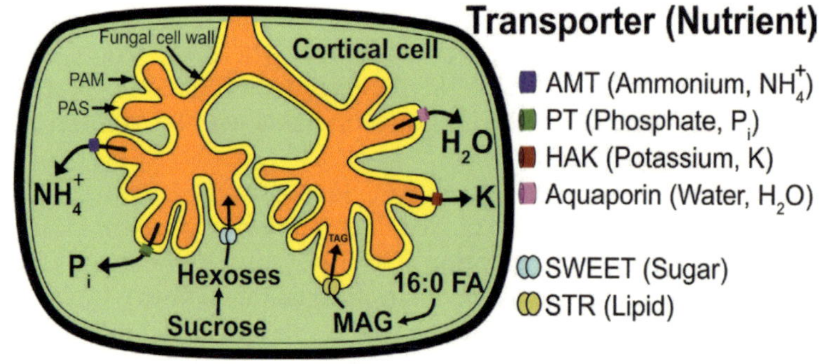

Figure 3.
Nutrient exchange in the arbuscular mycorrhiza symbiosis. Nutrients such as phosphate (P_i), ammonium (NH_4^+), potassium, and other resources such as water are captured by the AM fungus and transported to the arbusculated cells in the root cortex. Transporters in the periarbuscular membrane such as phosphate transporters (PT), ammonium transporters (AMT), high-affinity K^+ transporters (HAK), and aquaporins (AQP) facilitate the uptake by the plant from the mycorrhizal interface. In return, the plant rewards the fungus with carbohydrates and lipids transferred into the AM interface by the transporter STR.

or *Oryza sativa* (rice), respectively [61–63]. Most of these transporters are expressed explicitly after AM colonization and are exclusively located at the interface of the two symbionts in the PAM. Due to their unique expression pattern in response to AM fungal colonization, the expression of these AM-inducible P transporters can serve as a reliable indicator for AM colonization.

The PAM also contains P transporters that are essential for AM development but are not critical for symbiotic P_i uptake. For example, OsPT11 and OsPT13 are both critical for AM development in rice, and mutations in both transporters interfere with intraradical fungal development and growth of arbuscules, but only OsPT11 plays an active role in symbiotic P_i uptake. It has been suggested that OsPT13 could play a role as a P_i sensor at the PAM that regulates the development of arbuscules and thereby maximizes their symbiotic transport activity [64]. Similarly, AsPT1 and AsPT4, two AM-induced P_i transporters of *Astragalus sinicus*, that are also localized in the PAM, are essential for arbuscule development, but only AsPT4 is involved in symbiotic P_i transport [65]. AsPT1 could encode a functional transceptor, a protein with both solute transport and receptor-like signaling activity that could mediate symbiotic P_i uptake. Some AM-inducible P_i transporters are also expressed in nonmycorrhizal roots. For example, the AM-induced P_i transporter LjPT4 is essential for functional AM symbiosis but is also expressed in nonmycorrhizal root tips. It has been suggested that in nonmycorrhizal roots, LjPT4 integrates signals of the plant's P_i status and regulates P_i-dependent developmental programs, such as root branching [62].

3.2 Nitrogen uptake and transport

Nitrogen (N) is a major driver for crop yield, and its availability significantly impacts agricultural productivity. Since N can constitute in plant tissues between 1 and 5% of the total dry matter, plants require N in large quantities, and N is the nutrient that most often limits plant growth [66]. In soils, N is present in inorganic forms, such as nitrate (NO_3^-) or ammonium (NH_4^+), and in organic form, such as urea, free amino acids, and short peptides. The availability of these N pools, however, varies considerably due to soil heterogeneity and dynamic microbial conversions of these different N forms (e.g. nitrification and denitrification) [67]. In most aerobic soils,

NO_3^- is the primary type of N, whereas NH_4^+ can be the dominant form of N in acidic and/or anaerobic soils [68]. Plants can take up NH_4^+, NO_3^-, and organic N, but due to the strong competition with microorganisms for organic N uptake, they take up NH_4^+ and NO_3^- in larger quantities [69]. NO_3^- is very mobile in soils and readily leached, while NH_4^+ adsorbs onto the cation exchange complex of many soils and is only slowly released. Because plants have to also compete for NH_4^+ with soil microorganisms, which use NH_4^+ not only as an N source but also as an energy source and convert NH_4^+ into NO_3^- via nitrification, NO_3^- is a major N source for most higher plants [70].

The ERM of fungi can take up NH_4^+ and NO_3^-, but due to its higher energy efficiency, AM fungi typically prefer the uptake of NH_4^+ over NO_3^- [71]. AM fungi have high-affinity and low-affinity uptake systems for NH_4^+. Fungal high-affinity uptake systems for NH_4 have lower K_m values than plants, what allows AM fungi to take up NH_4^+ from soils with very low concentrations [72]. When NO_3^- is available, a high-affinity NO_3^- transporter is upregulated in the extraradical mycelium (ERM) of the AM fungus *Rhizophagus irregularis* [73]. However, the expression of this transporter is repressed by an external supply of NH_4^+ or a downstream metabolite such as glutamine [74]. Through a process that is known as N catabolite repression, NO_3^- transporters in many organisms are suppressed or degraded, when more preferred N sources, such as NH_4^+ become available, or when the concentrations of downstream metabolites in the cells increase [75]. Due to high mobility of NO_3^- in soils, the positive impact of AM fungi in N nutrition of plants is still under debate [76]. However, in *Medicago truncatula*, AM mycorrhizal growth responses can be tightly linked to mycorrhizal P but also N benefits [77]. In *MtPT4* mutants, in which the AM-induced P_i transporter is not expressed, intraradical root colonization is reduced, arbuscules prematurely degenerate, and the symbiosis fails [78]. Interestingly, the premature arbuscular degeneration (PAD) is suppressed when the plant is under N starvation, indicating that AM fungi can play an important role for the N nutrition of their host [71, 79].

Labeling, enzymatic measurements, and gene expression data suggest that N that is taken up by the AM fungus is first assimilated and converted to the amino acid arginine in hyphae of the ERM, then transferred via polyphosphates (polyP) to the intraradical mycelium (IRM), and in the IRM, arginine is converted back to NH_4^+ via the urea cycle before it is released into the AM interface [71, 74, 80]. In several plant species, NH_4^+ transporters of the AMT2 family are specifically expressed during AM symbiosis and are located in the PAM [81–83]. In *Medicago truncatula*, three AMT2 family NH_4^+ transporters are induced during the AM symbiosis. AMT2;4 is a functional NH_4^+ transporter, while AMT2;3 is unable to restore growth of a yeast high-affinity NH_4^+ mutant on low N media, indicating that AMT2;3 plays more a sensor or signaling role [83, 84]. Interestingly, only AMT2;3 was able to suppress the premature degeneration of arbuscules in *MtPT4* or *MtPT8* knockout mutants under N deficiency.

LjAMT2;2, an NH_4^+ transporter in the PAM of *Lotus japonicus*, transfers NH_3 instead of NH_4^+ [83]. The expression of LjAMT2;2 is induced under severe N starvation, and LjAMT2;2 overexpression increased the biomass and N content of roots and shoots and promoted plant growth under N deficiency. Like LjAMT2;2, the AM-inducible NH_4^+ transporter ZmAMT3;1 is able to recruit NH_4^+ but transports uncharged NH_3 rather than ionic NH_4^+ [85]. It has been estimated that the ZmAMT3;1-dependent mycorrhizal N uptake pathway is responsible for 68–74% of the symbiotic N uptake of AM maize plants in pot experiments and for >30% of the postsilking N uptake of field-grown maize. Recent evidence of the putative AM-inducible nitrate transporter OsNPF4.5, which is exclusively expressed in arbusculated cells and crucial for arbuscule development and symbiotic N uptake in rice, indicates that also NO_3^- can be transferred across the AM interface [86]. In knockout mutants of *OsNPF4.5*, a 45% decrease in symbiotic N uptake and AM colonization was observed. The expression of the AM-inducible plant lysine-histidine transporter *LjLHT1.2* in arbusculated cells of *Lotus japonicus* could indicate that N, also in the form of amino acids, can be transferred across the AM interface. A significant amino acid transport across the interface, however, would be costly for the AM fungus because plant-derived carbon skeletons that were used for the biosynthesis of amino acids in the ERM would be returned to the host plant. Labeling studies, however, confirmed that there is no significant carbon flux from the fungus to the host, that amino acids are first broken down in the IRM, and that N is transferred primarily in inorganic form across the mycorrhizal interface [80]. *LjLHT1.2* is also expressed in non-arbusculated cortical cells, which suggests a more general role of LjLHT1.2 in the reuptake and recycling of amino acids in the root cortex [87].

3.3 Potassium uptake and transport

Potassium (K) is a necessary macronutrient that plays a key role in enzyme activation, osmotic regulation, plant cell turgor generation, cell expansion, control of membrane electric potential, and pH homeostasis [66]. Plants have evolved a variety of transporters that differ in their structure and their transport mechanisms for the uptake of K from the soil, including voltage-gated K-channels, the carrier-like families KT/HAK/KUP, HKT uniporters and symporters, and cation-proton antiporters [for review see 88]. AM fungi have also been shown to have a positive impact on the K nutrition of their host plant, but the molecular mechanisms of fungal K transport are not well understood [89, 90]. The AM symbiosis triggers transcriptional responses in *Medicago* roots under K deprivation, and these responses, including, for example, the upregulation of a K^+/H^+ exchanger, could activate mechanisms that help mycorrhizal plants to tolerate long-term K deprivation or facilitate the uptake of K via the MP [90]. AtCHX20, an ortholog of the K^+/H^+ exchanger (*Medtr7g099800.1*), plays a critical role in K homeostasis and osmoregulation [91]. In *Lotus japonicus*, the potassium transporter LjHAK is highly upregulated in AM roots [92]. A LjHAK ortholog in tomato, SlHAK10 (high-affinity potassium transporter 10), is explicitly expressed in arbusculated cells and plays a role in the mycorrhizal K uptake pathway [93]. Under low-K supply conditions, loss of function of SlHAK10 decreased mycorrhizal K uptake and AM colonization rate but did not have a negative effect on arbuscule development. By contrast, SlHAK10 overexpression increased AM colonization, plant growth, K uptake, and the soluble sugar accumulation in roots under K deficiency. An increase in K uptake could promote the carbohydrate allocation to the roots and thereby stimulate the AM colonization of the roots.

3.4 Water uptake and transport

In addition to elemental nutrients, water is essential to drive plant growth and nutrient uptake. It is long known that AM fungi can have a positive impact on plant-water relations and can significantly improve the tolerance of plants to drought [94]. The positive impact of AM fungi on drought tolerance has been attributed to a variety of effects, including effects on stomatal conductance, an increase in water use efficiency, reductions of the oxidative damage under drought stress, modifications in the contents of plant hormones, such as strigolactones, jasmonic acid, and abscisic acid, improvements in plant water status by effects on hydraulic conductivity, and the activation of functional proteins, such as aquaporins [95]. For example, the AM symbiosis induced strigolactone biosynthesis under drought conditions and conferred drought tolerance in lettuce and tomatoes [96, 97]. Using a two-compartment system, it has been estimated that fungal water transport from the hyphal compartment can account for more than 30% of the water transpired by AM host plants. In addition, AM fungal hyphae were able to transport water along hyphae outside of hyphal cell membranes [98]. It has also been suggested that AM fungi increase the water uptake of plants by their positive effect on the expression of plant aquaporins [99, 100]. Aquaporins are membrane proteins that facilitate the transport of water and small solutes following an osmotic gradient. In *L. japonicus*, two putative aquaporin genes (*LjNIP1* and *LjXIP1*) are upregulated in AM plants [101]. The expression of *LjNIP1* is correlated with the expression of the AM-induced P_i transporter LjPT4 in arbusculated cells. Consistent with its function as aquaporin, LjNIP1 increased the water membrane permeability, when it was expressed in yeast protoplasts.

4. Carbon transport in root symbioses

4.1 Carbon transport in the arbuscular mycorrhizal symbiosis

The fact that AM fungi are obligate biotrophs and cannot complete their life cycle without the carbon supply from their host and the observation that host plants suppress the AM colonization of their root system when nutrients are readily available has led to the overall assumption that the host plant is in control of the symbiosis [102]. However, this phytocentric view disregards the long co-evolution of both partners in the AM symbiosis (~450 million years) that allowed both partners to improve their bargaining power in the symbiosis and contributed to the evolutionary stability of mutualism in the AM symbiosis [103, 104]. In CMNs, in which multiple host plants are interconnected and share the same mycorrhizal network, AM fungi preferentially allocate nutrient resources to host plants that provide more carbon benefits [105]. The carbon supply of the host is an important trigger for nutrient uptake and transport by the AM fungus [74, 106, 107]. Similarly, host plants are simultaneously colonized by communities of AM fungi that compete with their nutrient benefits for carbon from the mycorrhizal host. Despite the fact that different AM fungi can colonize the same root on a very small spatial scale, host plants are able to distinguish between AM fungi and preferentially allocate carbon resources to AM fungi that provide higher nutrient benefits [103]. This requires that host plants can fine-tune their carbon flux to different AM fungi on a very small spatial scale and at each individual AM interface. However, our current understanding of these processes is still very limited, particularly considering that the host plant exchanges carbon with the fungus for a variety of AM benefits.

In plants, sugar transporters are classified into monosaccharide transporters (MSTs), sucrose uptake transporters (SUTs), and SWEETs (Sugars Will Eventually be Exported Transporters). Both MSTs and SUTs contain 12 transmembrane α-helices, while SWEETs are characterized by seven transmembrane domains [108]. Plant MSTs show high expression levels in AM roots, and it has been suggested that MSTs could play an important role in funneling host plant carbon to the mycorrhizal interface [108, 109]. For example, the promoter of the *Medicago* hexose transporter *MtHex1* is particularly active in inner cortical cells that are adjacent to arbusculated cells [109]. In addition to monosaccharide transporters, there is increasing evidence that SUTs or SWEETs also play a role in the carbon allocation in mutualistic and pathogenic plant-microbe interactions. In split-root experiments with rhizobia- or AM-colonized root halves of *Medicago truncatula*, the expression levels of *MtSUT2* and *MtSUT4–1* were correlated with the amount of carbon that was allocated to the different root halves. This could indicate that both transporters play a role in the carbon transport regulation to different root symbionts [110]. High soil P_i availabilities reduce the AM colonization of potato plants, but when *SoSUT1* is overexpressed the response to high P_i availabilities is reduced [111]. By contrast, transgenic tomato plants, in which *SlSUT2* is downregulated, showed an increased AM colonization [112]. SlSUT2 is localized in the PAM but is presumably involved in the reuptake of sucrose from the AM interface into the host plant. A high expression of *SlSUT2* would reduce the carbon flux to the AM fungus and thereby reduce AM colonization.

While SUTs and monosaccharide transporters are symporters and require energy for the transmembrane transport of sugars, SWEETs are uniporters that can facilitate transmembrane transport in two directions and promote the diffusion of sugars along a concentration gradient. In soybeans, 52 different SWEET genes were identified [113]. Members of the SWEET transporter family show a distinct expression profile in plants and are involved in a variety of different physiological processes in plants, such as phloem transport, grain filling, floral transition, and the abiotic and biotic stress response of plants [114, 115]. In potato plants, the SWEETs *StSWEET2c, StSWEET7a,* and *StSWEET12a* are explicitly upregulated in arbusculated cells [116]. An overexpression of *SWEET7a* in potato plants increased the carbon sink strength of the roots, and these plants showed a faster colonization by AM fungi and the hemibiotrophic pathogen *Fusarium oxysporum f. sp. tuberosi* [117]. In *Medicago truncatula*, the two orthologs *MtSWEET1b* and *MtSWEET6* show high transcript levels in AM roots [110]. In addition, when the AM fungus had access to N and the carbon allocation to the AM root systems increased, *MtSWEET12*, *MtSWEET15c*, and *MtSWEET15d* showed high expression levels in AM roots but were downregulated in AM roots, when the host plant was directly supplied with N. However, many of these SWEETs have so far not been functionally characterized, and the fact that different SWEET transporters show similar expression profiles in AM roots indicates that there is some level of redundancy in the function of these transporters [110]. This redundancy likely also explains why loss of function mutants did not show an impairment in nodular or mycorrhizal function [118, 119]. For example, *MtSWEET1b* of *Medicago truncatula* is strongly upregulated in arbusculated cells and localizes to the PAM, and the overexpression of *MtSWEET1b* promoted the intraradical colonization of the host root by AM fungal hyphae, but loss of function mutants showed no impairment in the AM symbiosis [120].

Based on earlier findings, glucose seemed to be the most likely form in which the fungus takes up carbon from the AM interface [121]. The increased activities of an apoplastic acid invertase and sucrose synthase in AM roots, which could facilitate the conversion of sucrose into the glucose and fructose in the interfacial apoplast or in the cortical cytoplasm, seemed to confirm this view [122]. This is also consistent with the expression of the fungal monosaccharide transporter *MST2* in arbusculated cells of AM roots [123]. The fact that the expression of both, *MST2* and the AM-induced P_i transporter *MtPt4* is correlated, suggests that the exchange of carbon for P_i is tightly linked. By contrast, a reduction in the MST2 expression by gene silencing led to malformed arbuscules and reduced *MtPt4* expression.

AM fungi store carbon mainly in the form of lipids, and here, in particular, in the form of triacylglycerol and fatty acids (FAs). It has long been assumed that AM fungi use plant-derived sugars as precursors for lipid biosynthesis, and that the fungus synthesizes FAs exclusively in the intraradical mycelium [124]. However, there is growing evidence that AM fungi are FA auxotrophs, and that the plant must transfer FAs to the fungus to sustain the AM symbiosis [125, 126]. In plants, *de novo* FA biosynthesis takes place inside plastids and starts with the biosynthesis of long-chain FAs from malonyl-ACP and acetyl-CoA. Catalyzed by the acyl-ACP thioesterase-like proteins *FatA, FatB,* and *FatM, de novo* FA biosynthesis is terminated, and free FAs are exported out of the plastid. In AM roots, the expression of *FatM* is induced. Mutants in which *FatM* is impaired can still get colonized by AM fungi, but the AM development is significantly reduced [127]. In the endoplasmic reticulum, free FAs are bound to CoA to create acyl-CoA and then converted to sn-2 monoacylglycerol (ßMAGs) via RAM2. RAM2 encodes a glycerol-3-phosphate acyl transferase, and AM colonization is significantly impaired in RAM2 knockouts [128]. It has been suggested that FatM increases the release of 16:0 FAs from the plastid that are subsequently used by RAM2 to produce 16:0 ßMAG [129]. Host plants preferentially allocate carbon resources to beneficial AM fungi [103], and the finding that RAM2 expression is particularly induced in roots that are colonized with a beneficial AM fungus supports the view that RAM2 is a key element in the FA transport to the AM fungus [130]. It has been suggested that the ABCG half-transporters STR and STR2 are involved in releasing plant-derived lipids into the interfacial apoplast [129, 131]. STR and STR2 are essential for arbuscule formation, and silencing of STR/STR2 results in a stunted arbuscular phenotype [132, 133]. Both are localized in the PAM and function as heterodimers. Another transporter that is likely involved in lipid transport in *Medicago truncatula*, is ABCG3/WBC5. It has been suggested that ABCG3/WBC5 could be part of the RAM1-regulated lipid export pathway [125]. Together with *RAM2*, *MtABCG3* is strongly induced in mycorrhizal roots and regulated by the GRAS-domain transcription factor RAM1 (required for Arbuscular Mycorrhization 1) [125, 134].

4.2 Carbon transport in the symbiosis of legumes and rhizobia

Biological nitrogen fixation (BNF) by rhizobia is an energetically costly process. At least 16 molecules of ATP and 8 low potential electrons are necessary to convert N_2 into NH_3 [135]. Plants must provide rhizobia with a constant flow of energy in the form of reduced carbon compounds to maintain BNF activity. Root nodules constitute only a small fraction of the total biomass of a legume plant, but they can consume more than 25% of the total photosynthates of the plant. Sucrose is the main transport sugar that is translocated through the phloem to root nodules and to the other sink organs of the plant. However, how carbon is funneled to the root nodules and how this process is regulated is still largely unknown. For example, the nodule-specific sucrose transporter *MtSWEET11* of *Medicago truncatula* is expressed in infected root hair cells and in the meristem, invasion zone, and vascular system of nodules. However, although *MtSWEET11* seems to play a role in the sucrose distribution in root nodules, it is not essential for BNF [118]. The sucrose uptake transporter *GmSUT1* shows a high expression in the nodules of soybeans and is particularly localized in the peripheral fixation zone and the vascular bundles [136]. Since *GmSUT1* overexpression increased nodule number and plant N content, it has been suggested that GmSUT1 could play a role in funneling sucrose to the root nodules.

The symbiosome membrane (SYM) is the site of nutrient exchange between plants and rhizobia and is energized through the activities of H^+-ATPases that pump protons into the symbiosome space and provide both the plant and bacteroids with the necessary proton motive force to take up nutrients from the symbiosome space [137]. Among 197 proteins that were identified in the SYM of soybeans, were proteins involved in metabolism, protein folding and degradation, membrane trafficking, and solute transport, such as putative transporters for sulfate, calcium, hydrogen ions, peptide/dicarboxylate, and nitrate [138]. Mutations that interfere with dicarboxylate transport across the SYM interrupt N_2 fixation, and it has been suggested that C_4-dicarboxylates are the primary carbon source for bacteroids in active root nodules. The SYM has a well-characterized dicarboxylate transport system (Dct) encoded by the three gene loci *dctA*, *dctB*, and *dctD* [139]. While *dctB* and *dctD* encode a two-component regulatory system that responds to the presence of C_4-dicarboxylates, *dctA* encodes the structural protein for C_4-dicarboxylate transport. Succinate, malate, fumarate, aspartate, and oxaloacetate have been discussed as the most important substrates for rhizobial Dct systems.

Bacteroids transfer N primarily in the form of NH_4^+ to the plant but also secrete significant amounts of the amino acids aspartate and alanine [140]. NH_3, the product of BNF, is either transferred via passive diffusion across the bacteroid membrane or through the nodulin 26 channel [141]. Nodulin 26 is a member of the aquaporin superfamily but also shows an ammonia permease activity that is favored over its aquaporin activity [141]. Due to the low pH in the symbiosome space, NH_3 is protonated here to NH_4^+ and then taken up by the host via a cation channel that is permeable to NH_4^+ [142].

4.3 Regulation of nodulation and mycorrhizal colonization

Due to the high carbon costs of both root symbioses, plants are under high selective pressure to tightly control nodulation and AM colonization. N and P deficiency in plants stimulates the colonization of plants by rhizobial or AM fungal partners, while a high availability of N or P, or the inability of partners to provide benefits to the host plant, promotes the premature senescence of symbiotic organs [143, 144]. The interaction between the partners in the RL symbiosis is based on carbon-nitrogen trade-offs. Since the plant cannot determine the BNF efficiency of rhizobia during the recognition and root entry stages, the plant needs to sanction low-benefit rhizobia later [145, 146]. Legumes have control over the carbon supply and can influence the success of their symbiotic partners. In natural soils, non-fixing rhizobia strains are rare, but intermediate fixers are common. When the plant can choose between an intermediate fixer and a more effective strain, the plant will "sanction" the intermediate fixer by limiting its carbon supply, what results in smaller nodules with fewer viable bacteria. By contrast, when the only alternative is a non-fixing rhizobia strain, plants will not sanction intermediate fixers [147]. Environmental variation can help to explain why low- and high-benefit rhizobia still coexist. In an N-rich environment, the host might be less strict, thus allowing low-benefit strains to proliferate [148]. Low-benefit rhizobia might also be able to escape host plant sanctions by forming mixed nodules with high mutualistic rhizobia strains [149].

Arbuscules have only a short lifespan and are only functional for 2–3 days, before they become senescent and collapse [150]. This short life cycle seems like a waste of limited resources but allows plants to regulate their AM colonization. Plants can also induce a premature arbuscular degeneration (PAD) [84]. PAD has been discussed as one mechanism by which host plants can sanction low-benefit AM fungi or are able to reduce the AM colonization under high P conditions [79, 151]. Host plants can distinguish between high- and low-benefit AM fungi and allocate carbon resources accordingly [103, 130]. However, low-benefit AM fungi are able to persist in natural environments, and host plants tolerate the infection by low-benefit AM strains, to reduce their dependency on a single AM fungus for nutrient uptake.

However, plants not only control the colonization with low-benefit rhizobia but also limit nodule formation with high-benefit partners by the long-distance (systemic) autoregulation of nodulation (AON) pathway [152]. In response to rhizobial infection, a subset of genes encoding CLAVATA3/Embryo Surrounding Region (CLE) peptides are activated [153]. CLE peptides consist of 12–13 amino acids, secreted as signaling peptides from the C-terminal region of preproproteins. It has been reported that AON-related CLE peptides from the root negatively affect the number of nodules by inhibiting a leucine-rich repeat receptor-like-kinase (LRR-RLK) in the shoot, known as SUPER NUMERIC NODULES (*SUNN*) in *Medicago truncatula*, HYPERNODULATION ABERRANT ROOT FORMATION (*HAR1*) in *Lotus japonicus*, NODULE AUTOREGULATION RECEPTOR KINASE (*NARK*) in soybean, and SYMBIOSIS 29 (*SYM29*) in peas [154–157]. The perception of the root-derived signals by LRR-RLK in the shoot generates a shoot to root signal that inhibits further colonization. It has been suggested that this shoot-derived inhibitor could be a cytokinin [158, 159]. Currently, our understanding of the downstream targets of the AON pathway in the roots is limited. Compared to the wild type, supernodulating SUNN mutants, for example, showed an increase in polar auxin transport from the shoot to the root, and it has been suggested that higher auxin levels could be responsible for the interruption of AON [160]. In transgenic roots of *Medicago truncatula* that ectopically express either *MtCLE12* or *MtCLE13*, the early nodulation marker *MtENOD11* was not activated, suggesting that the systemic AON pathway could inhibit Nod factor signaling [153].

A prior AM colonization of one root half in a split-root system significantly reduces the AM colonization of the second root half, and there is evidence that the systemic autoregulation of mycorrhization (AOM) has similarities to the AON pathway [161, 162]. For example, hypernodulating mutants such as sunn, nark, har1, and sym29, also show a hypermycorrhizal colonization phenotype. However, whether other AON genes or mobile signals (CLE) are also involved in AOM is not yet known. However, the fact that prior nodulation in split-root studies can systemically suppress AM colonization and vice versa suggests that the same shoot-derived inhibitor is involved in AON and AOM.

5. Tripartite interactions of legumes

Our current understanding of the RL or AM symbiosis is mainly based on single plant/single symbiont studies, but field grown legumes form tripartite interactions, and are simultaneously colonized by both AM fungi and rhizobia [110, 163]. The inoculation with both root symbionts can lead to synergistic benefits, and plants can gain more from tripartite interactions than from single inoculations with either symbiont [163–165]. The rhizobial nitrogenase complex requires 16 ATP to fix one N_2 molecule, and consequently rhizobia require an adequate P supply for efficient BNF. Nodules act as strong P sinks in legume root systems to provide sufficient P resources for optimum BNF [110, 166]. Nonmycorrhizal soybean plants have lower nodule numbers and weights and show particularly under low P supply lower N fixation rates [163, 167]. The positive effect of the AM symbiosis on the P uptake of the plant has therefore been discussed as the primary reason for the stimulation of BNF in AM legumes [163]. In addition, AM fungi can also supply the plant with other microelements that are essential for N_2 fixation, including zinc, iron, manganese, and molybdenum [168, 169].

However, antagonistic responses have also been described, and the prior inoculation with either rhizobia or AM fungi can also suppress the root colonization by the other partner [170]. Whether the plant shows antagonistic or synergistic growth responses after dual inoculation depends on the environmental context [164], and the compatibility between symbiotic partners [165, 171]. As long as the root symbionts provide complementary rewards to the host plant, and the benefits outweigh the costs of these interactions, synergistic growth responses are more likely [172]. Rhizobia and AM fungi, however, are also competitors for the same resource from the host plant, and since both interactions are costly, the plant must be able to allocate carbon resources to both root symbionts according to their benefits. For example, an AM fungal partner is a stronger competitor for host plant carbon when it is able to provide N to the host, or when the plant is grown under high N but low P availabilities [110]. Our current understanding of the molecular mechanisms that control the carbon allocation from the host plant to individual root symbionts is very limited. Considering, the role that host plant carbon plays as an important trigger for symbiotic functioning, a better understanding of these processes is critical, because it may be key to improve the resource exchange between plants and symbionts and ultimately enhance productivity of agronomically important legumes [74, 105, 173, 174].

Conflict of interest

Authors declare no conflicts of interest in this work.

Author details

Debatosh Das, Salina Tripathi, Prema Mutyala, Dhruv Aditya Srivastava and Heike Bücking*
Division of Plant Science and Technology, University of Missouri, College of Agriculture, Food and Natural Resources, Columbia, MO, USA

*Address all correspondence to: heike.buecking@missouri.edu

References

[1] Turner TR, James EK, Poole PS. The plant microbiome. Genome Biology. 2013;**14**:209. DOI: 10.1186/gb-2013-14-6-209

[2] Herridge DF, Peoples MB, Boddey RM. Global inputs of biological nitrogen fixation in agricultural systems. Plant and Soil. 2008;**311**:1-18. DOI: 10.1007/s11104-008-9668-3

[3] Smith SE, Read DJ. Mycorrhizal Symbiosis. 3rd ed. New York: Academic Press; 2008. p. 787. DOI: 10.1016/B978-0-12-370526-6.X5001-6

[4] Parniske M. Arbuscular mycorrhiza: The mother of plant root endosymbioses. Nature Reviews Microbiology. 2008;**6**:763-775. DOI: 10.1038/nrmicro1987

[5] Strullu-Derrien C, Selosse M-A, Kenrick P, Martin FM. The origin and evolution of mycorrhizal symbioses: From palaeomycology to phylogenomics. New Phytologist. 2018;**220**:1012-1030. DOI: 10.1111/nph.15076

[6] Wang B, Qiu Y-L. Phylogenetic distribution and evolution of mycorrhizae in land plants. Mycorrhiza. 2006;**16**:299-363. DOI: 10.1007/s00572-005-0033-6

[7] Ryan MH, Graham JH. Is there a role for arbuscular mycorrhizal fungi in production agriculture? Plant and Soil. 2002;**244**:263-271. DOI: 10.1023/A:1020207631893

[8] Cameron DD. Arbuscular mycorrhizal fungi as (agro)ecosystem engineers. Plant and Soil. 2010;**333**:1-5. DOI: 10.1007/s11104-010-0361-y

[9] Smith SE, Jakobsen I, Grønlund M, Smith FA. Roles of arbuscular mycorrhizas in plant phosphorus nutrition: Interactions between pathways of phosphorus uptake in arbuscular mycorrhizal roots have important implications for understanding and manipulating plant phosphorus acquisition. Plant Physiology. 2011;**156**:1050-1057. DOI: 10.1104/pp.111.174581

[10] Smith SE, Smith FA, Jakobsen I. Functional diversity in arbuscular mycorrhizal (AM) symbioses: The contribution of the mycorrhizal P uptake pathway is not correlated with mycorrhizal responses in growth or total P uptake. New Phytologist. 2004;**162**:511-524. DOI: 10.1111/j.1469-8137.2004.01039.x

[11] Babikova Z, Gilbert L, Bruce TJA, Birkett M, Caulfield JC, Woodcock C, et al. Underground signals carried through common mycelial networks warn neighbouring plants of aphid attack. Ecology Letters. 2013;**16**:835-843. DOI: 10.1111/ele.12115

[12] Gorzelak MA, Asay AK, Pickles BJ, Simard SW. Inter-plant communication through mycorrhizal networks mediates complex adaptive behaviour in plant communities. AoB Plants. 2015:plv050. DOI: 10.1093/aobpla/plv050

[13] Figueiredo A, Boy J, Guggenberger G. Common mycorrhizae network: A review of the theories and mechanisms behind underground interactions. Frontiers in Fungal Biology. 2021;**2**:735299. DOI: 10.3389/ffunb.2021.735299

[14] Akiyama K, Matsuzaki KI, Hayashi H. Plant sesquiterpenes induce hyphal branching in arbuscular mycorrhizal fungi. Nature. 2005;**435**:824-827. DOI: 10.1038/nature03608

[15] Besserer A, Puech-Pagès V, Kiefer P, Gomez-Roldan V, Jauneau A, Roy S, et al. Strigolactones stimulate arbuscular mycorrhizal fungi by activating mitochondria. PLOS Biology. 2006;**4**:1239-1247. DOI: 10.1371/journal.pbio.0040226

[16] Bücking H, Abubaker J, Govindarajulu M, Tala M, Pfeffer PE, Nagahashi G, et al. Root exudates stimulate the uptake and metabolism of organic carbon in germinating spores of *Glomus intraradices*. New Phytologist. 2008;**180**:684-695. DOI: 10.1111/j.1469-8137.2008.02590.x

[17] Mostofa MG, Li W, Nguyen KH, Fujita M, Tran L-SP. Strigolactones in plant adaptation to abiotic stresses: An emerging avenue of plant research. Plant, Cell & Environment. 2018;**41**:2227-2243. DOI: 10.1111/pce.13364

[18] Das D, Paries M, Hobecker K, Gigl M, Dawid C, Lam H-M, et al. Phosphate starvation response transcription factors enable arbuscular mycorrhiza symbiosis. Nature Communications. 2022;**13**:477. DOI: 10.1038/s41467-022-27976-8

[19] Genre A, Chabaud M, Timmers T, Bonfante P, Barker DG. Arbuscular mycorrhizal fungi elicit a novel intracellular apparatus in Medicago truncatula root epidermal cells before infection. Molecular Plant Microbe Interactions. 2005;**17**:3489-3499. DOI: 10.1105/tpc.105.035410

[20] Coronado C, Zuanazzi J, Sallaud C, Quirion JC, Esnault R, Husson HP, et al. Alfalfa root flavonoid production is nitrogen regulated. Plant Physiology. 1995;**108**:533-542. DOI: 10.1104/pp.108.2.533

[21] Giraud E, Moulin L, Vallenet D, Barbe V, Cytryn E, Avarre J-C, et al. Legumes symbioses: Absence of *nod* genes in photosynthetic Bradyrhizobia. Science. 2007;**316**:1307-1312. DOI: 10.1126/science.1139548

[22] Madsen EB, Madsen LH, Radutoiu S, Olbryt M, Rakwalska M, Szczyglowski K, et al. A receptor kinase gene of the LysM type is involved in legume perception of rhizobial signals. Nature. 2003;**425**:637-640. DOI: 10.1038/nature02045

[23] Radutoiu S, Madsen LH, Madsen EB, Felle HH, Umehara Y, Grønlund M, et al. Plant recognition of symbiotic bacteria requires two LysM receptor-like kinases. Nature. 2003;**425**:585-592. DOI: 10.1038/nature02039

[24] Fisher RF, Long SR. Rhizobium–plant signal exchange. Nature. 1992;**357**:655-660. DOI: 10.1038/357655a0

[25] Liu C-W, Murray J. The role of flavonoids in nodulation host-range specificity: An update. Plants. 2016;**5**:33. DOI: 10.3390/plants5030033

[26] Wang Q, Yang S, Liu J, Terecskei K, Ábrahám E, Gombár A, et al. Host-secreted antimicrobial peptide enforces symbiotic selectivity in Medicago truncatula. Proceedings of the National Academy of Sciences, USA. 2017;**114**:6854-6859. DOI: 10.1073/pnas.1700715114

[27] Roy S, Liu W, Nandety RS, Crook A, Mysore KS, Pislariu CI, et al. Celebrating 20 years of genetic discoveries in legume nodulation and symbiotic nitrogen fixation. The Plant Cell. 2020;**32**:15-41. DOI: 10.1105/tpc.19.00279

[28] Downie JA. The roles of extracellular proteins, polysaccharides and signals in the interactions of rhizobia with legume roots. FEMS Microbiology Reviews. 2010;**34**:150-170. DOI: 10.1111/j.1574-6976.2009.00205.x

[29] Truchet G, Roche P, Lerouge P, Vasse J, Camut S, de Billy F, et al. Sulphated lipo-oligosaccharide signals of *Rhizobium meliloti* elicit root nodule organogenesis in alfalfa. Nature. 1991;**351**:670-673. DOI: 10.1038/351670a0

[30] Harris JM, Wais R, Long SR. Rhizobium-induced calcium spiking in *Lotus japonicus*. Molecular Plant Microbe Interactions. 2003;**16**:335-341. DOI: 10.1094/MPMI.2003.16.4.335

[31] Popp C, Ott T. Regulation of signal transduction and bacterial infection during root nodule symbiosis. Current Opinion in Plant Biology. 2011;**14**:458-467. DOI: 10.1016/j.pbi.2011.03.016

[32] Udvardi MK, Day DA. Metabolite transport across symbiotic membranes of legume nodules. Annual Review of Plant Physiology and Plant Molecular Biology. 1997;**48**:493-523. DOI: 10.1146/annurev.arplant.48.1.493

[33] Maillet F, Poinsot V, André O, Puech-Pagès V, Haouy A, Gueunier M, et al. Fungal lipochitooligosaccharide symbiotic signals in arbuscular mycorrhiza. Nature. 2011;**469**:58-63. DOI: 10.1038/nature09622

[34] Sprent JI. Evolving ideas of legume evolution and diversity: A taxonomic perspective on the occurrence of nodulation. New Phytologist. 2007;**174**:11-25. DOI: 10.1111/j.1469-8137.2007.02015.x

[35] Brundrett MC, Tedersoo L. Evolutionary history of mycorrhizal symbioses and global host plant diversity. New Phytologist. 2018;**220**:1108-1115. DOI: 10.1111/nph.14976

[36] Cope KR, Bascaules A, Irving TB, Venkateshwaran M, Maeda J, Garcia K, et al. The ectomycorrhizal fungus *Laccaria bicolor* produces lipochitooligosaccharides and uses the common symbiosis pathway to colonize *Populus* roots. The Plant Cell. 2019;**31**:2386-2410. DOI: 10.1105/tpc.18.00676

[37] Kistner C, Winzer T, Pitzschke A, Mulder L, Sato S, Kaneko T, et al. Seven *Lotus japonicus* genes required for transcriptional reprogramming of the root during fungal and bacterial symbiosis. The Plant Cell. 2005;**17**:2217-2229. DOI: 10.1105/tpc.105.032714

[38] Broghammer A, Krusell L, Blaise M, Sauer J, Sullivan JT, Maolanon N, et al. Legume receptors perceive the rhizobial lipochitin oligosaccharide signal molecules by direct binding. Proceedings of the National Academy of Sciences USA. 2012;**109**:13859-13864. DOI: 10.1073/pnas.1205171109

[39] Moling S, Pietraszewska-Bogiel A, Postma M, Fedorova E, Hink MA, Limpens E, et al. Nod factor receptors form heteromeric complexes and are essential for intracellular infection in *Medicago* nodules. The Plant Cell. 2014;**26**:4188-4199. DOI: 10.1105/tpc.114.129502

[40] He J, Zhang C, Dai H, Liu H, Zhang X, Yang J, et al. A LysM receptor heteromer mediates perception of arbuscular mycorrhizal symbiotic signal in rice. Molecular Plant. 2019;**12**:1561-1576. DOI: 10.1016/j.molp.2019.10.015

[41] Miyata K, Kozaki T, Kouzai Y, Ozawa K, Ishii K, Asamizu E, et al. The bifunctional plant receptor, OsCERK1, regulates both chitin-triggered immunity and arbuscular mycorrhizal symbiosis in rice. Plant Cell Physiology. 2014;**55**:1864-1872. DOI: 10.1093/pcp/pcu129

[42] Antolín-Llovera M, Ried Martina K, Parniske M. Cleavage of the SYMBIOSIS RECEPTOR-LIKE KINASE ectodomain

promotes complex formation with nod factor receptor 5. Current Biology. 2014;**24**:422-427. DOI: 10.1016/j.cub.2013.12.053

[43] Feng Y, Wu P, Liu C, Peng L, Wang T, Wang C, et al. Suppression of LjBAK1-mediated immunity by SymRK promotes rhizobial infection in *Lotus japonicus*. Molecular Plant. 2021;**14**:1935-1950. DOI: 10.1016/j.molp.2021.07.016

[44] Charpentier M, Bredemeier R, Wanner G, Takeda N, Schleiff E, Parniske M. *Lotus japonicus* CASTOR and POLLUX are ion channels essential for perinuclear calcium spiking in legume root endosymbiosis. The Plant Cell. 2008;**20**:3467-3479. DOI: 10.1105/tpc.108.063255

[45] Chen C, Fan C, Gao M, Zhu H. Antiquity and function of CASTOR and POLLUX, the twin ion channel-encoding genes key to the evolution of root symbioses in plants. Plant Physiology. 2008;**149**:306-317. DOI: 10.1104/pp.108.131540

[46] Groth M, Takeda N, Perry J, Uchida H, Dräxl S, Brachmann A, et al. NENA, a *Lotus japonicus* homolog of Sec13, is required for rhizodermal infection by arbuscular mycorrhiza fungi and rhizobia but dispensable for cortical endosymbiotic development. The Plant Cell. 2010;**22**:2509-2526. DOI: 10.1105/tpc.109.069807

[47] Kanamori N, Madsen LH, Radutoiu S, Frantescu M, Quistgaard EMH, Miwa H, et al. A nucleoporin is required for induction of Ca^{2+} spiking in legume nodule development and essential for rhizobial and fungal symbiosis. Proceedings of the National Academy of Sciences, USA. 2006;**103**:359-364. DOI: 10.1073/pnas.0508883103

[48] Singh S, Parniske M. Activation of calcium- and calmodulin-dependent protein kinase (CCaMK), the central regulator of plant root endosymbiosis. Current Opinion in Plant Biology. 2012;**15**:444-453. DOI: 10.1016/j.pbi.2012.04.002

[49] Singh S, Katzer K, Lambert J, Cerri M, Parniske M. CYCLOPS, a DNA-binding transcriptional activator, orchestrates symbiotic root nodule development. Cell Host & Microbe. 2014;**15**:139-152. DOI: 10.1016/j.chom.2014.01.011

[50] Gutjahr C, Parniske M. Cell and developmental biology of arbuscular mycorrhiza symbiosis. Annual Review in Cell Developmental Biology. 2013;**29**:593-617. DOI: 10.1146/annurev-cellbio-101512-122413

[51] Pimprikar P, Carbonnel S, Paries M, Katzer K, Klingl V, Bohmer Monica J, et al. A CCaMK-CYCLOPS-DELLA complex activates transcription of RAM1 to regulate arbuscule branching. Current Biology. 2016;**26**:987-998. DOI: 10.1016/j.cub.2016.01.069

[52] Yu N, Luo D, Zhang X, Liu J, Wang W, Jin Y, et al. A DELLA protein complex controls the arbuscular mycorrhizal symbiosis in plants. Cell Research. 2014;**24**:130-133. DOI: 10.1038/cr.2013.167

[53] Marsh JF, Rakocevic A, Mitra RM, Brocard L, Sun J, Eschstruth A, et al. *Medicago truncatula* NIN is essential for rhizobial-independent nodule organogenesis induced by autoactive calcium/calmodulin-dependent protein kinase. Plant Physiology. 2007;**144**:324-335. DOI: 10.1104/pp.106.093021

[54] Andriankaja A, Boisson-Dernier A, Frances L, Sauviac L, Jauneau A, Barker DG, et al. AP2-ERF transcription

factors mediate nod factor dependent Mt ENOD11 activation in root hairs via a novel cis-regulatory motif. The Plant Cell. 2007;**19**:2866-2885. DOI: 10.1105/tpc.107.052944

[55] Kaló P, Gleason C, Edwards A, Marsh J, Mitra RM, Hirsch S, et al. Nodulation signaling in legumes requires NSP2, a member of the GRAS family of transcriptional regulators. Science. 2005;**308**:1786-1789. DOI: 10.1126/science.1110951

[56] Smit P, Raedts J, Portyanko V, Debelle F, Gough C, Bisseling T, et al. NSP1 of the GRAS protein family is essential for rhizobial nod factor-induced transcription. Science. 2005;**308**:1789-1791. DOI: 10.1126/science.1111025

[57] Jin Y, Liu H, Luo D, Yu N, Dong W, Wang C, et al. DELLA proteins are common components of symbiotic rhizobial and mycorrhizal signalling pathways. Nature Communications. 2016;**7**:12433. DOI: 10.1038/ncomms12433

[58] Hirsch S, Kim J, Muñoz A, Heckmann AB, Downie JA, Oldroyd GE. GRAS proteins form a DNA binding complex to induce gene expression during nodulation signaling in *Medicago truncatula*. The Plant Cell. 2009;**21**:545-557. DOI: 10.1105/tpc.108.064501

[59] Schachtman DP, Reid RJ, Ayling SM. Phosphorus uptake by plants: From soil to cell. Plant Physiology. 1998;**116**:447-453. DOI: 10.1104/pp.116.2.447

[60] Bucher M. Functional biology of plant phosphate uptake at root and mycorrhiza interfaces. New Phytologist. 2007;**173**:11-26. DOI: 10.1111/j.1469-8137.2006.01935.x

[61] Paszkowski U, Kroken S, Roux C, Briggs SP. Rice phosphate transporters include an evolutionarily divergent gene specifically activated in arbuscular mycorrhizal symbiosis. Proceedings of the National Academy of Sciences, USA. 2002;**99**:13324-13329. DOI: 10.1073/pnas.202474599

[62] Volpe V, Giovannetti M, Sun X-G, Fiorilli V, Bonfante P. The phosphate transporters LjPT4 and MtPT4 mediate early root responses to phosphate status in non mycorrhizal roots. Plant, Cell & Environment. 2016;**39**:660-671. DOI: 10.1111/pce.12659

[63] Harrison MJ, Dewbre GR, Liu J. A phosphate transporter from *Medicago truncatula* involved in the acquisition of phosphate released by arbuscular mycorrhizal fungi. The Plant Cell. 2002;**14**:2413-2429. DOI: 10.1105/tpc.004861

[64] Yang S-Y, Grønlund M, Jakobsen I, Grotemeyer MS, Rentsch D, Miyao A, et al. Nonredundant regulation of rice arbuscular mycorrhizal symbiosis by two members of the PHOSPHATE TRANSPORTER1 gene family. The Plant Cell. 2012;**24**:4236-4251. DOI: 10.1105/tpc.112.104901

[65] Xie XA, Huang W, Liu FC, Tang NW, Liu Y, Lin H, et al. Functional analysis of the novel mycorrhiza-specific phosphate transporter AsPT1 and PHT1 family from *Astragalus sinicus* during the arbuscular mycorrhizal symbiosis. New Phytologist. 2013;**198**:836-852. DOI: 10.1111/nph.12188

[66] Hawkesford MJ, Cakmak I, Coskun D, De Kok LJ, Lambers H, Schjoerring JK, et al. Chapter 6 - functions of macronutrients. In: Rengel Z, Cakmak I, White PJ, editors. Marschner's Mineral Nutrition of Plants. San Diego: Academic Press; 2023. pp. 201-281

[67] Bloom AJ. The increasing importance of distinguishing among plant nitrogen

sources. Current Opinion in Plant Biology. 2015;**25**:10-16. DOI: 10.1016/j.pbi.2015.03.002

[68] Miller AJ, Cramer MD. Root nitrogen acquisition and assimilation. Plant and Soil. 2004;**274**:1-36. DOI: 10.1007/s11104-004-0965-1

[69] Kuzyakov Y, Xu X. Competition between roots and microorganisms for nitrogen: Mechanisms and ecological relevance. New Phytologist. 2013;**198**:656-669. DOI: 10.1111/nph.12235

[70] Rengel Z, Cakmak I, White PJ. Marschner's Mineral Nutrition of Plants. San Diego: Academic Press; 2023. p. 795p

[71] Bücking H, Kafle A. Role of arbuscular mycorrhizal fungi in the nitrogen uptake of plants: Current knowledge and research gaps. Agronomy. 2015;**5**:587-612. DOI: 10.3390/agronomy5040587

[72] Pérez-Tienda J, Valderas A, Camañes G, García-Agustín P, Ferrol N. Kinetics of NH_4^+ uptake by the arbuscular mycorrhizal fungus *Rhizophagus irregularis*. Mycorrhiza. 2012;**22**:485-491. DOI: 10.1007/s00572-012-0452-0

[73] Tian C, Kasiborski B, Koul R, Lammers PJ, Bücking H, Shachar-Hill Y. Regulation of the nitrogen transfer pathway in the arbuscular mycorrhizal symbiosis: Gene characterization and the coordination of expression with nitrogen flux. Plant Physiology. 2010;**153**:1175-1187. DOI: 10.1104/pp.110.156430

[74] Fellbaum CR, Gachomo EW, Beesetty Y, Choudhari S, Strahan GD, Pfeffer PE, et al. Carbon availability triggers fungal nitrogen uptake and transport in the arbuscular mycorrhizal symbiosis. Proceedings of the National Academy of Sciences, USA. 2012;**109**:2666-2671. DOI: 10.1073/pnas.1118650109

[75] Cooper TG, Sumrada RA. What is the function of nitrogen catabolite repression in *Saccharomyces cerevisiae*? Journal of Bacteriology. 1983;**155**:623-627. DOI: 10.1128/jb.155.2.623-627.1983

[76] Smith SE, Smith FA. Roles of arbuscular mycorrhizas in plant nutrition and growth: New paradigms from cellular to ecosystem scales. Annual Review Plant Biology. 2011;**62**:227-250. DOI: 10.1146/annurev-arplant-042110-103846

[77] Mensah JA, Koch AM, Antunes PM, Hart MM, Kiers ET, Bücking H. High functional diversity within arbuscular mycorrhizal fungal species is associated with differences in phosphate and nitrogen uptake and fungal phosphate metabolism. Mycorrhiza. 2015;**25**:533-546. DOI: 10.1007/s00572-015-0631-x

[78] Javot H, Penmetsa RV, Terzaghi N, Cook DR, Harrison MJ. A *Medicago truncatula* phosphate transporter indispensable for the arbuscular mycorrhizal symbiosis. Proceedings of the National Academy of Sciences, USA. 2007;**104**:1720-1725. DOI: 10.1073/pnas.060813610

[79] Javot H, Penmetsa RV, Breuillin F, Bhattarai KK, Noar RD, Gomez SK, et al. *Medicago truncatula* Mtpt4 mutants reveal a role for nitrogen in the regulation of arbuscule degeneration in arbuscular mycorrhizal symbiosis. Plant Journal. 2011;**68**:954-965. DOI: 10.1111/j.1365-313X.2011.04746.x

[80] Govindarajulu M, Pfeffer PE, Jin HR, Abubaker J, Douds DD, Allen JW. Nitrogen transfer in the arbuscular mycorrhizal symbiosis. Nature. 2005;**435**:819-823. DOI: 10.1038/nature03610

[81] Kobae Y, Tamura Y, Takai S, Banba M, Hata S. Localized expression of arbuscular mycorrhiza-inducible ammonium transporters in soybean. Plant and Cell Physiology. 2010;**51**:1411-1415. DOI: 10.1093/pcp/pcq099

[82] Koegel S, Lahmidi NA, Arnould C, Chatagnier O, Walder F, Ineichen K, et al. The family of ammonium transporters (AMT) in *Sorghum bicolor*: Two AMT members are induced locally, but not systemically in roots colonized by arbuscular mycorrhizal fungi. New Phytologist. 2013;**198**:853-865. DOI: 10.1111/nph.12199

[83] Guether M, Neuhauser B, Balestrini R, Dynowski M, Ludewig U, Bonfante P. A mycorrhizal-specific ammonium transporter from *Lotus japonicus* acquires nitrogen released by arbuscular mycorrhizal fungi. Plant Physiology. 2009;**150**:73-83. DOI: 10.1104/pp.109.136390

[84] Breuillin-Sessoms F, Floss DS, Gomez SK, Pumplin N, Ding Y, Levesque-Tremblay V, et al. Suppression of arbuscule degeneration in *Medicago truncatula* phosphate transporter 4 mutants is dependent on the ammonium transporter 2 family protein AMT2;3. The Plant Cell. 2015;**27**:1352-1366. DOI: 10.1105/tpc.114.131144

[85] Hui J, An X, Li Z, Neuhäuser B, Ludewig U, Wu X, et al. The mycorrhiza-specific ammonium transporter ZmAMT3;1 mediates mycorrhiza-dependent nitrogen uptake in maize roots. The Plant Cell. 2022;**34**:4066-4087. DOI: 10.1093/plcell/koac225

[86] Wang S, Chen A, Xie K, Yang X, Luo Z, Chen J, et al. Functional analysis of the OsNPF4.5 nitrate transporter reveals a conserved mycorrhizal pathway of nitrogen acquisition in plants. Proceedings of the National Academy of Science, USA. 2020;**117**:16649-16659. DOI: 10.1073/pnas.2000926117

[87] Guether M, Volpe V, Balestrini R, Requena N, Wipf D, Bonfante P. LjLHT1.2 - a mycorrhiza-inducible plant amino acid transporter from *Lotus japonicus*. Biology and Fertility of Soils. 2011;**47**:925-936. DOI: 10.1007/s00374-011-0596-7

[88] Ragel P, Raddatz N, Leidi EO, Quintero FJ, Pardo JM. Regulation of K^+ nutrition in plants. Frontiers in Plant Science. 2019;**2019**:10. DOI: 10.3389/fpls.2019.00281

[89] Garcia K, Ané J-M. Polymorphic responses of *Medicago truncatula* accessions to potassium deprivation. Plant Signaling & Behavior. 2017;**12**:e1307494. DOI: 10.1080/15592324.2017.1307494

[90] Garcia K, Chasman D, Roy S, Ané J-M. Physiological responses and gene co-expression network of mycorrhizal roots under K deprivation. Plant Physiology. 2017;**173**:1811-1823. DOI: 10.1104/pp.16.01959

[91] Padmanaban S, Chanroj S, Kwak JM, Li X, Ward JM, Sze H. Participation of endomembrane cation/H^+ exchanger AtCHX20 in osmoregulation of guard cells. Plant Physiology. 2007;**144**:82-93. DOI: 10.1104/pp.106.092155

[92] Guether M, Balestrini R, Hannah M, He J, Udvardi MK, Bonfante P. Genome-wide reprogramming of regulatory networks, transport, cell wall and membrane biogenesis during arbuscular mycorrhizal symbiosis in *Lotus japonicus*. New Phytologist. 2009;**182**:200-212. DOI: 10.1111/j.1469-8137.2008.02725.x

[93] Liu J, Liu J, Liu J, Cui M, Huang Y, Tian Y, et al. The potassium transporter SlHAK10 is involved in mycorrhizal potassium uptake. Plant Physiology.

2019;**180**:465-479. DOI: 10.1104/pp.18.01533

[94] Augé RM. Water relations, drought and vesicular-arbuscular mycorrhizal symbiosis. Mycorrhiza. 2001;**11**:3-42. DOI: 10.1007/s005720100097

[95] Bahadur A, Batool A, Nasir F, Jiang S, Mingsen Q, Zhang Q, et al. Mechanistic insights into arbuscular mycorrhizal fungi-mediated drought stress tolerance in plants. International Journal of Molecular Sciences. 2019;**20**:4199. DOI: 10.3390/ijms20174199

[96] Ruiz-Lozano JM, Aroca R, Zamarreño ÁM, Molina S, Andreo-Jiménez B, Porcel R, et al. Arbuscular mycorrhizal symbiosis induces strigolactone biosynthesis under drought and improves drought tolerance in lettuce and tomato. Plant, Cell & Environment. 2016;**39**:441-452. DOI: 10.1111/pce.12631

[97] Visentin I, Vitali M, Ferrero M, Zhang Y, Ruyter-Spira C, Novák O, et al. Low levels of strigolactones in roots as a component of the systemic signal of drought stress in tomato. New Phytologist. 2016;**212**(4):954-963. DOI: 10.1111/nph.14190

[98] Kakouridis A, Hagen JA, Kan MP, Mambelli S, Feldman LJ, Herman DJ, et al. Routes to roots: Direct evidence of water transport by arbuscular mycorrhizal fungi to host plants. New Phytologist. 2022;**236**:210-221. DOI: 10.1111/nph.18281

[99] Quiroga G, Erice G, Ding L, Chaumont F, Aroca R, Ruiz-Lozano JM. The arbuscular mycorrhizal symbiosis regulates aquaporins activity and improves root cell water permeability in maize plants subjected to water stress. Plant, Cell & Environment. 2019;**42**:2274-2290. DOI: 10.1111/pce.13551

[100] Ruiz-Lozano JM, del Mar AM, Bárzana G, Vernieri P, Aroca R. Exogenous ABA accentuates the differences in root hydraulic properties between mycorrhizal and non mycorrhizal maize plants through regulation of PIP aquaporins. Plant Molecular Biology. 2009;**70**:565-579. DOI: 10.1007/s11103-009-9492-z

[101] Giovannetti M, Balestrini R, Volpe V, Guether M, Straub D, Costa A, et al. Two putative-aquaporin genes are differentially expressed during arbuscular mycorrhizal symbiosis in *Lotus japonicus*. BMC Plant Biology. 2012;**12**:14. DOI: 10.1186/1471-2229-12-186

[102] Smith SE, Smith FA. Fresh perspectives on the roles of arbuscular mycorrhizal fungi in plant nutrition and growth. Mycologia. 2012;**104**:1-13. DOI: 10.3852/11-229

[103] Kiers ET, Duhamel M, Beesetty Y, Mensah JA, Franken O, Verbruggen E, et al. Reciprocal rewards stabilize cooperation in the mycorrhizal symbiosis. Science. 2011;**333**:880-882. DOI: 10.1126/science.1208473

[104] Bücking H, Mensah JA, Fellbaum CR. Common mycorrhizal networks and their effect on the bargaining power of the fungal partner in the arbuscular mycorrhizal symbiosis. Communicative & Integrative Biology. 2016;**9**:e1107684. DOI: 10.1080/19420889.2015.1107684

[105] Fellbaum CR, Mensah JA, Cloos AJ, Strahan GD, Pfeffer PE, Kiers ET, et al. Fungal nutrient allocation in common mycelia networks is regulated by the carbon source strength of individual host plants. New Phytologist.

2014;**203**:645-656. DOI: 10.1111/nph.12827

[106] Bücking H, Shachar-Hill Y. Phosphate uptake, transport and transfer by the arbuscular mycorrhizal fungus *Glomus intraradices* is stimulated by increased carbohydrate availability. New Phytologist. 2005;**165**:899-912. DOI: 10.1111/j.1469-8137.2004.01274.x

[107] Hammer EC, Pallon J, Wallander H, Olsson PA. Tit for tat? A mycorrhizal fungus accumulates phosphorus under low plant carbon availability. FEMS Microbiology Ecology. 2011;**76**:236-244. DOI: 10.1111/j.1574-6941.2011.01043.x

[108] Harrison MJ. A sugar transporter from *Medicago truncatula*: Altered expression pattern in roots during vesicular-arbuscular (VA) mycorrhizal associations. The Plant Journal. 1996;**9**:491-503. DOI: 10.1046/j.1365-313x.1996.09040491.x

[109] Gaude N, Bortfeld S, Duensing N, Lohse M, Krajinski F. Arbuscule-containing and non-colonized cortical cells of mycorrhizal roots undergo extensive and specific reprogramming during arbuscular mycorrhizal development. Plant Journal. 2012;**69**:510-528. DOI: 10.1111/j.1365-313X.2011.04810.x

[110] Kafle A, Garcia K, Wang X, Pfeffer PE, Strahan GD, Bücking H. Nutrient demand and fungal access to resources control the carbon allocation to the symbiotic partners in tripartite interactions of *Medicago truncatula*. Plant, Cell & Environment. 2019;**42**:270-284. DOI: 10.1111/pce.13359

[111] Gabriel-Neumann E, Neumann G, Leggewie G, George E. Constitutive overexpression of the sucrose transporter SoSUT1 in potato plants increases arbucular mycorrhiza fungal root colonization under high, but not under low, soil phosphorus availability. Journal of Plant Physiology. 2011;**168**:911-919. DOI: 10.1016/j.jplph.2010.11.026

[112] Bitterlich M, Krügel U, Boldt-Burisch K, Franken P, Kühn C. The sucrose transporter SlSUT2 from tomato interacts with brassinosteroid functioning and affects arbuscular mycorrhiza formation. Plant Journal. 2014;**78**:877-889. DOI: 10.1111/tpj.12515

[113] Patil G, Valliyodan B, Deshmukh R, Prince S, Nicander B, Zhao M, et al. Soybean (*Glycine max*) SWEET gene family: Insights through comparative genomics, transcriptome profiling and whole genome re-sequence analysis. BMC Genomics. 2015;**16**:520. DOI: 10.1186/s12864-015-1730-y

[114] Gautam T, Dutta M, Jaiswal V, Zinta G, Gahlaut V, Kumar S. Emerging roles of SWEET sugar transporters in plant development and abiotic stress response. Cell. 2022;**11**:1303. DOI: 10.3390/cells11081303

[115] Ji J, Yang L, Fang Z, Zhang Y, Zhuang M, Lv H, et al. Plant SWEET family of sugar transporters: Structure, evolution and biological functions. Biomolecules. 2022;**12**:205. DOI: 10.3390/biom12020205

[116] Manck-Götzenberger J, Requena N. Arbuscular mycorrhiza symbiosis induces a major transcriptional reprogramming of the potato SWEET sugar transporter family. Frontiers in Plant Science. 2016;**2016**:7. DOI: 10.3389/fpls.2016.00487

[117] Tamayo E, Figueira-Galán D, Manck-Götzenberger J, Requena N. Overexpression of the potato monosaccharide trransporter *StSWEET7a* promotes root colonization by symbiotic and pathogenic fungi by increasing root sink strength. Frontiers in

Plant Science. 2022;**2022**:13. DOI: 10.3389/fpls.2022.837231

[118] Kryvoruchko IS, Sinharoy S, Torres-Jerez I, Sosso D, Pislariu CI, Guan D, et al. MtSWEET11, a nodule-specific sucrose transporter of *Medicago truncatula*. Plant Physiology. 2016;**171**:554-565. DOI: 10.1104/pp.15.01910

[119] Sugiyama A, Saida Y, Yoshimizu M, Takanashi K, Sosso D, Frommer WB, et al. Molecular characterization of LjSWEET3, a sugar transporter in nodules of *Lotus japonicus*. Plant and Cell Physiology. 2016;**58**:298-306. DOI: 10.1093/pcp/pcw190

[120] An J, Zeng T, Ji C, de Graaf S, Zheng Z, Xiao TT, et al. A *Medicago truncatula* SWEET transporter implicated in arbuscule maintenance during arbuscular mycorrhizal symbiosis. New Phytologist. 2019;**224**:396-408. DOI: 10.1111/nph.15975

[121] Pfeffer PE, Douds DD, Bécard G, Shachar-Hill Y. Carbon uptake and the metabolism and transport of lipids in an arbuscular mycorrhiza. Plant Physiology. 1999;**120**:587-598. DOI: 10.1104/pp.120.2.587

[122] Schaarschmidt S, Kopka J, Ludwig-Müller J, Hause B. Regulation of arbuscular mycorrhization by apoplastic invertases: Enhanced invertase activity in the leaf apoplast affects the symbiotic interaction. Plant Journal. 2007;**51**:390-405. DOI: 10.1111/j.1365-313X.2007.03150.x

[123] Helber N, Wippel K, Sauer N, Schaarschmidt S, Hause B, Requena N. A versatile monosaccharide transporter that operates in the arbuscular mycorrhizal fungus *Glomus* sp. is crucial for the symbiotic relationship with plants. The Plant Cell. 2011;**23**:3812-3823. DOI: 10.1105/tpc.111.089813

[124] Trépanier M, Bécard G, Moutoglis P, Willemot C, Gagné S, Avis TJ, et al. Dependence of arbuscular-mycorrhizal fungi on their plant host for palmitic acid synthesis. Applied and Environmental Microbiology. 2005;**71**:5341-5347. DOI: 10.1128/aem.71.9.5341-5347.2005

[125] Luginbuehl LH, Menard GN, Kurup S, van Erp H, Radhakrishnan G, Breakspear A, et al. Fatty acids in arbuscular mycorrhizal fungi are synthesized by the host plant. Science. 2017;**356**:1175-1178. DOI: 10.1126/science.aan0081

[126] Keymer A, Pimprikar P, Wewer V, Huber C, Brands M, Bucerius SL, et al. Lipid transfer from plants to arbuscular mycorrhiza fungi. eLife. 2017;**6**:e29107. DOI: 10.7554/eLife.29107

[127] Bravo A, York T, Pumplin N, Mueller LA, Harrison MJ. Genes conserved for arbuscular mycorrhizal symbiosis identified through phylogenomics. Nature Plants. 2016;**2**:15208. DOI: 10.1038/nplants.2015.208

[128] Wang E, Schornack S, Marsh JF, Gobbato E, Schwessinger B, Eastmond PJ, et al. A common signaling process that promotes mycorrhizal and oomycete colonization in plants. Current Biology. 2012;**22**:2242-2246. DOI: 10.1016/j.cub.2012.09.043

[129] Bravo A, Brands M, Wewer V, Dörmann P, Harrison MJ. Arbuscular mycorrhiza-specific enzymes FatM and RAM2 fine-tune lipid biosynthesis to promote development of arbuscular mycorrhiza. New Phytologist. 2017;**214**:1631-1645. DOI: 10.1111/nph.14533

[130] Cope KR, Kafle A, Yakha JK, Pfeffer PE, Strahan GD, Garcia K, et al. Physiological and transcriptomic response of *Medicago truncatula* to colonization by high- or low-benefit arbuscular mycorrhizal fungi. Mycorrhiza. 2022;**32**:281-303. DOI: 10.1007/s00572-022-01077-2

[131] Banasiak J, Jamruszka T, Murray JD, Jasiński M. A roadmap of plant membrane transporters in arbuscular mycorrhizal and legume–rhizobium symbioses. Plant Physiology. 2021;**187**:2071-2091. DOI: 10.1093/plphys/kiab280

[132] Zhang Q, Blaylock LA, Harrison MJ. Two *Medicago truncatula* half-ABC transporters are essential for arbuscule development in arbuscular mycorrhizal symbiosis. The Plant Cell. 2010;**22**:1483-1497. DOI: 10.1105/tpc.110.074955

[133] Radhakrishnan GV, Keller J, Rich MK, Vernié T, Mbadinga Mbadinga DL, Vigneron N, et al. An ancestral signalling pathway is conserved in intracellular symbioses-forming plant lineages. Nature Plants. 2020;**6**:280-289. DOI: 10.1038/s41477-020-0613-7

[134] Hogekamp C, Arndt D, Pereira PA, Becker JD, Hohnjec N, Kuster H. Laser microdissection unravels cell-type-specific transcription in arbuscular mycorrhizal roots, including CAAT-box transcription factor gene expression correlating with fungal contact and spread. Plant Physiology. 2011;**157**:2023-2043. DOI: 10.1104/pp.111.186635

[135] Kafle A, Garcia K, Peta V, Yakha JK, Soupir A, Bücking H. Beneficial plant microbe interactions and their effect on nutrient uptake, yield, and stress resistance of soybeans. In: Kasai M, editor. Soybean: Biomass, Yield and Productivity. London: Intechopen; 2018. DOI: 10.5772/intechopen.81396

[136] Deng L, Zhao S, Yang G, Zhu S, Tian J, Wang X. Soybean GmSUT1 transporter participates in sucrose transport to nodules during rhizobial symbiosis. Plant Growth Regulation. 2022;**96**:119-129. DOI: 10.1007/s10725-021-00764-y

[137] Catalano CM, Lane WS, Sherrier DJ. Biochemical characterization of symbiosome membrane proteins from *Medicago truncatula* root nodules. Electrophoresis. 2004;**25**:519-531. DOI: 10.1002/elps.200305711

[138] Clarke VC, Loughlin PC, Gavrin A, Chen C, Brear EM, Day DA, et al. Proteomic analysis of the soybean symbiosome identifies new symbiotic proteins. Molecular Cell Proteomics. 2015;**14**:1301-1322. DOI: 10.1074/mcp.M114.043166

[139] Yurgel SN, Kahn ML. Dicarboxylate transport by rhizobia. FEMS Microbiology Reviews. 2004;**28**:489-501. DOI: 10.1016/j.femsre.2004.04.002

[140] White J, Prell J, James EK, Poole P. Nutrient sharing between symbionts. Plant Physiology. 2007;**144**:604-614. DOI: 10.1104/pp.107.097741

[141] Hwang JH, Ellingson SR, Roberts DM. Ammonia permeability of the soybean nodulin 26 channel. FEBS Letters. 2010;**584**:4339-4343. DOI: 10.1016/j.febslet.2010.09.033

[142] Schwember AR, Schulze J, Del Pozo A, Cabeza RA. Regulation of symbiotic nitrogen fixation in legume root nodules. Plants (Basel). 2019;**8**:333. DOI: 10.3390/plants8090333

[143] Lambert I, Pervent M, Le Queré A, Clément G, Tauzin M, Severac D, et al. Responses of mature symbiotic nodules to the whole-plant systemic nitrogen signaling. Journal of

Experimental Botany. 2020;**71**:5039-5052. DOI: 10.1093/jxb/eraa221

[144] Luginbuehl LH, Oldroyd GED. Understanding the arbuscule at the heart of endomycorrhizal symbioses in plants. Current Biology. 2017;**27**:R952-R963. DOI: 10.1016/j.cub.2017.06.042

[145] Westhoek A, Field E, Rehling F, Mulley G, Webb I, Poole PS, et al. Policing the legume-rhizobium symbiosis: A critical test of partner choice. Scientific Reports. 2017;7:1419. DOI: 10.1038/s41598-017-01634-2

[146] Kiers ET, Rousseau RA, West SA, Denison RF. Host sanctions and the legume-rhizobium mutualism. Nature. 2008;**425**:78-81. DOI: 10.1038/nature01931

[147] Westhoek A, Clark LJ, Culbert M, Dalchau N, Griffiths M, Jorrin B, et al. Conditional sanctioning in a legume-rhizobium mutualism. Proceedings of the National Academy of Sciences, USA. 2021;**2021**:118. DOI: 10.1073/pnas.2025760118

[148] Vuong HB, Thrall PH, Barrett LG. Host species and environmental variation can influence rhizobial community composition. Journal of Ecology. 2017;**105**:540-548. DOI: 10.1111/1365-2745.12687

[149] Checcucci A, Azzarello E, Bazzicalupo M, Galardini M, Lagomarsino A, Mancuso S, et al. Mixed nodule infection in *Sinorhizobium meliloti–Medicago sativa* symbiosis suggest the presence of cheating behavior. Frontiers in Plant Science. 2016;**2016**:7. DOI: 10.3389/fpls.2016.00835

[150] Kobae Y, Hata S. Dynamics of periarbuscular membranes visualized with a fluorescent phosphate transporter in arbuscular mycorrhizal roots of rice. Plant and Cell Physiology. 2010;**51**:341-353. DOI: 10.1093/pcp/pcq013

[151] Breuillin F, Schramm J, Hajirezaei M, Ahkami A, Favre P, Druege U, et al. Phosphate systemically inhibits development of arbuscular mycorrhiza in *Petunia hybrida* and represses genes involved in mycorrhizal functioning. Plant Journal. 2010;**64**:1002-1017. DOI: 10.1111/j.1365-313X.2010.04385.x

[152] Suzaki T, Nishida H. Autoregulation of legume nodulation by sophisticated transcriptional regulatory networks. Molecular Plant. 2019;**12**:1179-1181. DOI: 10.1016/j.molp.2019.07.008

[153] Mortier V, Den Herder G, Whitford R, Van de Velde W, Rombauts S, D'Haeseleer K, et al. CLE peptides control *Medicago truncatula* nodulation locally and systemically. Plant Physiology. 2010;**153**:222-237. DOI: 10.1104/pp.110.153718

[154] Schnabel E, Journet E-P, de Carvalho-Niebel F, Duc G, Frugoli J. The *Medicago truncatula* SUNN gene encodes a CLV1-like leucine-rich repeat receptor kinase that regulates nodule number and root length. Plant Molecular Biology. 2005;**58**:809-822. DOI: 10.1007/s11103-005-8102-y

[155] Searle IR, Men AE, Laniya TS, Buzas DM, Iturbe-Ormaetxe I, Carroll BJ, et al. Long-distance signaling in nodulation directed by a CLAVATA1-like receptor kinase. Science. 2003;**299**:109-112. DOI: 10.1126/science.1077937

[156] Krusell L, Madsen LH, Sato S, Aubert G, Genua A, Szczyglowski K, et al. Shoot control of root development and nodulation is mediated by

a receptor-like kinase. Nature. 2002;**420**:422-426. DOI: 10.1038/nature01207

[157] Suzaki T, Yoro E, Kawaguchi M. Leguminous plants: Inventors of root nodules to accommodate symbiotic bacteria. International Review of Cell and Molecular Biology. 2015;**316**:111-158. DOI: 10.1016/bs.ircmb.2015.01.004

[158] Sasaki T, Suzaki T, Soyano T, Kojima M, Sakakibara H, Kawaguchi M. Shoot-derived cytokinins systemically regulate root nodulation. Nature Communications. 2014;**5**:4983. DOI: 10.1038/ncomms5983

[159] Tiwari M, Pandey V, Singh B, Yadav M, Bhatia S. Evolutionary and expression dynamics of LRR-RLKs and functional establishment of KLAVIER homolog in shoot mediated regulation of AON in chickpea symbiosis. Genomics. 2021;**113**:4313-4326. DOI: 10.1016/j.ygeno.2021.11.022

[160] van Noorden GE, Ross JJ, Reid JB, Rolfe BG, Mathesius U. Defective long-distance auxin transport regulation in the *Medicago truncatula* super numeric nodules mutant. Plant Physiology. 2006;**140**:1494-1506. DOI: 10.1104/pp.105.075879

[161] Wang C, Reid JB, Foo E. The art of self-control - autoregulation of plant-microbe symbioses. Frontiers in Plant Science. 2018;**9**:988. DOI: 10.3389/fpls.2018.00988

[162] Meixner C, Vegvari G, Ludwig-Müller J, Gagnon H, Steinkellner S, Staehelin C, et al. Two defined alleles of the LRR receptor kinase GmNARK in supernodulating soybean govern differing autoregulation of mycorrhization. Physiologia Plantarum. 2007;**130**:261-270. DOI: 10.1111/j.1399-3054.2007.00903.x

[163] Püschel D, Janouskova M, Voriskova A, Gryndlerova H, Vosatka M, Jansa J. Arbuscular mycorrhiza stimulates biological nitrogen fixation in two *Medicago* spp. through improved phosphorus acquisition. Frontiers in Plant Science. 2017;**2017**:8. DOI: 10.3389/fpls.2017.00390

[164] Larimer AL, Clay K, Bever JD. Synergism and context dependency of interactions between arbuscular mycorrhizal fungi and rhizobia with a prairie legume. Ecology. 2014;**95**:1045-1054. DOI: 10.1890/13-0025.1

[165] Bournaud C, James EK, de Faria SM, Lebrun M, Melkonian R, Duponnois R, et al. Interdependency of efficient nodulation and arbuscular mycorrhization in *Piptadenia gonoacantha*, a Brazilian legume tree. Plant, Cell & Environment. 2017;**41**:2008-2020. DOI: 10.1111/pce.13095

[166] Vadez V, Beck DP, Lasso JH, Drevon JJ. Utilization of the acetylene reduction assay to screen for tolerance of symbiotic N2 fixation to limiting P nutrition in common bean. Physiologia Plantarum. 1997;**99**:227-232. DOI: 10.1111/j.1399-3054.1997.tb05406.x

[167] Mortimer PE, Perez-Fernandez MA, Valentine AJ. The role of arbuscular mycorrhizal colonization in the carbon and nutrient economy of the tripartite symbiosis with nodulated *Phaseolus vulgaris*. Soil Biology & Biochemistry. 2008;**40**:1019-1027. DOI: 10.1016/j.soilbio.2007.11.014

[168] Chen BD, Li XL, Tao HQ, Christie P, Wong MH. The role of arbuscular mycorrhiza in zinc uptake by red clover growing in a calcareous soil spiked with various quantities of zinc. Chemosphere. 2003;**50**:839-846. DOI: 10.1016/s0045-6535(02)00228-x

[169] Ibiang YB, Mitsumoto H, Sakamoto K. Bradyrhizobia and arbuscular mycorrhizal fungi modulate manganese, iron, phosphorus, and polyphenols in soybean (*Glycine max* (L.) Merr.) under excess zinc. Environmental and Experimental Botany. 2017;**137**:1-13. DOI: 10.1016/j.envexpbot.2017.01.011

[170] Sakamoto K, Ogiwara N, Kaji T. Involvement of autoregulation in the interaction between rhizobial nodulation and AM fungal colonization in soybean roots. Biology and Fertility of Soils. 2013;**49**:1141-1152. DOI: 10.1007/s00374-013-0804-8

[171] Meghvansi MK, Prasad K, Harwani D, Mahna SK. Response of soybean cultivars toward inoculation with three arbuscular mycorrhizal fungi and *Bradyrhizobium japonicum* in the alluvial soil. European Journal of Soil Biology. 2008;**44**:316-323. DOI: 10.1016/j.ejsobi.2008.03.003

[172] Afkhami ME, Almeida BK, Hernandez DJ, Kiesewetter KN, Revillini DP. Tripartite mutualisms as models for understanding plant–microbial interactions. Current Opinion in Plant Biology. 2020;**56**:28-36. DOI: 10.1016/j.pbi.2020.02.003

[173] Konvalinková T, Jansa J. Lights off for arbuscular mycorrhiza: On its symbiotic functioning under light deprivation. Frontiers in Plant Science. 2016;**2016**:7. DOI: 10.3389/fpls.2016.00782

[174] Kleinert A, Venter M, Kossmann J, Valentine A. The reallocation of carbon in P deficient lupins affects biological nitrogen fixation. Journal of Plant Physiology. 2014;**171**:1619-1624. DOI: 10.1016/j.jplph.2014.07.017

Chapter 8

Mycorrhizal Symbiosis for Sustainable Optimization of Tropical Agriculture: A Review of Research

Ricardos M. Aguégué, Corentin Akpode, Tomabu Adjobimey, Haziz Sina, Sylvestre A. Assogba, Abdel D. Koda, Nadège A. Agbodjato, Nestor R. Ahoyo Adjovi, Adolphe Adjanohoun, Olubukola O. Babalola and Lamine Baba-Moussa

Abstract

Excessive application of chemical fertilizers and other agrochemicals can cause significant imbalances in soils and agricultural ecosystems. To minimize these impacts, biofertilizers and organic fertilizers are needed to maintain a sustainable production system. The use of subterranean microorganisms in agriculture to stimulate plant growth and improve yields has recently received increasing interest. In this context, mycorrhizae represent a viable solution to mitigate these adverse effects. Mycorrhizal fungi are able to form a symbiotic relationship with the roots of plants in the environment. Mycorrhizal fungus helps the plant to absorb nutrients and water. In addition, mycorrhizal fungi play a crucial role in storing carbon (C) in the soil. Most previous studies have just considered the effects of AMF species on a specific crop in one particular area but have not assessed the balance of AMF in production systems in tropical agriculture. This consideration should allow for the optimization of cropping practices through a review of the work on the use of AMF in tropical agriculture production systems. In this paper, we will discuss, through different examples of experiments carried out in the tropics, the performance of different strategies for managing the potential of AMF to maintain a sustainable production system.

Keywords: bioinoculation, SDG#2, sustainable agriculture, symbiosis, tropical environment

1. Introduction

In Sub-Saharan Africa, the population increase is estimated at 2 billion in 2050. This is more than double what it was in 2010 (800 million inhabitants) [1, 2]. Food needs will evolve considerably, even four times those of 2010 [1, 3]. Indeed, the technical performance of African agriculture, particularly crop yields, which are generally low, would not be sufficient to cover such changes in needs [1] without an effective alternative in new agricultural technologies. It is, therefore, necessary to expand the cultivated land in Africa on a large scale by more than 122 million ha, that is, a growth of more than 47% compared to the initial situation in 2010.

In this context, it is essential to have a better yield of corn grains for the self-sufficiency of the population. Thus, in most cases, producers use chemical agricultural inputs (pesticides and mineral fertilizers), which, in the long run, degrade the soil and pollute waterways and the environment. Indeed, the contribution of organic fertilizer is necessary for these soils on which chemical fertilizers are used to avoid their rapid degradation by water and wind currents [4]. Establishing a reliable and sustainable agricultural technology without adverse effects on soil health and the environment to meet food needs remains the major concern of researchers in the agricultural field. Indeed, the organic farming system remains the best option because it obliges farmers to use organic amendment resources of the remains from livestock, green organic matter, and other organic manures [5, 6]. Nevertheless, these inputs are costly, and their forms remain relatively insoluble. Also, their effects on phosphorus availability in soils in organic agriculture are often limited [5, 7].

Techniques for regenerating the health of cultivated soils that incorporate a judicious combination of organic matter of different C/N ratios, on the one hand, and the use of the soil microflora and soil fauna, on the other, are needed to reduce the rate of mineralization of organic matter supplied to soils, thereby maintaining microbial life in these soils for long periods.

Beneficial microorganisms include arbuscular mycorrhizal fungi (AMF) that are associated with at least 90% of terrestrial plants [8, 9]. A majority of terrestrial plants, especially grapevine [10], are colonized by about 400 arbuscular mycohizal fungi that allow them better nutrient uptake and stress tolerance [11]. The interactions between plant, soil, and mycorrhizal fungi are ecologically and agriculturally beneficial systems. In nature, most plants are in association with mycorrhizae. Thanks to this association, the plant increases its water and nutrient-absorption surface. Also, they have better accessibility to the elements available in the soil but not accessible. In addition, mycorrhizae contribute to the mobilization of carbon (C) in the soil.

In this review, we will synthesize research on the performance of different AMF potential management strategies to maintain a sustainable production system.

2. Functional diversity of arbuscular mycorrhizal fungi in the rhizosphere

The first classification was established by Taxter [12] and later modified by many mycorrhizologists [13, 14]. The classical identification of arbuscular mycorrhizal fungi is based on the structural morphology of the spore. Those that have similar morphology and form a single type of spore wall are phylogenetically related. However, populations of spores isolated and identified from rhizophytic soils do not always constitute communities of arbuscular mycorrhizal fungal species that infect roots [15]. Nevertheless, the low morphological variation among spores often makes their identification difficult.

Schwarzott et al. [16] describe the molecular classification based on in vitro DNA amplification of variable regions of the fungal genome from spores or from mycorrhized roots [17]. Indeed, multigene analysis has been used to identify arbuscular mycorrhizal fungi [18] to confirm the phylogenetic structure proposed by Schüßler

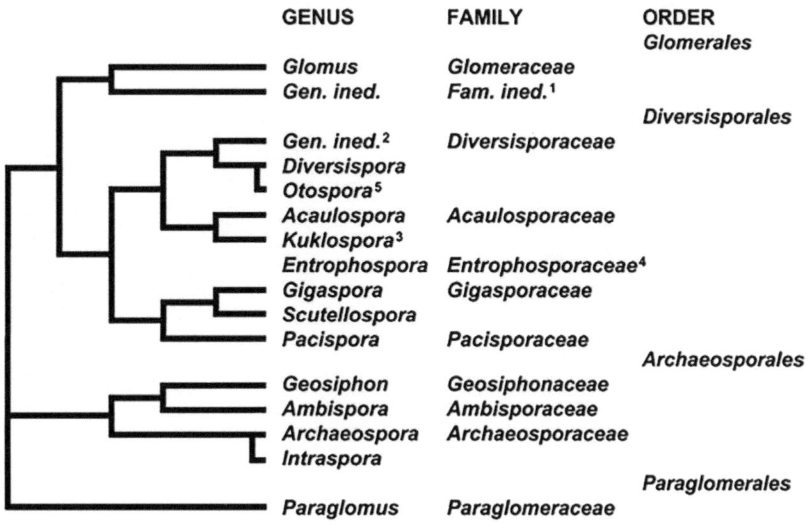

Figure 1.
Phylogenetic relationships of Glomeromycetes taxa [13, 24].

et al. [13]. Furthermore, biochemical approaches based on the use of specific antibodies [19] and lipid profiles [20] are also used for the identification of arbuscular mycorrhizal fungi.

Morton and Benny [21], who classified arbuscular mycorrhizal fungi in the order Glomales, subdivided them into two suborders, Glominae (vesicles forming) and Gigasporinae. The Glominae include two large families, the Glomaceae represented by the genera Glomus and Sclerocystys, a genus eliminated by Redecker et al. [22], and the Acaulosporaceae by Acaulospora and Entrophospora. The Gigasporinae include three families, the Gigasporaceae represented by Gigaspora and Scutellospora, the Archaeosporaceae represented by the genus Archaeospora, and the Paraglomaceae by Paraglomus [23].

Figure 1 shows the phylogenetic classification of the kingdom Mycota (Fungi) based on the analysis of 18S rRNA nucleotide sequences [13]. [1]Species currently named *Glomus* [2]contains *Glomus fulvum, Gl. megalocarpum, Gl. pulvinatum*; [3]contains *Kuklospora colombiana* and *Ku. kentinensis* (formerly *Entrophospora*) [25]; [4]contains one genus with *Entrophospora infrequent* and *A. baltica* [25], neither of which are phylogenetically characterized; [5]*Otospora* [26] contains *Otospora bareai* [27].

3. Role of arbuscular mycorrhizal fungi in promoting crop growth

The success of a production technique could depend on controlling the factors that influence the development of arbuscular mycorrhizal fungi (AMF). AMF are another group of symbiotic soil microbes capable of directly or indirectly influencing soil properties, affecting plant growth and community structure [28, 29]. Their multiple beneficial attributes allow them to be involved in many processes, such as biofertilization, biostimulation, and bioprotection [30]. Indeed, AMF confer many positive effects on host plants, including promoting plant growth, stabilizing soil aggregation, maintaining soil moisture, improving tolerance to abiotic and biotic

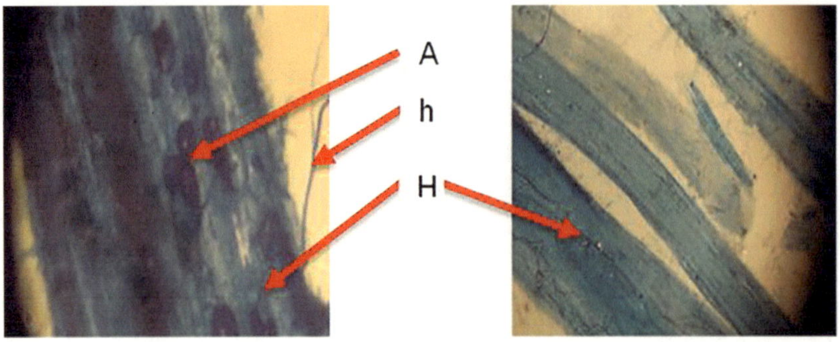

Figure 2.
Infected corn roots colonized by shrub-shaped mycorrhizal fungi structures. (A: vesicular; h: intra-root hyphae; H: extra-root hyphae (×180) [38].

stress, and increasing plant biodiversity [30–33]. In addition, these beneficial fungi have shown a potential ability to increase crop resilience and performance [34] and are known to affect their host plants differently. AMF are present in the soil in the form of spores and extra radicular hyphae, as well as in plant roots, with which they form a mutualistic association, and are considered an essential component of various ecosystems [35, 36]. The most important contribution of AMF fungi to plant growth is due to the uptake of phosphorus and other elements by the extra-radical hyphae and their transfer to root tissues. Indeed, intra-root hyphae, vesicles, and intra- and extra-root spores are structures capable of AMF propagation [37]. In addition to being a propagule, vesicles (**Figure 2**) are a reserve organ; their production is related to the stage of AMF development, and their presence can vary with carbon allocation by the host plant [39, 40]. However, factors such as low light intensity and defoliation, factors that limit photosynthesis and thus the carbon content of the plant, significantly reduce sporulation as well as colonization of new roots [37]. AMF provides nutrients, primarily phosphorus (P) and nitrogen (N), to host plants in exchange for carbon (C). The increase in available phosphorus and exchangeable potassium and magnesium levels in the soil of AMF-treated plants would certainly mean more leaves, branches, and biomass in AMF-treated plants than in control [41]. Mycorrhization of tissue-grown propagules can produce plants with increased levels of biologically active secondary metabolites [42]. The exploration of a larger volume of soil and the possibility of alteration of primary minerals by mycorrhizal fungi improve the phosphate nutrition of plants [43, 44]. This improved acquisition of inorganic nutrients by fungal symbionts also concerns other macro- (N, K, Mg, Na, S) and micro- (B, Br, Cl, Cu, Cr, Cs, Co, Fe, Mo, Mn, Ni, Si, Zn) soil nutrients [37, 44]. Mycorrhizal associations play a potential role in the decomposition and mineralization of plant organic matter and the mobilization of nutrients to the host plant [44, 45].

4. Arbuscular mycorrhizal fungi as biofertilizers to increase crop yield

The work of Xie et al. [46] and Battini et al. [47] on tomato showed that AMFs were responsible for P mobilization in inoculated tomato plants. In addition, Balliu et al. [48] on tomato plants inoculated with AMF observed a significant increase in N uptake. Campo et al. [49] showed the importance of mycorrhizal fungi on growth, productivity, and disease resistance in the rice crop. In addition, in the pasture, the dry season appears to favor AMF species diversity. It is possible that the lower humidity of this season influences the sporulation of more species [50, 51]. Studies conducted by Nekou et al. [52] and Ngakou et al. [53] in Cameroon in 5 agro-ecological zones showed that the use of commercial mycorrhizal fungi on cowpea plants improved their growth by 17–46%. In Togo, inoculation of soybean plants with mycorrhizae resulted in an improvement in plant height from 14 to 32 cm [54]. Also, the work of Ngakou et al. [55] on garlic plants inoculated with mycorrhizae under greenhouse conditions observed an improvement in plant growth from 4.85 cm to 6.28 cm. Ogou et al. [54] obtained in Togo an average mycorrhizae-induced pod gain on soybean of +126.83% compared to the control. Moussa et al. [56] reported an average mycorrhizal inoculation-induced improvement of 23.4% on *Vigna subterranea* plants. Hemissi et al. [57], in Tunisia, obtained a gain of 20 kg/ha more wheat in mycorrhized plants than in control plants. The role of AMF on *Drymaria cordata* is indicative of the apparent potential of this association to improving production [41]. Similar efficacy of G. fasciculatum on the growth and performance of normal and regenerated *Andrographis paniculata* plants has been studied [58]. Numerous works have shown the effectiveness of endomycorrhizal symbiosis in improving maize productivity [59–63].

These improvements in plant growth and yield by arbuscular fungi are explained by the fact that mycorrhizae degrade and mineralize soil organic matter and mobilize the resulting nutrients to the plant [64, 65]. In addition, mycorrhizae develop extra-radical mycelial hyphae that explore a larger volume of soil not accessible to plant roots [66]. *Solanum lycopersicum*, inoculated with *G. fasciculatu*m significantly improved morphological characteristics [67].

Another important point regarding plant health is the presence of toxic metals such as Cu, Cd, Zn, and Pb in the soil. It has been suggested that the most appropriate use for these elements is potentially toxic elements (PTEs) [68, 69]. AMF are capable of absorbing Cu, and 1 g of AMF hyphae has a Cu content of 3–14 mg [70].

5. Role of glomalin secreted by mycorrhizal fungi

The growth of mycorrhizal hyphae in the soil is accompanied by the production of glomalin (a glycoprotein), which improves the aggregation of soil particles [44, 71]. Glomalin is a hydrophobic protein. Glomalin, an N-linked glycoprotein [72], which is considered to be an AMF gene product, is defined as a protein secreted by AMF hyphae and spores [73]. It contains iron (2–5%), oxygen (4–6%), phosphorus (0:03–0.1%), carbon (36–59%), hydrogen (33–49%), and nitrogen (3–5%) [73–75]. The reddish-brown appearance of glomalin extracts is due to the iron content [71, 75]. According to Wright [76], the concentration of glomalin in soils is very high compared to humic acid, especially in the presence of insoluble humus or minerals in soils treated with sodium hydroxide. The glomalin secreted by the mycorrhizae in the soil allows the stabilization and optimization of the PTE. Moreover, in addition to its beneficial roles for the soil, in the rhizosphere of plants, it plays a protective role for the microorganism and the plant roots against toxins [73]. There is a strong relationship between glomalin concentration and soil aggregate stability [75, 77]. Because of this role, glomalin preserves unstable compounds in soil aggregates and thus reduces the degradation of soil organic matter [78]. N-linked glomalin [72], which is considered the gene product of AMF, is defined as a glycoprotein secreted by AMF hyphae and spores [73]. Glomalin, plays an excellent protective role for hyphae as well as promotes soil aggregation.

Glomalin is measured from the soil as Glomalin-Related Soil Protein (GRSP) [78–80]. Among soil microorganisms, AMF have the ability to stabilize soil structure, which reduces stress at the plant level [81, 82]. Indeed, glomalin produced by arbuscular mycorrhizal fungi is one of the key factors of soil quality. It is essential for the formation of soil aggregates because it is one of the components of soil organic matter [75, 83]. Also, it contributes to the nutrient storage capacity and water holding capacity [83]. Land use change can alter the abundance and diversity of arbuscular mycorrhizal fungi species and the content of glomalin-related soil protein (GRSP) in the soil [84–86]. Glomalin from the decomposition of hyphae wall and spores of mycorrhizal fungi even after their death is quantified in soil as GRSP [87, 88]. GRSP is considered an essential component of the soil organic carbon (SOC) pool in terrestrial ecosystems [80, 89]. Indeed, GRSP is a key link in appreciating soil fertility, its water holding capacity, and aeration and nutrient content for better plant productivity [90].

GRSP is composed of two soil proteins: the easily extractable protein (EE-GRSP) and the total protein (T-GRSP). Assuming the C content of glomalin to be 32% [91], concerning equivalent dry mass of soil. The contribution of TG to SOC varies from season to season in agroforestry and forest [92]. Indeed, it is higher during the dry season compared to the rainy season, when it is 5.69% in soils under pasture, followed by AS3 (4.5%) and AS2 (4.31%). The work of Driver et al. [93] on *Rhizophagus intraradices* (N.C. Schenck & G.S. Sm). C. Walker & A. Schüßler showed that 80% of the glomalin secretion came from the wall of hyphae and spores.

The glomalin secreted by the mycorrhizae in the soil allows the stabilization and optimization of the PTE. In addition, glomalin plays a protective role for microorganisms and plant roots against toxins [73].

6. Mechanisms employed by arbuscular mycorrhizal fungi for salt stress amelioration

Soil salinity is a global problem because it negatively affects plant productivity and yield, especially in arid and semi-arid regions of the world. Under salt stress, plant growth and biomass have suffered a setback. The reasons may be the unavailability of nutrients and the expenditure of energy to counteract the toxic effects of NaCl. This is because excess salt decreases the availability of soil water to plants, inhibits plant metabolism and nutrient uptake, and is also responsible for osmotic imbalance [94]. Among the various biotechnological techniques used to combat the harmful effects of salt stress, the use of arbuscular mycorrhizal fungi (AMF) is considered an effective approach for bioenhancement of salt stress [95]. Possible mechanisms for salinity stress mitigation by AMF include: (1) enhancing plant nutrient uptake, particularly P; (2) elevating the K/Na ratio; (3) providing higher osmosolute accumulation; and (4) maintaining higher antioxidant enzyme activities [94]. AMF adjust all physiological and biochemical properties of the host plant (**Figure 3**) [96–98]. The mycorrhizal symbiosis with the roots of plants includes several stages, namely, (i) the formation of the appessoriun (ap), (ii) the penetration and development of hyphae into the root cortex (a), and (iii) the formation of vesicles in the root cortex (v). The salinity has a very negative impact on the productivity of the plant. Indeed, it inhibits the fundamental needs of the plant in water and nutrients, which leads to its physiological dryness and the decrease of the osmotic potential. Nevertheless, with the symbiosis of arbuscular mycorrhizal fungi, the plant resists saline stresses by a better absorption of water and nutrients. The action of mycorrhizae can be summarized as: (i) increased accumulation

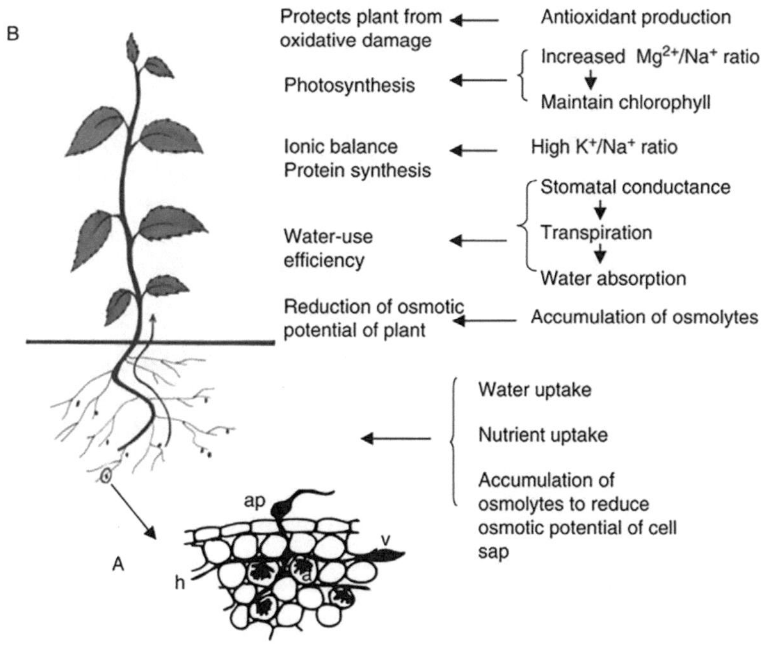

Figure 3.
The intricate functioning of arbuscular mycorrhizal fungi in ameliorating salt stress in plants [96].

of osmolytes, which allows a decrease in osmotic potential; (ii) efficient management of water and photosynthesis; and (iii) production of antioxidants to scavenge ROS [96].

AMF can protect cucumber growth from salt stress [99]. However, AMF inoculation ameliorated the negative effects by increasing biomass; pigment synthesis; antioxidant enzyme activity, including superoxide dismutase, catalase, ascorbate peroxidase, and glutathione reductase; and ascorbic acid content, which might be the result of lower levels of lipid peroxidation and electrolyte leakage. Mycorrhization has been found to increase the capacity of the host plant by improving its growth and biomass [96]. Colla et al. [100] reported improved growth, yield, water status, nutrient content, and fruit quality of *Cucurbita pepo* plants colonized by Glomus intraradices when exposed to salinity stress. Under NaCl stress, the combination of mycorrhizal fungi with compost significantly improved plant growth; P, K+, N, and Ca^{2+} uptake; leaf water potential; stomatal conductance; all antioxidant enzyme activities; and proline and soluble sugar content [101]. The application of anti-salinity increased the yield of green bean pods under all levels of salinity stress, especially with AMF followed by B. Megatherium, compared to non-inoculated plants [102].

The positive action of AMF under salt stress may be due to higher concentration of osmolytes (glycine-betaine, sugars) and polyamines and more and larger plastoglobules (higher concentration of α-tocopherol) in AMF-inoculated plants compared to in non-AMF-inoculated plants. While lower Na + and Cl—ions provide less ion toxicity, higher osmolytes and tocopherols provide osmotic adjustment and better ability to scavenge free radicals generated by salt stress, respectively [103]. Landwehr et al. [43] reported abundant AMF spores in extremely alkaline soils with a pH as high as 11, independent of soil types and NaCl, Na_2CO_3, Na_2SO_4, or $CaSO_4$ salt types, although the degree of colonization varied among individuals. Saint-Etienne et al. [104] reported significant negative correlations between salt levels and soil mycorrhizal infection (measured as most probable number values); that is, as soil salinity increased from 5 to 22%, the level of infection decreased from 301 to 20 most probable numbers per 100 g of soil. Under salt stress conditions, a beneficial effect of AMF symbiosis has been observed on water status, osmolyte accumulation, and plant growth of *Phragmites australis* [105].

However, it should be noted that in the presence of NaCl, colonization of plant roots by some AMF is reduced [106], probably due to the suppression of arbuscular mycorrhizae formation due to the effect of NaCl on its fungi [95, 106, 107]. In addition, in the presence of NaCl, spore germination is delayed [107].

7. Interactions between mycorrhizae and beneficial rhizobacteria

Plant growth and development can be promoted or inhibited by microorganisms in the rhizosphere [108]. The synergistic action between plant growth promoting rhizobacteria (PGPR) and mycorrhizae (AMF) is of great importance for the improvement of productivity and sustainability of agricultural and natural ecosystems [109–111]. The beneficial effect of the interaction between RMPs and MFAs is demonstrated by several authors [108, 112, 113]. Indeed, the phytohormones produced by PGPRs allow a good development of the plant roots, which favors the colonization of the latter by AMF [112, 114]. According to Hodge [113], the accumulation of photosynthetic products is affected by the activities of AMF on PGPR in the mycorrhizosphere. As a result of increasing mycorrhizal colonization by PGPR, glomalin production increases [115].

The latest studies have shown various abiotic factors that have had marked effects on plant growth and development. Among these, N soil, pH, temperature, erosion, waterlog, salinity, heavy metals, fungicides, and drought are the major factors that affect plant growth [116, 117]. Hence, plant growth and yield will be affected [118]. However, diverse research [119, 120] have explored the significance of beneficial rhizobacteria and mycorrhizae in the growth and development of several plants in stress environments (**Figure 4**).

This positive interaction mainly occurs among PGP rhizobacteria, mycorrhizae, and plants [122]. The addition of a single bacterial inoculum might have core effects on the rhizosphere structure. Still, it will depend on whether the fresh inoculum is already a part of that bacterial population or not [123]. Similarly, it can be more effective when a combination of PGPR is used [124]. The inoculum composed of the mixture of PGPR and AMF, due to the synergistic effect of its two microorganisms, allows for better colonization and nutrient uptake by plant roots [122]. In the same way, the rise in root exudates by the microbes triggers the fungus growth and hence increases the rate of root colonization BN [108]. After the inoculation of 20 different *Medicago truncatula* assents with Funneliformis mosseae strains, several interesting results on physiology and gene expression at the level of individual plants were observed [125]. Studying the *P. fluorescens* C7R12, it was shown as an operative biocontrol mediator in comparison to the Fusarium species [126]. Antifungal metabolites produced by Pseudomonas, instead of inhibiting the action of the G. mosseae strain, allowed the colonization of the roots by the hyphae of the latter [127]. In addition, the synthesis of Rhizobia exopolysaccharides (EPSs) improved the synergy between bacterial strains and mycorrhizal structure [128]. Thus, the combination of the rhizobacteria can raise the activity of arbuscular mycorrhizae in the symbiotic association [129]. The occurrence of PGP rhizobacteria, as well as mycorrhizae in the rhizosphere, can encourage the growth of fungus hyphae by enhancing cell permeability and help the roots in penetration to fungus [130], and fungus hyphae can raise the activities of bacterial strains, which solubilize the phosphorus [62, 131–134].

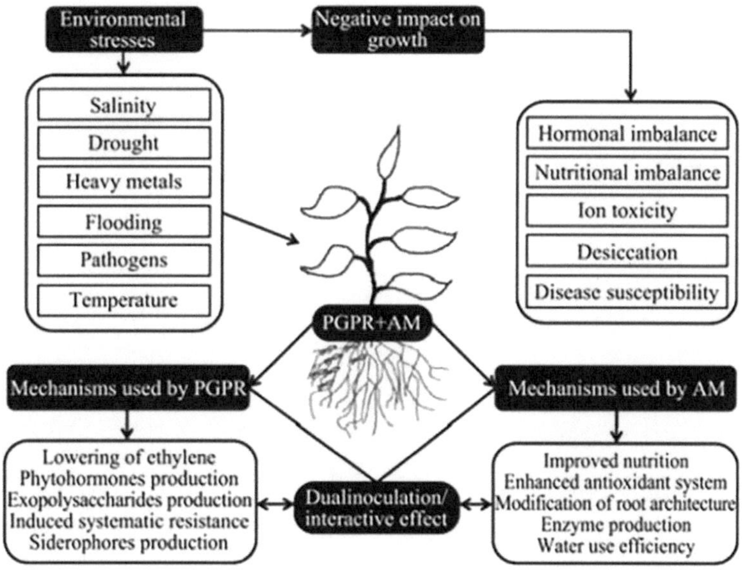

Figure 4.
Potential mechanisms used by PGP Rhizobacteria and mycorrhizae for improving plant growth under stress circumstances [121].

In addition, AMF increase plant uptake of water and mineral elements, especially phosphorus, to the plant [135]. This is because the elongation of the extra-radical mycelium increases the exchange surface between soil minerals and plant roots. Thus, areas inaccessible to the plant are explored by the extra-root mycelia to collect water and nutrients and transfer them to the host plant, allowing for improved growth, yield, and quality of plant production [134]. Moreover, Raklami et al. noted that the field inoculation with PGPR-Rhizobia-mycorrhizae improved growth, nutrition, and productivity of bean and wheat plants compared to the uninoculated control and other treatments based on PGP rhizobacteria as well as mycorrhizae. All the treatments were beneficent for *Vicia faba* L. and *Triticum durum* L. plants. The best treatment was the inoculation with PGPR- Rhizobia-mycorrhizae. Plant growth under normal or stressful situations could be improved by the use of PGP rhizobacteria and mycorrhizal strains, alone or in combination [118].

Authors' contributions

RMA, CA, TA: Conceptualization, methodology, writing—original draft; RMA, CA, TA; HS, SAA, ADK and NAA: Writing—Revision; NRAA, AA, OOB, and LB-M: Acquisition of financing; OOB and LB-M: Writing and editing. All authors have read and approved the published version of the manuscript. All authors participated equally in the work and approved the final submission.

Declaration of competing interest

The authors declare that they have no financial interests or personal relationships that might have appeared to influence the work reported in this article.

Author details

Ricardos M. Aguégué[1], Corentin Akpode[1], Tomabu Adjobimey[2], Haziz Sina[1], Sylvestre A. Assogba[1], Abdel D. Koda[1], Nadège A. Agbodjato[1,3], Nestor R. Ahoyo Adjovi[4], Adolphe Adjanohoun[4], Olubukola O. Babalola[3]* and Lamine Baba-Moussa[1]*

1 Laboratoire de Biologie et de Typage Moléculaire en Microbiologie, Département de Biochimie et de Biologie Cellulaire, Faculté des Sciences et Techniques, Université d'Abomey-Calavi, Cotonou, Bénin

2 Unité de Biochimie et de Biologie Moléculaire, Faculté des Sciences et Techniques, Université d'Abomey-Calavi, Bénin

3 Food Security and Safety Focus Area, Faculty of Natural and Agricultural Sciences, North-West University, Mmabatho, South Africa

4 Institut National des Recherches Agricoles du Bénin, Cotonou, Bénin

*Address all correspondence to: olubukola.babalola@nwu.ac.za and laminesaid@yahoo.fr

References

[1] Schmitt B, Forslund A, Tibi A, Guyomard H, Debaeke P. Comment assurer les disponibilités alimentaires du continent africain en 2050 ? Relecture africaine de l'étude INRAE (Place des agricultures européennes dans le monde) l'horizon 2050 conduite par INRAE à la demande de Pluriagri. France: Rapport de l'étude; 2021. 62 p

[2] Undp A. Africa human development report 2012. In: Towards a Food Secure Future, 267,636. United Nations Development Program (UNDP); 2012

[3] Paillard S. Agrimonde. Agricultures et alimentation du monde en 2050 : scénarios et défis pour un développement durable. 2009

[4] Alalaoui AC. Fertilisation minérale des cultures: les éléments fertilisants majeurs (Azote, Potassium et Phosphore). Bulletin Mensuel d'Information et de Liaison. 2007;**155**:1-4

[5] Colomb B, Carof M, Aveline A, Bergez JE. Stockless organic farming: Strengths and weaknesses evidenced by a multicriteria sustainability assessment model. Agronomy for Sustainable Development. 2013;**33**(3):593-608

[6] Watson CA, Atkinson D, Gosling P, Jackson L, Rayns FW. Managing soil fertility in organic farming systems. Soil Use & Management. 2002;**18**:239-247

[7] Morel C, Le Clech B, Linères M, Pellerin S. Gare à la baisse de la biodisponibilité du phosphore. AlterAgri. 2006;**79**:21-23

[8] Cairney JWG, Sawyer NA, Sharples JM, Meharg AA. Intraspecific variation in nitrogen source utilisation by isolates of the ericoid mycorrhizal fungus Hymenoscyphus ericae (Read) Korf and Kernan. Soil Biology and Biochemistry. 2000;**32**(8-9):1319-1322

[9] Hibbett DS, Gilbert LB, Donoghue MJ. Evolutionary instability of ectomycorrhizal symbioses in basidiomycetes. Nature. 2000;**407**(6803):506-508

[10] Trouvelot S, Bonneau L, Redecker D, Van Tuinen D, Adrian M, Wipf D. Arbuscular mycorrhiza symbiosis in viticulture: A review. Agronomy for Sustainable Development. 2015a;**35**(4):1449-1467

[11] Heckman DS, Geiser DM, Eidell BR, Stauffer RL, Kardos NL, Hedges SB. Molecular evidence for the early colonization of land by fungi and plants. Science. 2001;**293**(5532):1129-1133. DOI: 101,126/science.1061457

[12] Taxter RA. Revision of Ondogonaceae. Proceedings of the American Academic Arts Science. 1922;**57**:291-351

[13] Schüßler A, Schwarzott D, Walker C. A new fungal phylum, the Glomeromycota: Phylogeny and evolution. Mycological Research. 2001;**105**:1413-1421

[14] Sidhoum W. Diversité des mycorhizes arbusculaires chez la variété, Sigoise, d'olivier (Olea europea L.) : étude de leurs efficacités sur la croissance des plants. Université d'Oran Es-Senia; 2011 Mémoire de Magister en Biotechnologie, 182

[15] Sanders IR. Plants and arbuscular mycorrhizal fungal diversity: Are we looking at the relevant levels of diversity and are we using the right techniques? New Phytologist. 2004:415-418

[16] Schwarzott D, Walker C, Schüßler A. Glomus, the largest genus

of the arbuscular mycorrhizal fungi (Glomales), is nonmonophyletic. Molecular Phylogenetics and Evolution. 2001;**21**(2):190-197

[17] Alguacil MM, Roldán A, Torres MP. Complexity of semi-arid gypsophilous shrub communities mediates the AMF biodiversity at the plant species level. Microbial Ecology. 2009;**57**(4):718-727

[18] Redecker D, Raab P. Phylogeny of the Glomeromycota (arbuscular mycorrhizal fungi): Recent developments and new gene markers. Mycologia. 2006;**98**(6):885-895

[19] Hahn A, Bonfante P, Horn K, Pausch F, Hock B. Production of monoclonal antibodies against surface antigens of spores from arbuscular mycorrhizal fungi by an improved immunization and screening procedure. Mycorrhiza. 1993;**4**(2):69-78

[20] Madan R, Pankhurst C, Hawke B, Smith S. Use of fatty acids for identification of AM fungi and estimation of the biomass of AM spores in soil. Soil Biology and Biochemistry. 2002;**34**(1):125-128

[21] Morton JB, Benny GL. Revised classification of arbuscular mycorrhizal fungi (Zygonycetes): A new order, Glomales, two new busorders, Glomineae and Gigasporineae, with an emendation of Glomaceae. Mycotaxon. 1990;**37**:471-491

[22] Redecker D, Morton JB, Bruns TD. Ancestral lineages of arbuscular mycorrhizal fungi (Glomales). Molecular Phylogenetics and Evolution. 2000;**14**(2):276-284

[23] Morton JB, Redecker D. Two new families of Glomales, Archaeosporaceae and Paraglomaceae, with two new genera Archaeospora and Paraglomus, based on concordant molecular and morphological characters. Mycologia. 2001;**93**(1):181-195

[24] Walker C, Vestberg M, Demircik F, Stockinger H, Saito M, Sawaki H, et al. Molecular phylogeny and new taxa in the Archaeosporales (Glomeromycota): Ambispora fennica gen. sp. nov., Ambisporaceae fam. nov., and emendation of Archaeospora and Archaeosporaceae. Mycological Research. 2007;**111**(2):137-153

[25] Sieverding E, Oehl F. Revision of Entrophospora and description of Kuklospora and Intraspora, two new genera in the arbuscular mycorrhizal Glomeromycetes. Journal of Applied Botany and Food Quality. 2006;**80**:69-81

[26] Palenzuela J, Ferrol N, Boller T, Azcon-Aguilar C, Oehl F. Otospora bareai, a new fungal species in the Glomeromycetes from a dolomitic shrub land in sierra de Baza National Park (Granada, Spain). Mycologia. 2008;**100**:296-305

[27] Krüger M, Stockinger H, Krüger C, Schüßler A. DNA-based species level detection of Glomeromycota: One PCR primer set for all arbuscular mycorrhizal fungi. New Phytologist. 2009;**183**(1):212-223

[28] Chouyia FE, Fiorentino N, Rouphael Y, Ventorino V, Fechtali T, Visconti D, et al. Assessing the effect of P-solubilizing bacteria and mycorrhizal fungi on tomato yield and quality under different crop rotations. Scientia Horticulturae. 2022;**293**(110):740

[29] Idbella M, Bonanomi G, De Filippis F, Amor G, Chouyia FE, Fechtali T, et al. Contrasting effects of Rhizophagus irregularis versus bacterial and fungal seed endophytes on Trifolium repens plant-soil feedback. Mycorrhiza. 2021;**31**(1):103-115

[30] Rouphael Y, Franken P, Schneider C, Schwarz D, Giovannetti M, Agnolucci M, et al. Arbuscular mycorrhizal fungi act as biostimulants in horticultural crops. Scientia of Horticulturae. 2015;**196**:91-108

[31] Gashgari R, Selim S, Abdel-Mawgoud M, Warrad M, Habeeb TH, Saleh AM, et al. Arbuscular mycorrhizae induce a global metabolic change and improve the nutritional and health benefits of pennyroyal and parsley. Acta Physiologiae Plantarum. 2020;**42**(6):1-11

[32] Keymer A, Pimprikar P, Wewer V, Huber C, Brands M, Bucerius SL, et al. Lipid transfer from plants to arbuscular mycorrhiza fungi. eLife. 2017;**6**:e29107

[33] Parihar M, Rakshit A, Meena VS, et al. The potential of arbuscular mycorrhizal fungi in C cycling: A review. Archives of Microbiology. 2020;**202**:1581-1596

[34] Zhu C, Tian G, Luo G, Kong Y, Guo J, Wang M, et al. N-fertilizer-driven association between the arbuscular mycorrhizal fungal community and diazotrophic community impact wheat yield. Agriculture, Ecosystems and Environment. 2018;**254**:191-201

[35] Hodge A, Campbell CD, Fitter AH. An arbuscular mycorrhizal fungus accelerates decomposition and acquires nitrogen directly from organic material. Nature. 2001;**413**(6853):297-299

[36] Zhang F, Zou YN, Wu QS, Kuˇca K. Arbuscular mycorrhizas modulate root polyamine metabolism to enhance drought tolerance of trifoliate orange. Environmental and Experimental Botany. 2020;**171**(103):962

[37] Smith SE, Read DJ. Mycorrhizal Symbiosis. 3rd ed. Great Britain: Academic Press; 2008. pp. 11-145

[38] AguéguéMR, NestorRAA, NadègeAA, Pâcome NA, Marcellin A, Nicodeme WC, et al. Greenhouse evaluation of the growth of *Zea mays* L. inoculated by arbuscular mycorrhizal fungi strains in native arbuscules on ferrous soil. Journal of Agricultural and Crop Research. 2020;**8**:55-63

[39] Dodd JC, Boddington CL, Rodriguez A, Gonzalez-Chavez C, Mansur I. Mycelium of arbuscular mycorrhizal fungi (AMF) from different genera: Form, function and detection. Plant and Soil. 2000;**226**(2):131-151

[40] Timonen S, Smith FA, Smith SE. Microtubules of mycorrhizal fungus Glomus intraradices in symbiosis with tomato roots. Canadian Journal of Botany. 2001;**79**:307-313

[41] Indira MN, Tejavathi DH. Growth response of in vitro regenerated *Drymaria cordata* (L.) Willdex Roem. & Schult. to inoculation with Arbuscular Mycorrhizal fungi. Current Trends in Biotechnology and Pharmacy. 2021;**15**(5):452-458

[42] Kapoor R, Sharma D, Bhatnagar AK. Arbuscular mycorrhizae in micropropagation systems and their potential applications. Scientia Horticulturae. 2008;**116**(3):227-239

[43] Landwehr M, Hildebrandt U, Wilde P, Nawrath K, Tóth T, Biró B, et al. The arbuscular mycorrhizal fungus Glomus geosporum in European saline, sodic and gypsum soils. Mycorrhiza. 2002;**12**(4):199-211

[44] Duponnois R, Hafidi M, Wahbi S, Sanon A, Galiana A, Baudoin E, et al. La symbiose mycorhizienne et la fertilité des sols dans les zones arides: Un outil biologique sous-exploité dans la gestion des terres de la zone sahélo-saharienne. La Grande Muraille Verte: Capitalisation

des recherches et valorisation des savoirs locaux, édité par A. Dia et Robin Duponnois, Synthèses. Marseille: IRD; 2012:351-369

[45] Gobat JM, Aragno M, Matthey W. The living soil: Basic pedology-soil biology. The living soil: Basic pedology-soil biology; 2003

[46] Xie Y, Wang F, Wang K, Yue H, Lan X. Responses of bacterial phoD gene abundance and diversity to crop rotation and feedbacks to phosphorus uptake in wheat. Applied Soil Ecology. 2020;**154**:103604

[47] Battini F, Grønlund M, Agnolucci M, Giovannetti M, Jakobsen I. Facilitation of phosphorus uptake in maize plants by mycorrhizosphere bacteria. Scientific Reports. 2017;**7**(1):1-11

[48] Balliu A, Sallaku G, Rewald B. AMF inoculation enhances growth and improves the nutrient uptake rates of transplanted, salt-stressed tomato seedlings. Sustainability. 2015;**7**(12):15967-15981

[49] Campo S, Martín-Cardoso H, Olivé M, Pla E, Catala-Forner M, Martínez-Eixarch M, et al. Effect of root colonization by arbuscular mycorrhizal fungi on growth, productivity and blast resistance in rice. Rice. 2020;**13**(1):1-14

[50] Caproni AL, Granha JRDDO, Fornaciari AJ, Nobre CP, Mendonça LP, Berbara RLL. Diversity of arbuscular mycorrhizal fungi in an Amazon environment after mining. Floresta e Ambiente. 2018;**25**

[51] Coutinho ES, Barbosa M, Beiroz W, Mescolotti DL, Bonfim JA, Berbara RLL, et al. Soil constraints for arbuscular mycorrhizal fungi spore community in degraded sites of rupestrian grassland: Implications for restoration. European Journal of Soil Biology. 2019;**90**:51-57

[52] Nekou GN, Sontsa-Donhoung AM, Bahdjolbe M, Tobolbaï R, Nwaga D. Response of the Leek (Allium porrum)-Mycorrhizal fungus Symbiosis to cutting levels, light exposure and seedling density. European Journal of Biology and Biotechnology. 2021;**2**(3):1-6

[53] Ngakou A, Nwaga D, Ntonifor NN, Tamo M, Nebane CLN, Parh IA. Contribution of Arbuscular Mycorrhizal Fungi (AMF), rhizobia and Metarhizium anisopliae to cowpea production in Cameroon. International Journal of Agricultural Research. 2007;**2**(9):754-764

[54] Ogou A, Tchabi A, Tounou AK, Agboka K, Sokame BM. Effet de quatre souches de champignons mychoriziens arbusculaires sur Meloidogyne spp. principal nématode parasitaire du soja (*Glycine max*, L.) au Togo. Journal of Applied Biosciences. 2018;**127**:12758-12769

[55] Ngakou A, Takoukam TS, Ngo NL, Tobolbaï R, Maryamou I. Appropriate compost/soil ratios for sustainable production of garlic (*Allium sativum* L.) under Mycorrhization in pots experiment. Journal of Plant Science. 2020;**15**(3):64-73

[56] Moussa M, Albert N, Fernand-Nestor FT. Influence de l'interaction mycorhizes-insectes floricoles sur les rendements de Vigna subterranea variété rouge (Fabaceae) à Dang (Ngaoundéré, Cameroun). International Journal of Biological and Chemical Science. 2018;**12**(4):1897-1913

[57] Hemissi I, Labidi S, Dhifalli F, Hammami R, Hachana A, Hlel D, et al. Effet de la bio inoculation mycorhizienne et rhizobiale sur les composantes de rendement du blé dur (*Triticum durum* Desf.). In: Annales de l'INRA. Vol. 92. 2019

[58] Tejavathi DH, Anitha P, Murthy SM, Nijagunaiah R. Effect of

AM fungal association with normal and micropropagated plants of Andrographis paniculata Nees on biomass, primary and secondary metabolites. International Research Journal of Plant Science. 2011;**2**(12):338-348

[59] Aguegue RM, Assogba SA, Salami HA, Koda AD, Agbodjato NA, Amogou O, et al. Organic fertilizer based on Rhizophagus intraradices: Valorization in a farming environment for maize in the south, Center and north of Benin. Frontiers in Sustainable Food Systems. 2021;**5**:605-610

[60] Alori ET, Fawole OB, Afolayan AJ, Babalola OO. Arbuscular Mycorrhizal Fungi: Occurrence and factors affecting their distribution in selected soils of Southern Guinea Savanna of Nigeria. In: Dermesonlouoglou E, Giannou V, Gogou E, Taoukis P, editors. 29th Effost International Conference. Food Science Research and Innovation. Delivering sustainable solutions to the global economy and society. Food science and process engineering research applications contributing to food security and water conservation EFFOST2015_0755. Athens, Greece; 2015. pp. 2020-2026. Available from: http://www.effost2015.org/proceedings.php

[61] Fasusi O, Amoo A, Babalola OO. Propagation and characterization of viable arbuscular mycorrhizal fungal spores within maize plant (*Zea mays* L.). Journal of the Science of Food and Agriculture. 2021;**101**:5834-5841

[62] Koda AD, Dagbenonbakin G, Assogba F, Agbodjato NA, N'tcha C, Assogba S, et al. Impact of native arbuscular mycorrhizal fungi based fertilizers on to increase maize productivity in North Benin. African Journal of Agricultural Research. 2020;**16**(9):1298-1306

[63] Sylvestre AA, Nestor AAR, Saliou B, et al. Inoculation of native arbuscular mycorrhizal fungi based bio-fertilizers for improvement of maize productivity in Central Benin. African Journal of Agricultural Research. 2020;**16**(5):652-660

[64] Igiehon ON, Babalola OO. Rhizobium and mycorrhizal fungal species improved soybean yield under drought stress conditions. Current Microbiology. 2021;**78**(4):1615-1627

[65] Igiehon OO, Ademola EA, Anthony TI. A review on the consumption of vended fruits: Microbial assessment, risk, and its control. Food Quality and Safety. 2020;**4**(2):77-81

[66] Clark RÁ, Zeto SK. Mineral acquisition by arbuscular mycorrhizal plants. Journal of Plant Nutrition. 2000;**23**(7):867-902

[67] Kavatagi PK, Lakshman HC. Effect of arbuscular mycorrhizal fungi for their symbiotic efficiency on two varieties of Solanum lycopersicum. International Journal of Pharma and Biosciences. 2012;**3**(3):1007-1017

[68] Alloway BJ. Heavy Metals in the Soil. London: Chapman and Hall; 1995

[69] Gadd GM. Interactions of fungi with toxic metals. In: The Genus Aspergillus. Boston, MA: Springer; 1994. pp. 361-374

[70] Gonzalez-Chavez C, D'haen J, Vangronsveld J, Dodd JC. Copper sorption and accumulation by the extraradical mycelium of different Glomus spp. (arbuscular mycorrhizal fungi) isolated from the same polluted soil. Plant and Soil. 2002;**240**(2):287-297

[71] Rillig MC, Wright SF, Nichols KA, Schmidt WF, Torn MS. Large contribution of arbuscular mycorrhizal fungi to soil carbon pools in

tropical forest soils. Plants and Soil. 2001;**233**:167-177

[72] Schindler FV, Mercer ER, Rice JA. Chemical characteristics of glomalin-related soil protein (GRSP) extracted from soils of varying organic matter content. Soil Biology and Biochemistry. 2007;**39**:320-329

[73] Atakan A, Özkaya HO. Arbuscular Mycorrhizal Fungi and Glomalin. Turkish Journal of Agriculture—Food, Science and Technology. 2021;**9**(12):2371-2375

[74] Lovelock CE, Wright SF, Clark DA, Ruess RW. Soil stocks of glomalin produced by arbuscular mycorrhizal fungi across a tropical rain forest land scape. Journal of Ecology. 2004;**92**:278-287

[75] Wright SF, Upadhyaya A, Buyer JS. Comparison of N-linked oligo saccharides of glomalin from arbuscular mycorrhizal fungi and soils by capillary electrophoresis. Soil Biology and Biochemistry. 1998;**30**:1853-1857

[76] Wright SF, Nichols K. Glomalin: Hiding place for a third of the world's stored soil carbon. Agricultural Research. 2002;**50**(9):4

[77] Wright SF, Anderson RL. Aggregate stability andg lomalin in alternative crop rotations for the central Great Plains. Biology and Fertility of Soils. 2000;**31**:249-253

[78] Rillig MC. Arbuscular mycorrhizae and terrestrial ecosystem processes. Ecological Letter. 2004;**2004**:740-754

[79] Singh PK, Singh M, Tripathi BN. Glomalin: An arbuscular mycorrhizal fungal soil protein. Protoplasma. 2013;**250**:663-669

[80] Wright SF, Upadhyaya A. Extraction of an abundant and unusual protein from soil and comparison with hyphal protein of arbuscular mycorrhizal fungi. Soil Science. 1996;**161**:575-586

[81] Lehmann A, Rillig MC. Understanding mechanisms of soil biota involvement in soil aggregation: A way forward with saprobic fungi? Soil Biology and Biochemistry. 2015;**88**:298-302

[82] Ortiz N, Armada E, Duque E, Roldan A, Azcon R. Contribution of arbuscular mycorrhizal fungi and/or bacteria to enhancing plant drought tolerance under natural soil conditions: Effectiveness of auto chthonous strains. Journal of Plant Physiology. 2015;**174**:87-96

[83] Paul EA, Clark FE. Soil Biology and Biochemistry. San Diego, CA: Academic Press; 1989

[84] Nogueira LR, Silva CF, Pereira MG, Gaia-Gomes JH, Silva EMR. Biological properties and organic matter dynamics of soil in pasture and natural regeneration areas in the Atlantic forest biome. Rev. Bras. Ci^enc. Solo. 2016;**40**:1-13

[85] da Silva CF, Pereira MG, Feitosa JCF, Fonseca Júnior AM, Gaia-Gomes JH, Menezes CEG. Soil organic matter fractions, chemical attributes and aggregation under forestry and agricultural systems. Communication Science. 2018;**8**:459-468

[86] Silva CF, Pereira MG, Santos VL, Miguel DL, Silva EMR. Fungos Micorrízicos Arbusculares : Composição, Comprimento de Micélio Extrarradicular e Glomalina em Áreas de Mata Atlântica, Rio De Janeiro. Ciência Florest. 2016;**26**:419-433

[87] Gillespie AW, Farrell RE, Walley FL, Ross AR, Leinweber P, Eckhardt KU, et al. Glomalin-related soil protein

contains non-mycorrhizal-related heat-stable proteins, lipids and humic materials. Soil Biology and Biochemistry. 2011;**43**(4):766-777

[88] Moreira FMS, Siqueira JO. In: Lavras E, editor. Microbiologia e bioquímica do solo. 2. da Ufla. 2006. p. 729

[89] Jia X, Zhao Y, Liu T, Huang S, Chang Y. Elevated CO2 increases glomalin-related soil protein (GRSP) in the rhizosphere of *Robinia pseudoacacia* L. seedlings in Pb-and Cd-contaminated soils. Environmental Pollution. 2016;**218**:349-357

[90] Fokom R, Adamou S, Teugwa MC, Boyogueno AB, Nana WL, Ngonkeu MEL, et al. Glomalin related soil protein, carbon, nitrogen and soil aggregate stability as affected by land use variation in the humid forest zone of South Cameroon. Soil and Tillage Research. 2012;**120**:69-75

[91] Seguel A, Barea JM, Cornejo P, Borie F. Role of arbuscular mycorrhizal symbiosis in phosphorus-uptake efficiency and aluminium tolerance in barley growing in acid soils. Crop and Pasture Science. 2015;**66**(7):696-705

[92] Priscila SM, da Cristiane F, Marcos GP, Eliane MR, da Ricardo MT, André LCF, et al. Short-term modifications of mycorrhizal fungi, glomalin and soil attributes in a tropical agroforestry. Acta Oecologica. 2022;**114**(103):815

[93] Driver JD, Holben WE, Rillig MC. Characterization of glomalin as a hyphal wall component of arbuscular mycorrhizal fungi. Soil Biology and Biochemistry. 2005;**37**(1):101-106

[94] Hajiboland R. Role of arbuscular mycorrhiza in amelioration of salinity. In: Salt Stress in Plants. New York, NY: Springer; 2013. pp. 301-354

[95] Evelin H, Devi TS, Gupta S, Kapoor R. Mitigation of salinity stress in plants by arbuscular mycorrhizal symbiosis: Current understanding and new challenges. Frontiers in Plant Science. 2019;**10**:470

[96] Evelin H, Kapoor R, Giri B. Arbuscular mycorrhizal fungi in alleviation of salt stress: A review. Annals of Botany. 2009;**104**(7):1263-1280

[97] Bencherif K, Dalpé Y, Lounès Hadj-Sahraoui A. Arbuscular mycorrhizal fungi alleviate soil salinity stress in arid and semi-arid areas. In: Microorganisms in Saline Environments: Strategies and Functions. Cham: Springer; 2019. pp. 375-400

[98] Plenchette C. Utilisation des mycorhizes en agriculture et horticulture. In: Strullu DG, Garbaye J, Perrin R, Plenchette C, editors. Les mycorhizes des arbres et des plantes cultivées. Paris: Lavoisier; 1991. pp. 131-196

[99] Hashem A, Alqarawi AA, Radhakrishnan R, Al-Arjani ABF, Aldehaish HA, Egamberdieva D, et al. Arbuscular mycorrhizal fungi regulate the oxidative system, hormones and ionic equilibrium to trigger salt stress tolerance in *Cucumis sativus* L. Saudi Journal of Biological Sciences. 2018;**25**(6):1102-1114

[100] Colla G, Rouphael Y, Cardarelli M, Tullio M, Rivera CM, Rea E. Alleviation of salt stress by arbuscular mycorrhizal in zucchini plants grown at low and high phosphorus concentration. Biology and Fertility of Soils. 2008;**44**(3):501-509

[101] Ait-El-Mokhtar M, Baslam M, Ben-Laouane R, Anli M, Boutasknit A, Mitsui T, et al. Alleviation of detrimental effects of salt stress on date palm (*Phoenix dactylifera* L.) by the application of arbuscular mycorrhizal fungi and/or

compost. Frontiers in Sustainable Food Systems. 2020;**4**:131

[102] Abdel Motaleb NA, Abd Elhady SA, Ghoname AA. AMF and Bacillus Megatherium neutralize the harmful effects of salt stress on bean plants. Gesunde Pflanzen. 2020;**72**(1):29-39

[103] Evelin H, Giri B, Kapoor R. Ultrastructural evidence for AMF mediated salt stress mitigation in Trigonella foenum-graecum. Mycorrhiza. 2013;**23**(1):71-86

[104] Saint-Étienne L, Paul S, Imbert D, et al. Arbuscular mycorrhizal soil infectivity in a stand of the wetland tree *Pterocarpus officinalis* along a salinity gradient. Forest Ecology and Management. 2006;**232**:86-89

[105] Al-Garni SMS. Increasing NaCl-salt tolerance of a halophytic plant Phragmites australis by mycorrhizal symbiosis. American-Eurasian Journal of Agricultural and Environmental Science. 2006;**1**(2):119-126

[106] Sheng M, Tang M, Chan H, Yang B, Zhang F, Huang Y. Influence of arbuscular mycorrhizae on photosynthesis and water status of maize plants under salt stress. Mycorrhiza. 2008;**18**:287-296

[107] Juniper S, Abbott LK. Soil salinity delays germination and limits growth of hyphae from propagules of arbuscular mycorrhizal fungi. Mycorrhiza. 2006;**16**(5):371-379

[108] Barea JM, Pozo MJ, Azcon R, Azcon-Aguilar C. Microbial co-operation in the rhizosphere. Journal of Experimental Botany. 2005;**56**(417):1761-1778

[109] Adesemoye AO, Torbert HA, Kloepper JW. Enhanced plant nutrients use efficiency with PGPR and AMF in an integrated nutrient management system. Canadian Journal of Microbiology. 2008;**54**(10):876-886

[110] Requena N, Jimenez I, Toro M, Barea JM. Interactions between plant-growth-promoting rhizobacteria (PGPR), arbuscular mycorrhizal fungi and Rhizobium spp. in the rhizosphere of Anthyllis cytisoides, a model legume for re vegetation in mediterranean semi-arid ecosystems. New Phytologist. 1997;**136**:667-677

[111] Walley FL, Germida JJ. Response of spring wheat (*Triticum aestivum*) to interactions between Pseudomonas species and Glomus clarum NT4. Biology and Fertility of Soils. 1997;**24**:365-371

[112] Barea JM, Azcón R, Azcón-Aguilar C. Mycorrhizosphere interactions to improve plant fitness and soil quality. Antonie Van Leeuwenhoek. 2002;**81**(1):343-351

[113] Hodge A. Microbial ecology of the arbuscular mycorrhizal. Microbiology and Ecology. 2000;**32**:91-96

[114] Dwivedi D, Johri BN, Ineichen K, Wray V, Wiemken A. Impact of antifungals producing rhizobacteria on the performance of Vigna radiata in the presence of arbuscular mycorrhizal fungi. Mycorrhiza. 2009;**19**(8):559-570

[115] Purin S, Rillig MC. The arbuscular mycorrhizal fungal protein glomalin: Limitations, progress, and a new hypothesis for its function. Pedobiologia. 2007;**51**:123-130

[116] Ole, P., Pierdomenico, P., Laurentius, A.C.J., Voesenek, (2017). Flooding and low oxygen responses in plants. Functional Plant Biology, 44, 3-6.

[117] Sehgal A, Sita K, Bhandari K, Kumar S, Kumar J, Prasad PVV, et al.

Influence of drought and heat stress, applied independently or in combination during seed development, on qualitative and quantitative aspects of seeds of lentil (Lens culinaris Medikus) genotypes, differing in drought sensitivity. Plant Cell and Environment. 2018;**42**:198-211

[118] Khan A, Ding Z, Ishaq M, Khan I, Ahmed AA, Khan AQ, et al. Applications of beneficial plant growth promoting rhizobacteria and mycorrhizae in rhizosphere and plant growth: A review. International Journal of Agricultural and Biological Engineering. 2020;**13**(5):199-208

[119] Igiehon NO, Babalola OO. Biofertilizers and sustainable agriculture: Exploring arbuscular mycorrhizal fungi. Applied Microbiology and Biotechnology. 2017;**101**:4871-4881

[120] Igiehon NO, Babalola OO. Below-ground-above-ground plant-microbial interactions: Focusing on soybean, rhizobacteria, and Mycorrhizal fungi. The Open Microbiology Journal. 2018;**12**:261-279

[121] Nadeem SM, Ahmad M, Zahir ZA, Javaid A, Ashraf M. The role of mycorrhizae and plant growth promoting rhizobacteria (PGPR) in improving crop productivity under stressful environments. Biotechnology Advances. 2014;**32**:429-448

[122] Richardson AE, Barea J-M, McNeill AM, Prigent-Combaret C. Acquisition of phosphorus and nitrogen in the rhizosphere and plant growth promotion by microorganisms. Plant and Soil. 2009;**321**:305-339

[123] Gadhave KR, Devlin PF, Ebertz A, Ross A, Gange AC. Soil inoculation with Bacillus spp. modifies root endophytic bacterial diversity, evenness, and community composition in a context-specific manner. Microbial Ecology. 2018;**76**(3):741-750

[124] Gadhave KR, Hourston JE, Gange AC. Developing soil microbial inoculants for pest management: Can one have too much of a good thing? Journal of Chemical Ecology. 2016;**42**(4):348-356

[125] Watts-Wlliams SJ, Cavagnaro TR, Tyerman SD. Variable effects of arbuscular mycorrhizal fungal inoculation on physiological and molecular measures of root and stomatal conductance of diverse Medicago truncatula accessions. Plant Cell and Environment. 2018;**42**:285-294

[126] Pivato B, Offre P, Marchelli S, Barbonaglia B, Mougel C, Lemanceau P, et al. Bacterial effects on arbuscular mycorrhizal fungi and mycorrhiza development as influenced by the bacteria, fungi, and host plant. Mycorrhiza. 2009;**19**:81-90

[127] Barea JM, Jeffries P. Arbuscular mycorrhizas in sustainable soil-plant systems. Mycorrhiza. 1995:521-560

[128] Bianciotto V, Andreotti S, Balestrini R, Bonfante P, Perotto S. Extracellular polysaccharides are involved in the attachment of Azospirillum brasilense and Rhizobium leguminosarum to arbuscular mycorrhizal structures. European Journal of Histochemistry. 2001;**45**(1):39-50

[129] Artursson V, Finlay RD, Jansson JK. Interactions between arbuscular mycorrhizal fungi and bacteria and their potential for stimulating plant growth. Environmental Microbiology. 2006;**8**:1-10

[130] Jeffries P, Gianinazzi S, Perotto S, Turnau K, Barea JM. The contribution of arbuscular mycorrhizal fungi in sustainable maintenance of plant health and soil fertility. Biology and Fertility of Soils. 2003;**37**(1):1-16

[131] Linderman RG. Vesicular-arbuscular mycorrhizae and soil microbial interactions. Mycorrhizae in Sustainable Agriculture. 1992;**54**:45-70

[132] Agbodjato NA, Assogba SA, Babalola OO, Koda AD, Aguegue RM, Sina H, et al. Formulation of biostimulants based on arbuscular mycorrhizal fungi for maize growth and yield (research topic: Soil-plant-microbe interactions: An innovative approach towards improving soil health and plant growth). Frontiers in Agronomy. 2022;**4**:894489

[133] Aguégué MR, Ahoyo Adjovi NR, Agbodjato NA, Noumavo PA, Assogba S, Salami H, et al. Efficacy of native strains of arbuscular mycorrhizal fungi on maize productivity on ferralitic soil in benin. Agricultural Research. 2021;**2021**:1-15

[134] Barea JM, Azcón R, Azcón-Aguilar C. Mycorrhizosphere interactions to improve a sustainable production of legumes. In: Microbes for Legume Improvement. Cham: Springer; 2017. pp. 199-225

[135] Baslam M, Qaddoury A, Goicoechea N. Role of native and exotic mycorrhizal symbiosis to develop morphological, physiological and biochemical responses coping with water drought of date palm, Phoenix dactylifera. Trees. 2014;**28**(1):161-172

Chapter 9

Essential Role of Symbiotic Microorganisms Supporting Forests in East Asia under Changing Environment

Yannan Wang, Laiye Qu, Yutaka Tamai, Toshihiro Watanabe and Takayoshi Koike

Abstract

Regeneration success of forests is strongly dependent on symbiotic microorganisms, that is, arbuscular mycorrhiza (AM) and ectomycorrhiza (ECM). In the northeastern part of Asia, larch and birch are used as timber resources, and in the south, fir, cedar, cypress, and oak are used as timber resources. Planted forests have reached the time of harvesting and/or thinning, and after the forestry practices, it is expected that they will become mixed forests equipped with resistance to weather damage; that is, drought, heat, typhoons, etc. On the other hand, the physical production environment has changed greatly, therefore, we investigated the growth of the major trees and the role of mycorrhizal fungi in the northeastern Asia. Elevated O_3 decreased growth, colonization rates of ECM, and the biodiversity; however, elevated CO_2 moderated or increased them in larch. Except for disease of rot and damping off, we discuss wise use of symbiotic microbe in far East Asia.

Keywords: symbiotic microorganisms, conifers, regeneration, mixed forests, changing environment

1. Introduction

Various microorganisms grow in the soil [1, 2]. They are divided into parasitic fungi and symbiotic fungi that attach to living organisms, and saprophytic fungi that decompose and utilize dead things, based on how they acquire nutrients (e.g., [3, 4]). Most land plants form symbiotic associations with mycorrhizal fungi. These below-ground fungi act an essential role in terrestrial ecosystems as they regulate nutrient and carbon cycles [1]. They influence soil structure and ecosystem multifunctionality. Up to 80% of plant N and P is provided by mycorrhizal fungi and many plant species are depending on these symbionts for growth and survival under changing environments [5, 6].

Man-made forests in Japan are mainly composed of coniferous trees and are approaching the final harvest, but the production cost is very expensive and they are still in a state where they cannot be used. After harvesting, mixed reforestation is recommended considering the resistance to various damages [7–9]. On the other hand, since 1999, China has planted *Larix gmelinii* (Dahurian larch) in the north, *Populus* sp. (poplar) in the center, and *Cunninghamia lanceolata* (Chinese fir) [10, 11] in the south as forest rehabilitation as "Gain for Green Program". Recently planted area is the largest in the world, larger than the land area of Japan [12]. Also, in the southwest, *Eucalyptus* sp. (eucalyptus species) are mainly planted and harvested with a method of short rotation intensive culture of about less than 10 years to support the production of wood materials. However, in recent years, there have been cases of concern about the growth of third-generation afforestation areas [13].

Regeneration after wildfires, which frequently occur in the Russian Far East, is mainly carried out by natural regeneration of Dahurian larch, red pine (*Pinus sylvestris* var. *mongolica*), and birch (*Betula platyphylla*) [6, 14, 15]. In both cases, the importance of the role of ectomycorrhizal fungi and biochar has been pointed out (e.g., [16, 17]).

In this chapter, we reviewed briefly on basic information about AM (arbuscular mycorrhiza = vesicular and arbuscular mycorrhiza: VA) and ECM, essential role of symbionts in natural forests [18, 19], Sugi-cedar (*Cryptomeria japonica*: AM conifer) plantation in Japan [9], and regeneration success after forestry practices in northeast China under changing environment [5, 6].

2. Principal of plants and microbe interactions

Four major mycorrhizal types have been described based on their structure and function, namely arbuscular mycorrhiza (AM; formerly named as VA), ECM, orchid mycorrhiza, and ericoid mycorrhiza [18]. In addition, Fabaceae or Leguminosae trees (*Robinia* sp. [20]) and alder tree species (*Alnus* sp. [21]) are also introduced for restoration of degraded lands, especially under nutrient-poor conditions. Therefore, we discuss on effects of nitrogen and ECM colonization (e.g., [22, 23]). Mycorrhizal fungi are a type of fungi colonizing plant roots to form "mycorrhizae" and form symbiosis relations as "mycorrhizal fungi." In contrast to ECM, those whose hyphae enter the cell wall are collectively called "endomycorrhiza."

2.1 General information on AM, ECM, and Rhizobium

2.1.1 Evolution of AM and ECM

Fossil records and molecular clock dating suggest that the ancestors of the plants that entered the land were pteridophytes that appeared 400–300 million years ago and coexisted with AM [24–26] (**Figure 1**). At that time, the CO_2 concentration was about 20 times higher than today, and green plants were photosynthesizing.

Since lignin-decomposing fungi did not exist 400 to 300 million years ago, fossil fuels such as coal and petroleum were produced and accumulated, and then we are using them today (e.g., [27]). About 300 million years ago, gymnosperm (conifers) took precedence, ectomycorrhizal fungi established a symbiotic relationship, and about 150 million years ago angiosperms appeared, and the activities of symbiotic fungi are essential for more than 80% of plants [2, 26, 28].

ECM (about 200 million years ago) and ericoid mycorrhizas (about 100 million years ago) evolved subsequently as the organic matter content of some ancient soils increased and sclerophyllous vegetation arose as a response to nutrient-poor soils, respectively. This section discussed the function of ECM associated with reforestation

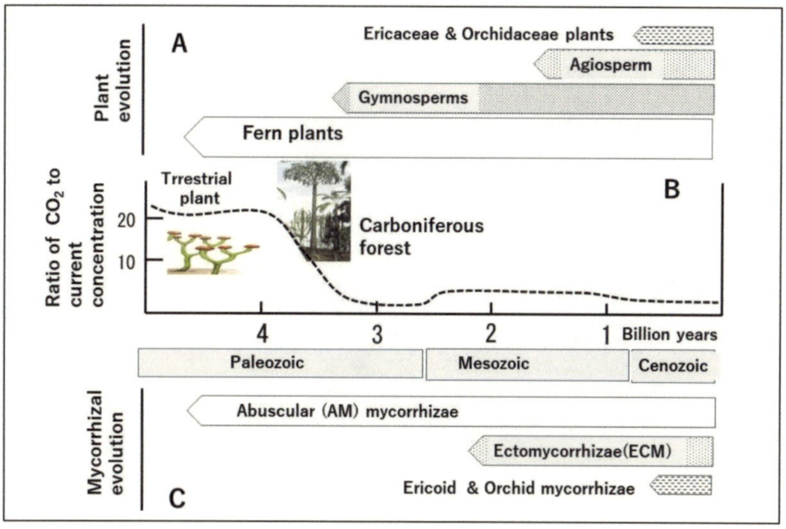

Figure 1.
Evolution of plant (A), atmospheric CO_2 concentration (B), and mycorrhizae (C) (Original figure was offered by M. Saito).

in Sugi-cedar (*Cryptomeria japonica*) and larch (*Larix* sp.) plantations under changing atmospheric environments.

2.1.2 AM

AM fungi have existed since ancient times when plants evolved from the sea to the land and took root on the ground (**Figure 1**). Colonization of plants with AM fungi does not result in significant changes in root external morphology. Inside the root, a unique structure with finely branched tips called "arbuscule" is formed. Fungi whose hyphae form highly branched (arbre: French; tree-shaped) growths (arbuscles) between the cell walls and cell membranes of plant root cells, thereby conveying phosphorous (P) and other materials to plants and taking up organic compounds from root (**Figure 2A2**). Mutualistic symbiotic relationships are maintained by the association of plants. Spores that are present in soil germinate, "infect" the root system, and form arbuscule structures inside the cells of root.

When AM fungi colonize the roots of plants, they absorb phosphoric acid from the hyphae spread in the soil and supply it to the symbiotic plants (hosts) through the dendrites. On the other hand, by receiving energy (carbohydrates) produced by photosynthesis, plants mutually benefit and symbiotically with AM, ECM, etc. [31]. There is a problem to overcome when using AM fungi, which are obligate symbiotic microorganisms, as an inoculation material.

2.1.3 ECM

Ectomycorrhiza (ECM) is a type of mycorrhiza that has hyphae on the surface of the root, mainly fine root and between epidermal cells and cortical cells, but does not enter the cells. The hyphae of ECM usually form a mesh-like structure that is stretched between the intercellular spaces of cortical cells, and this "mesh" is called the Hartig net (**Figure 2A3**, and **B**). In addition, the hyphae form a thick mycelial

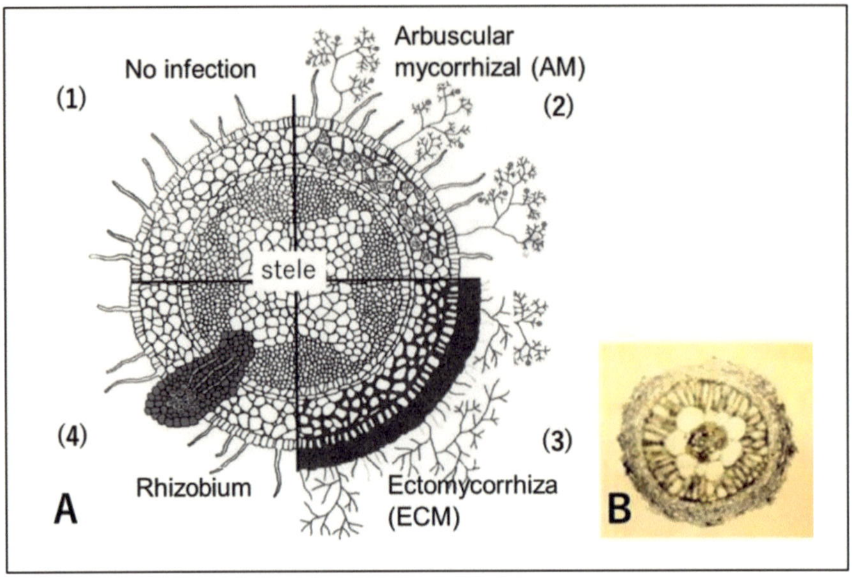

Figure 2.
Illustration of typical mycorrhizae in root (A) and Hartig net (B). A: Modification from Bücking H et al. [29], and B: Courtesy of Y. Matsuda [30].

layer called a fungal sheath (fungal mantle) around the rootlets. The presence of ECM increases the uptake of P, water, and nitrogen (N) by the roots, especially when the concentration of these nutrients in the soil is low [2, 18, 32]. The presence of ECM also contributes to the improvement of resistance to soil pathogens, and the presence of mycelium prevents invasion of other microorganisms and harmful substances.

2.1.4 Rhizobium

Rhizobium is a soil microorganism that performs so-called symbiotic N fixation, forming nodules on the roots of leguminous plants, in which atmospheric N is reduced by nitrogenase, converted to ammonia nitrogen, and supplied to the host (**Figure 2A4**). A symbiotic relationship is established within the nodule by supplying photosynthetic products from the host plant.

Frankia sp. is a genus of bacteria classified as actinomycetes. They have the ability to fix N and live symbiotically with a group of angiosperms called "actinorhizal plants," typically alder (*Alnus* sp.) trees (e.g., [21]). *Frankia* sp. live symbiotically in nodules, which form in the roots of actinorhizal plants. In the root nodule, *Frankia* sp. reduces N molecules to ammonia by N fixation and supplies it to the host plant.

2.2 Nitrogen environment

The activities of symbiotic fungi are enhanced in nutrient-poor environments (e.g., [18, 33]). Therefore, increased N deposition reduces the health and vitality of symbiotic forest ecosystems. N is usually insufficient n temperate and boreal forests [34], and according to Aber et al. [35], who first noted the effects of N deposition as four stages, growth is promoted in the initial stage (stages 0 and I). However, N addition experiments on nutrient-poor soil in North America began to decline (stages

I and II), and excess N flowed into the watershed (stage III). The activity of mycorrhizal fungi can be expected at Stage I or below.

In general, forests circulate nutrients through litterfall as a self-fertilizing function (e.g., [36]). Deciduous forests in Japan tended to be 10–30 kgN hr^{-1} yr^{-1} higher than coniferous forests in values before air pollution became serious in the 1960s. Therefore, in places where N deposition exceeds the amount of self-fertilization, attention must be paid to N runoff. In an experiment using tree seedlings [37], nitrogen (N) fertilizer was added up to about 80 kgN hr^{-1} yr^{-1} and found the N tolerance. in the order of *Larix gmelinii* x *kaempferi* (Hybrid F_1), *Cryptomeria japonica* (Sugi-cedar), *Abies firma* (Fir), *Picea asperata* (dragon spruce), and *Quercus acuta.* However, growth of following tree seedlings is suppressed with 50 ~ 80 kg N hr^{-1} yr^{-1}; *Pinus densiflora* (red pine), *P. tabutaeformls*, *Fagus crenata* (Siebold's beech), *Castanopsis sieboldii,* and *Q. glauca* (evergreen oak).

3. Essential role of symbiotic microbes in a mixed forest

Many man-made coniferous forests in Japan are approaching the harvest, but the cost of harvesting and transporting the timber is too expensive to make use of it and price of raw material (=timber) is too low [38, 39]. Tending the plantation is essential because delaying the harvesting period will result in strong wind, heavy snowfall, etc. Although there is an administrative guideline to delay the harvest time, if implemented, weather damage like typhoon will occur, therefore, forest tending is essential for conservation of land and biodiversity (**Figure 3**) [40].

A framework is proposed for considering how tree species and their mycorrhizal associates differentially couple carbon (C) and nutrient (mainly N) cycles in temperate forests [41]. In AM-dominated forest stands (**Figure 4A**), an inorganic nutrient economy is predicted as a result of the fast decomposition of high-quality litterfall (=high nitrogen content with low defense chemicals, such as phenolics including lignin) and enhanced mineralization rates of C and nitrogen (N) [42]. In contrast, an organic nutrient dynamic is estimated as a result of low-quality litterfall (=low N with high phenolics) and slow turnover rates of C and N in ECM-dominated stands (**Figure 4B**). These trends may indicate limited losses of inorganic nutrients.

Figure 3.
Sugi-cedar (Cryptomeria japonica) plantations with adequate thinning (A) and no thinning (B) in Southwestern Japan (after [40]). Forest floor is an aerobic condition with rich vegetation (A) vs. that an anaerobic condition with no ground vegetation (B). Photos are courtesy of E. Takahashi.

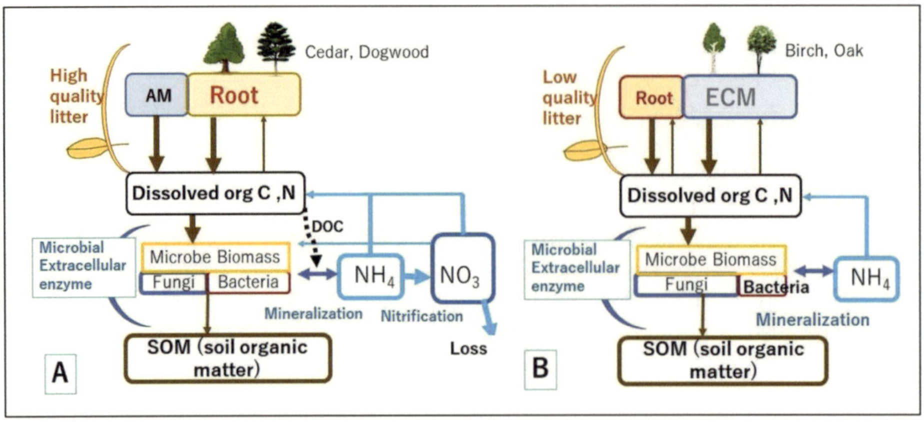

Figure 4.
Trees predominantly associate with a single type of mycorrhizal fungi AM or ECM, and the two types of fungi differ in their nutrient acquisition; abundance of AM (A) and ECM (B) trees may provide an integrated index of biogeochemical transformations relevant to C cycling and nutrient retention in a forest ecosystem. Within a mycorrhizal group, the size of the boxes and arrows indicates relative importance of pools and fluxes, respectively (Modification from Phillips et al. [41]).

In order to improve soil fertility and avoid weather damage, it is necessary to promote mixing with broad-leaved trees for biodiversity conservation [9]. This statement is applicable to the large-scale plantations in China and in South Korea. This section discussed the case of Sugi-cedar (AM conifer) plantation in the cool temperate zone in Japan for making a mixed forest and moderating methane (CH_4) emission in closed forests [43].

3.1 Conifer plantation to mixed broad-leaved forest by thinning

Thinning is an effective method to create mixed broad-leaved trees (hardwoods)—conifer forests with successful hardwood recruitment [44]. However, it remains unclear to what extent mycorrhizal association (AM vs. ECM) affects hardwood recruitment. In this section, we explored which mycorrhizal types of hardwood were more advantageous to recruit into a Sugi-cedar plantation in northern Japan [9, 45, 46].

We compared seedling recruitment success (i.e., seedling emergence, survival, and growth) between AM and ECM hardwoods during 5 years after thinning at two levels: 67% thinned (intensive), 33% thinned (weak), and un-thinned treatments (control). We also evaluated the colonization percent of mycorrhizal fungi for the seedlings of the most dominant AM (basswood: *Cornus controversa*) and ECM (*Quercus serrata*) hardwoods in both the Sugi-cedar plantation and adjacent forests dominated by ECM tree species (e.g., *Q. serrata, Alnus hirsuta* var. *sibirica, Castanea crenata*) [30, 45–48].

From 5 years of monitoring, rate of emergence and survival of seedlings was greater in AM-type broad-leaved tree species than in ECM-ones. Height growth of the seedlings was also greater for AM than ECM species, especially for early emerging cohorts in thinning treatments [45]. Colonization rate of AM fungi to AM seedlings was greater in the Sugi-cedar (AM-type) plantation than in the adjacent ECM-dominant forest stands. Conversely, colonization rate of ECM fungi to ECM-type seedlings was lower in the plantation than in broad-leaved forests. These results suggest that AM broad-leaved trees could easily be recruited to AM conifer plantations.

Growth of the AM seedlings was enhanced, which translocates nutrients by way of hyphal networks [18]. Recruitment of ECM-type seedlings was strongly inhibited; however, this may be due to the lack of association of the ECM fungi in the AM-type conifer (Sugi cedar) plantation. Consequently, mycorrhizal associations play an important role in recruitment success of mycorrhizal broad-leaved trees into AM conifer plantations.

3.2 Methane emission and the possible role of earthworms after thinning

Methane (CH_4) is the second strongest greenhouse gas after CO_2, excluding water vapor, and forests currently act as a source of CH_4 consumption (= CH_4 absorption; e.g., [34, 49]).

However, due to the increase in atmospheric CO_2 concentration, foliar density of canopy trees increases larger, and the incident light to the forest floor decreases (e.g., [50–52]), inducing the forest floor in a moist state (= anaerobic condition) [53, 54]. Furthermore, looking at individual leaves, stomata tend to close with increasing CO_2, so transpiration is suppressed [51, 52]. Therefore, the water content of the soil, in which the root system develops, is maintained at a high level. In the case of brown forest soil (Cambisols), CH_4 is released when the atmospheric CO_2 concentration is about 500 ppm and the soil moisture content exceeds 35% (**Figure 5A**) [55].

However, although thinning reduces the transpiration surface of the canopy, the rise in soil temperature promotes the activity of earthworms, and the earthworms eat fallen leaves or litterfall that cover the ground surface andprevent evaporation. These earthworm activities promote drying of soil surface on a forest floor (**Figure 6**) [56]. Increased soil porosity and near-surface aerobic conditions by earthworms [9, 56] can keep the methane-consuming capacity of forests. These tendencies were confirmed in AM-type cedar forests (**Figure 5B**) [43]. ECM-type mixed spruce-broad-leaved forests [56] and ECM-type birch-oak-beech stand with a free-air CO_2 enrichment (FACE) [49].

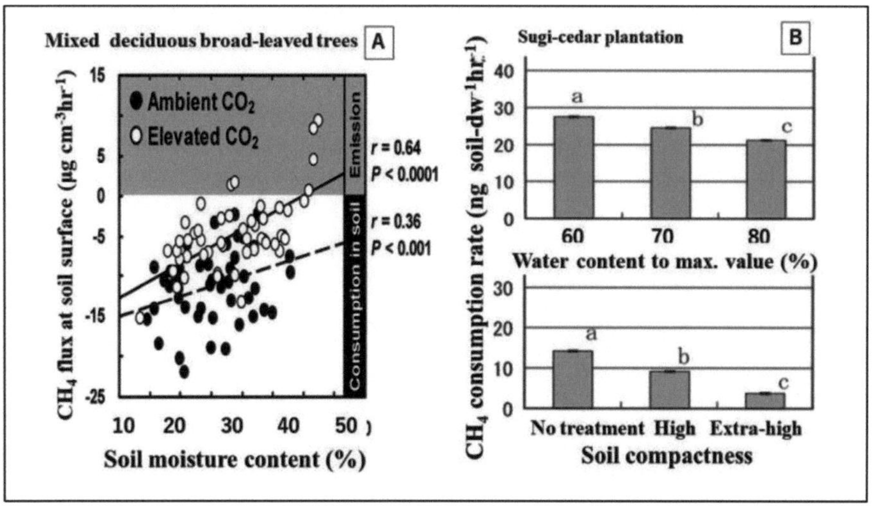

Figure 5.
Methane consumption rate under various conditions. Adopted from (A) Kim et al. [44, 50], (B) from Mametani et al. In B, different letters indicate significant differences (p < 0.05), [40].

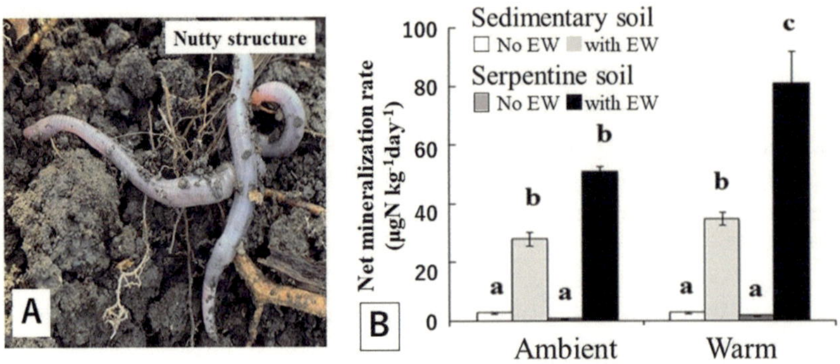

Figure 6.
Nutty soil structure by earthworm (A) and net mineralization without/with Earthworm in different soil fertility (rich-sedimentary soil; poor-serpentine soil under elevated Temperature (+3°C) (B). A: Courtesy of Makoto K, B: adopted from Makoto et al. In B, different letters indicate significant differences ($p < 0.05$), [45].

3.3 Reforestation at acidic and infertile sites with mycorrhizae

Seedlings of three kinds of evergreen *Fagaceae* species (*Quercus glauca, Q. salicina, Castanopsis cuspidate*) planted on acidic [57], nutrient-poor colluvial soil showed accelerated growth when inoculated with ECM fungi [58, 59]. Of course, differences in growth among the seedlings seem to be strongly related to differences in nutrient acquisition of species. Utility of ECM (*Astraeus hygrometricus* or *Scleroderma citrinum*) inoculation to the tree seedlings was examined for their growth in acidic soil conditions with no information on native bacterial flora of ECM. Six months after planting, seedlings of the three species inoculated with *A. hygrometricus* were growing well, especially, *Q. salicina* (**Figure** 7). The growth of seedlings inoculated with *S. citrinum* was inferior to seedlings inoculated with *A. hygrometricus*.

This difference in inoculation effect of two ECM species to *Q. salicina* seedlings may be attributed to an affinity between host-symbiont. Acidic soils are widespread in many parts of the world, turning them into wastelands. Phosphorus (P) is adsorbed on the soil and cannot be used by plants [18]. P is available to host plants through organic acids produced by AM and ECM. Helper bacteria perform important activities when the holobiont is formed [60, 61]. Furthermore, when soil pH < 4.5, toxic Al is liberated and inhibits root system function. However, some *Melastoma* sp. growing in tropical peatlands accumulate Al in their roots to prevent excessive accumulation of Fe, which inhibits growth in acidic environments [62].

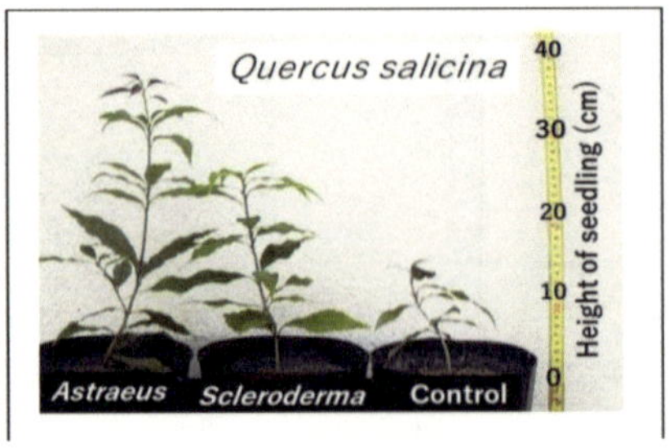

Figure 7.
Growth of an evergreen oak inoculated with different ECM (Astraeus hygrometricus vs. Scleroderma cepa) vs. no-ECM inoculation after 4 months in a phytotron (by courtesy of M. Kayama).

4. An application of symbiotic microbes in regeneration of the man-made forests

4.1 Principal idea of mycorrhizal technology

With the implementation of China's key forestry projects such as Natural Forest Conservation Program (NFCP) and conversion of farmlands with steep slopes to man-made forests, the area of zonal secondary broad-leaved forest has increased rapidly in subtropical areas [11]. Among them, afforestation projects are carried out on barren mountains, wastelands, and degraded lands. In order to improve the ability of trees to survive and grow under natural harsh conditions, attention should be paid to wise use of the symbiotic relationship between trees and ECM and AM, which is the key to the success of afforestation.

Mycorrhizal colonization with the seedlings has become the third factor of afforestation after "matching species with the site, improved variety, and strong seedling" (e.g., [58, 63]). China has been committed to the application and promotion of mycorrhizal technology. Namely, the Chinese Academy of Forestry led the implementation of forest mycorrhizal biotechnology project, which has played an important role in improving the quality and the survival rate of seedlings in afforestation.

4.2 Role of biochar in afforestation and regeneration after forest fire

Forest fires occur frequently in the Far East, especially on the Russian territory, where large areas are burning, but regeneration takes place relatively quickly outside the salt accumulation areas [14, 64, 65]. The burned forest floor is full of charcoal with few surviving individuals (**Figure 8**). If the thick-barked larch (*Larix gmelinii*) trees would not receive a canopy fire, the surviving individuals become mother trees to regenerate naturally. Experimentally, charcoal is effective for the growth of seedlings of larch and pine, in which it is located about 1 cm above the ground surface but not randomly distributed [66] and the charcoal combustion temperature is relatively low (<400~600°C) [17, 67].

As charcoal is a carbon-enriched material with a highly aromatic and porous structure, it is highly resistant to microbial decomposition and thus remains in soil for thousands of years. The abundant pores existing charcoal surfaces have powerful adsorption.

Figure 8.
Forest regeneration after forest fire in the Russian Far East (by courtesy of K. Makoto for A and C; of Y. Matsuura for B) (A) After stem fire, (B) Survived larch individuals with thick bark and regenerated seedlings, and (C) Larch seedlings with biochar.

Charcoal on soil-plant systems explains the complex direct and indirect pathways of these influences that occur during succession after fires in boreal ecosystems. It influences soil-plant systems related to the element composition and nutrient availability in soils [16, 68] and to the abundance of phenolics released from *Ericaceae* plants in the understory of boreal forests [69]. We found a strong bias in the studied processes toward nutrient mineralization rather than immobilization, which suggests that it is risky to draw general conclusions about the influence of charcoal on soil nutrient dynamics. Artificial forest fire experiments gave information about an increase incident light to forest floor and enhancement of litter and humus decomposition and moreover, given the possibility that charcoal accelerates CO_2 release in a postfire forest for relatively long-term [67].

4.3 Afforestation using seedlings associated with ECM

Some conifer species (e.g., pine) and broad-leaved tree species (e.g., oak) are tightly dependent on ECM fungi in their natural state (**Figure** 7), otherwise, they cannot survive [70]. In the early years, China had repeatedly failed to introduce foreign pines because of the lack of symbiotic mycorrhizal fungi, which made the introduced plants grow poorly or even unable to survive until the mycorrhizal fungi of origin of pines are introduced. Commercial ECM fungi: *Pisolithus tinctorius* and *Scleroderma polyrhizum* use practically for "greening material" in pine species and larch [63, 71] and eucalyptus and acacia (e.g., [72]), which can promote the growth of host plants and improve the stress resistance of host plants. At initial stage, ECM is hardly inoculated with host plants because ECM requires lots of photosynthates, therefore, we should introduce "helper" bacteria for accelerating symbiotic relations [73, 74].

After ECM inoculation, the seedlings had a positive effect on seedling height, ground diameter, total biomass, and root diameter. This was found in the studies on *Pinus tabulaeformis*, *Cunninghamia lanceolate*, *Taxus cuspidate,* and *Cyclobalanopsis gilva* [75]. The inoculation of ECM fungi can improve the quality of container larch F_1 seedlings [76, 77] and reduce the use of slow-release fertilizer, which is helpful to cultivate high-quality seedlings and save seedling costs [78] but not always true for larch seedlings [71, 79, 80].

AM fungi also play an important role in maintaining the stability of the southern coastal mangrove ecosystem. Plants in the mangrove ecosystem can form a symbiotic system with a variety of AM fungi under aerobic conditions. Inoculation with AM fungi can significantly increase plant height, stem diameter, and aboveground biomass of mangrove apples (*Sonneratia caseolaris*) and promote the uptake of phosphorus and potassium by roots [81]. Yuan et al. [82] found that AM fungi could promote the uptake and utilization of potassium nitrate and ammonium sulfate in NO_3-N fertilizer by oil trees (*Elaeagnus mollis*) seedlings. In the experiment of inoculation AM fungi and rhizobia on shrub legume (*Amorpha fruticosa*) found that there is an interactive recognition process between mycorrhizal fungi and rhizobia, that double inoculation could significantly increase the mycorrhizal colonization rate and nodule number [83–85].

The rate of mycorrhizal colonization and the number of nodules by double inoculation were significantly higher than those by other inoculation treatments [63, 71, 86, 87], which enhanced the nitrogen-fixing ability of rhizobia and promoted plant growth.

4.4 Afforestation in adversity and biological control

Mycorrhizal colonization of the plants (=**holobiont**; e.g., [88]) can improve both the survival rate of seedlings and the absorption and utilization of nutrients in soil by holobiont. A holobiont is defined as an assemblage of a host (plant) and the many other species living in or around it. Once holobiont is formed, symbionts promote the growth of host plants, and we expect an enhancement of the resistance of seedlings to diseases, drought, organic pollutants, and heavy metal stress [89]. In order to improve the survival rate of afforestation tree seedlings, researchers suggest using the mutualistic symbiotic relationship between mycorrhizal fungi and host trees to increase the success rate of afforestation under adverse conditions in non-forested areas and degraded lands where soil and vegetation have been destroyed (e.g., [30, 33, 63, 88, 90]).

Inoculation of ECM fungi in acidified soil in southeast Sichuan, China, could improve the viability and aluminum (Al) resistance [62, 91] of *Pinus massoniana* in poor soil, which increases the local coverage [92]. Under the conditions of copper (Cu) and cadmium (Cd) stress in soil, the inoculation of ECM fungi on *Pinus tabulaeformis* can promote not only the growth and development of host plants and increase the biomass but also significantly reduce the concentration of heavy metals. ECM-host plants (holobiont) moderate the transfer of heavy metals from roots to shoots, improved the resistance to heavy metal stress, and improved the survival rate of forestation [88, 93].

In the process of coal gangue hill reclamation, inoculated ash tree (*Fraxinus chinensis*) seedlings with AM fungi, survival rate of the transplanted seedlings was significantly improved [42, 94]. They also found that ECM (*Gomphidium viscidus*) can significantly alleviate effect of coal gangue on the growth of *Pinus bungeana* seedlings, and the number of lateral roots and dry weight of inoculated seedlings were significantly higher than those of uninoculated seedlings, so as to promote the ecological restoration of gangue areas.

Using mycorrhizal fungi for biological control of tree diseases has the advantages of low cost, high efficiency, avoiding environmental pollution caused by chemical control, and can promote the robust growth of seedlings. The ECM fungi (*Boletus edulis*, *Amanita pantherine, Pisolithus tinctorius*) could enhance the resistance of *Pinus massoniana* seedlings to pine dieback disease, and also enhance the resistance of pine (*P. thunbergii* and *P. massoniana*) to pine wood nematode disease, delaying or reducing the death of seedlings [95]. Meng et al. [96] inoculated ECM against the damping-off of Dahurian larch (*Larix gmelinii*) seedlings in the alpine mountains and found that, compared with the control area, the incidence of damping-off decreased by 14–22%, and that of European red pine (*P. sylvestris*) decreased by 12–20%.

4.5 Edible mushroom production

ECM has always been a keyword with high frequency in the field of mycorrhizal research in China, mainly because the symbiotic relationship between ECM fungi and gymnosperms has great guiding significance for afforestation [97]. In addition, the fruit bodies of various ECM fungi are edible mushrooms with high economic value [98]. With the deepening of people's understanding of the edible or medicinal value of ECM fungi, more and more scholars are committed to the research on the artificial cultivation technology of ECM fungi. China has rich in wild edible fungi resources [99]. At present, the production of ECM edible fungi abroad has formed a certain scale in China, which has played a significant role in the national economy and brought great wealth.

From the holobiont viewpoint, an increase in host plant photosynthesis by thinning may allocate more photosynthates to belowground associated with ECM, and then increase mushroom production [100, 101].

5. Regeneration of seedlings on fallen logs

In forests, there are symbiotic fungi that play an important role in the acquisition of nutrients and water after seedlings have settled. However, it has long been known that AM, ECM and some kinds of fungi damage and cause seeds to rot and damp off in seedlings [102, 103]. Here, we discuss the role of CWD in the regeneration of natural regeneration. At this time, we introduce the activity of fungi that decays the CWD, which acts as a seed and seedling bed and helps naturally regenerate the forest.

5.1 Role of CWD

Regeneration of spruce (*Picea* sp.) in high-elevation mountain or cool temperate forests often depends on the presence of coarse woody debris (CWD), as logs provide sites with more favorable conditions for spruce regeneration compared to the forest floor (**Figure 9**) [104, 105]. However, there is little quantitative knowledge on the factors that are conducive to or hindering spruce establishment on CWD.

Variables describing microsite conditions were measured, and fungi were isolated from wood samples. Seedling density depended on the type of rot, log inclination, and decay stage of logs. Sapling density depended mainly on light availability, the time of tree fall, and the distance between the log surface and the forest floor [105]. A total of 22 polypore fungi were isolated from the wood samples, four of them being threatened species as disease sources [104]. White- and brown-rot fungi were found in all decay stages. The visual assessment of the type of rot in the field corresponded in only 15% of cases to the wood rot type caused by the isolated fungi; we should pay attention to the presence of harmful microbes and make field assessments of rot type [99].

5.2 White rot vs. brown rot

Seedling and sapling densities followed negative binomial distributions, with many logs of all decay stages having low regeneration densities [105]. The degree of ground contact, white-rot-causing *Armillaria* spp. presence, white-rot-causing *Phellinus nigrolimitatus* presence, and log diameter were positively related to both seedling and sapling density [99]. Also, tree death as a result of wind uprooting was positively related to seedling density. Conversely, the presence of brown-rot-causing

Figure 9.
Natural regeneration from spruce seedlings on fallen trees with brown rot in temperate mixed conifer-deciduous broad-leaved forests in northern Japan.

Fomitopsis pinicola and tree death as a result of bark beetle attack was negatively related to regeneration densities. The low cover of vegetation from sides positively affected seedling density; however, heavily covered logs were less occupied by seedlings. Former studies provide evidence that large logs originating from wind uprooting or butt rot infection are most appropriate [104, 106].

6. Function of symbiotic microbes under changing environment

We have discussed the role of symbiotic fungi in forest regeneration and tending practices, therefore, we need to further discuss forest management that responds to the activities of symbiotic fungi in response to rapid environmental changes, including recent global warming by increasing CO_2 [90, 107, 108]. As most of them are experimental studies [22, 109], we will introduce examples of rehabilitation in acidic soils and salinized soils as well as physical environmental changes (e.g., elevated CO_2, ozoneO_3, N loading).

6.1 CO_2, O_3, and N loading

Several Pinaceae species (*Pinus* sp. and *Larix* sp.) in northeastern Asia, especially Korea, show declining symptoms due to soil acidification and air pollution [63]. In the industrial area of northwest Korea, growth and physiological activities of *Pinus koraiensis* and *P. rigida* are adversely influenced by increased deposition of fluorine (F), chlorine (Cl), and manganese (Mn), involving both wet and dry deposition; these also have harmful effects on the ectomycorrhizal status [110]. The ECM colonization level at the damaged site was 30% less than at a control site. To determine the role of ECM with seedlings of *P. densiflora, P. koraiensis, P. rigida,* and *Larix kaempferi* during progressive soil acidification, the relation between relative total dry mass (TDM) and the molar ratio of base cation (BC) to aluminum (Al) (BC/Al) BC to manganese (Mn) or BC/(Al + Mn)) was examined [63].

The TDM of *P. koraiensis* and *L. kaempferi* seedlings was approximately 40% at a BC/Al ratio of 1.0, but that of *P. rigida* and *P. densiflora* was approximately 50% at the same BC/Al ratio. With an increasing amount of proton (H^+) added to the soil, the water-soluble concentrations of Ca, Mg, K, Al, and Mn increased, especially below a soil pH of 3.8. Development of ectomycorrhizae, that is, *Pisolithus tinctorius* (*Pt*), Ectodrench (EC), or *Cenococcum geophilum* (*Cg*) in all four test species was significantly greater in the 10 and 30 $mmolH^+ \cdot kg^{-1}$ treatments than in control treatments [63]. ECM development in the 60 and 90 $mmolH^+ \cdot kg^{-1}$ treatments was significantly lower than in the other treatments. Atmospheric CO_2 increases yearly, which may affect mycorrhizal growth and diversity in forest ecosystems through host plant physiology.

The ECM development of each species was significantly higher under elevated CO_2 concentration (720 ppm) than at ambient levels (360 ppm) in the study at the time [63]. The P concentration in needles and roots following inoculation with *Pt,* EC, or *Cg* was significantly higher in all species than without inoculation, at both CO_2 concentrations. Furthermore, no photosynthetic down-regulation was found in any species when inoculated with *Pt,* EC, or *Cg,* and photosynthesis was significantly higher than in non-inoculated seedlings at both CO_2 concentrations [111]. Stomatal limitation in each species at each CO_2 concentration was less than in non-inoculated seedlings [71].

Simulated nitrogen (N) deposition (50 kg N $hr^{-1}yr^{-1}$) increased photosynthesis and growth of hybrid larch F_1 for the initial 2 years. No significant effects were

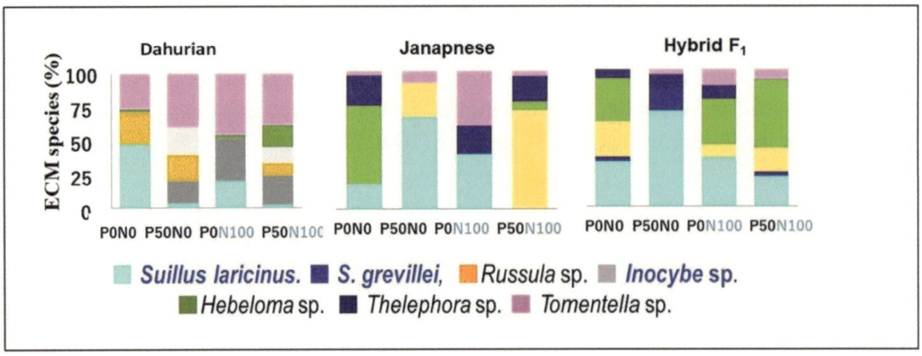

Figure 10.
Ectomycorrhizae (ECM) colonization of three larch species grown under different N and P (as solution of ammonium nitrate (NH_4NO_3) and potassium phosphate (KH_2PO_4)). The P50N100 means 50 kgP ha^{-1} yr^{-1}, 100 kgN ha^{-1} yr^{-1} for four growth seasons, and zero indicates no nutrient treatment. Suillus larcicinus, S. grevillea, and Inocybe sp. are specialist ECM to larch. However, Suillus grevillea and Inocybe sp. were only found in F_1 and Dahurian larch (DL: Larix gemelinii), respectively [23].

found for the next 3–8 years on brown forest soil (Dystric Cambisols). Among three larch species (Japanese, Dahurian, and F_1 hybrid larch), the dominant ECM species preferred N (**Figure 10**) [23].

Elevated O_3 decreased growth of larch seedlings, ECM colonization rates, and ECM species diversity; however, elevated CO_2 moderated or increased them in hybrid larch F_1. After elevated CO_2 and O_3 fumigation on F_1 seedlings, we found *Suillus* sp., a larch specialist ECM in F_1 [112], which means these ECM will be effective materials for larch plantations [97].

6.2 Combination effect of salt stresses

Soil salinization area is estimated to increase due to global warming [113], especially low precipitation region in the northeastern part of Asia. Inoculation of mycorrhizal fungi can often promote the growth of plants in a saline environment, increase the tolerance of plants to excessive sodium (Na) ions in the environment, and reduce the toxicity of mycorrhizal fungi to plants. Mycorrhizal fungi increase the ability of plants to resist salt stress by promoting the absorption of mineral elements, promoting the balance of elements in plants, and improving the water state of plants [95]. In addition, due to rapid development of industrialization and urbanization, emissions of ozone (O_3) precursors have been increasing rapidly, which has led to pronounced air pollution of ground-level O_3 [5, 114]. The effects of elevated ozone (eO_3) and soil salinization with alkaline salts in northeastern (NE) China is a serious concern affecting the success of the national replanting project [115]. As planted areas exceed 4 million hectares in China, we must consider future afforestation efforts after thinning and harvesting.

We evaluate the effects of elevated O_3 and alkaline salt on Dahurian larch F_1 (DL: *Larix gmelinii* var. *japonica*) and Japanese larch (JL: *L. kaempferi*) seedlings, and their hybrid larch F_1 (hereafter F_1, *L. gmelinii* var. *japonica* x *L. kaempferi*). As a result, effect of elevated O_3 on three larch species has hardly been found to interact with salinity treatments [116]. However, based on a leaf nutrient analysis, DL and JL have certain negative influences by the salinity treatment. The relationship between needle Na and other mineral contents indicated that both maintained K contents even with excess Na contents in needles. DL showed relatively lower reduction of other mineral contents, indicating higher salt tolerance of needle element homeostasis than JL. Low salinity stress did not significantly inhibit the growth of F_1 and even promoted it to some extent. There were no interaction effects of eO_3 and salt stress on both species. These results indicated that DL seedlings may be more suitable than JL seedlings as a future afforestation species under eO_3 levels of about <70 ppb at saline soil conditions [115].

In northern Japan, the growth of *Picea abies* and *P. glehnii*, which have been planted along the highways, is often suppressed due to several environmental stresses [89]. To examine the adverse effects of deicing salt (mainly NaCl), we measured ECM colonization, needle lifespan, and physiological capacity of the two spruce species at a site with damaged trees, near the roadside, and a site with healthy trees. These elements (Na and Cl) accumulated in spruce needles from the soil and snow at the damaged site. However, no ECM colonization was found in both spruce on the damaged site. Poor physiological traits might be attributed to an accumulation of deicing salt in the needles, resulting in the suppression of tree growth.

7. Conclusion

Fungi, which have coexisted with land plants since ancient times, play an important role in regenerating forests and maintaining and improving ecosystem services. In high CO_2 environments, commensal fungi act as sinks and play an important role in maintaining ecosystems. Methane (CH_4) is the second largest greenhouse gas after CO_2. At present, forests are methane sinks, but as the CO_2 concentration increases, the leaves in the upper canopy grow thicker, suppressing transpiration and creating an anaerobic state, turning them into sources of methane emissions. To avoid this, and maintain forests as a source of methane consumption, it is necessary to keep the forest floor dry (aerobic conditions). In addition, since monoculture forests are not resistant to various damages; therefore, mixed forests are required. In order to maintain a bright forest floor, thinning should be carried out intensively and direct sunlight should be introduced to raise the soil temperature, and the activity of earthworms would make the forest floor an aerobic condition. For these, we should introduce broad-leaved trees with AM and/or ECM to expect natural regeneration after thinning or harvesting conifer trees. We would like to expect charcoal to enhance the activities of symbiotic fungi that are involved in every aspect of forest regeneration, even in the rapidly changing physical environment.

Acknowledgements

We thank Prof. Masanori Saito for offering original illustration of evolution of plant and symbiotic microorganism with CO_2 concentration. Thanks are also due to Dr. Tatsu Ezawa for offering information on AM and plants; to financial support in part by Development Program of China 2017YFE0127700; LY. Qu, JST (No. JPMJSC18HB: representative researcher, M. Watanabe of TUAT and T. Watanabe of HU). The third author thanks for kind consideration and support of The National Federation of Land Improvement Associations of Shimane for field studies.

Author details

Yannan Wang[1], Laiye Qu[2*], Yutaka Tamai[3], Toshihiro Watanabe[3] and Takayoshi Koike[2,3*]

1 Jiangxi Academy of Forestry, Nanchang, China

2 Research Center for Eco-Environmental Sciences, CAS; University of Chinese Academy of Sciences, Beijing, China

3 Research Faculty of Agriculture, Hokkaido University, Sapporo, Japan

*Address all correspondence to: lyqu@rcees.ac.cn and tkoike@agr.hokudai.ac.jp

References

[1] Paul FA. Soil Microbiology, Ecology, and Biochemistry. 4th ed. Amsterdam, Elsevier; 2014

[2] Saito M. Mycorrhizal World. Tokyo: Tsukiji-shokan; 2020 (in Japanese)

[3] Averill C, Bhatnagara JM, Dietzeb MC, Pearsed WD, Kivline SN. Global imprint of mycorrhizal fungi on whole-plant nutrient economics. PNAS (Proceedings of the National Academy of Sciences). 2019;**116**:23163-23168. DOI: /10.1073/pnas.1906655116

[4] Tedersoo L, Bahram M, Zobel M. How mycorrhizal associations drive plant population and community biology. Science. 2020;**367**:6480. DOI: 10.1126/science.aba1223

[5] Qu LY, Wang XN, Mao QZ, Agathokleous E, Choi DS, Tamai Y, et al. Responses of ectomycorrhizal diversity of larch and its hybrid seedlings and saplings to elevated CO_2, O_3, and high nitrogen loading. Eurasian Journal of Forest Research. 2022;**22**:23-27. DOI: 10.14943/EJFR

[6] Qu LY, Wang YN, Masyagina O, Kitaoka S, Fujita S, Kita K, et al. Conifers—Recent Advances; Larch: A Promising Deciduous Conifer as an Eco-Environmental Resource. London, UK: IntechOpen; 2022. www.intechopen.com/chapters/80260

[7] Koike T, Nakamura M, Miyamoto T. Basic Textbook for Forest Protection. Tokyo: Nobunkyo-Publisher; 2021 (in Japanese)

[8] Seiwa K, Kunii D, Masaka K, Hayashi S, Tada C. Hardwood mixture enhances soil water infiltration in a conifer plantation. Forest Ecology and Management. 2021;**498**(6):119508. DOI: 10.1016/j.foreco.2021.119508

[9] Seiwa K. Mixed Sugi-cedar and Broadleaved Forests- revival of Ecosystem Services. Tokyo: Nobunkyo Publisher; 2022 (in Japanese)

[10] Wang YN. On the topics of Chinese fir (*Cunninghamia lanceolata*). Forest Technology. 2019;**931**:24-26 (in Japanese)

[11] Zhang P, Shao G, Zhao G, le Master DC, Parker GR, Dunning JB, et al. China's forest policy for the 21st century. Science. 2000;**288**:2135-2136. DOI: 10.1126/science.288.5474.2135

[12] Liang B, Wang J, Zhang Z, Jia Zhang J, Zhang J, Cressey EL, et al. Planted forest is catching up with natural forest in China in terms of carbon density and carbon storage. Fundamental Research. 2022;**2**:688-696. DOI: 10.1016/j.fmre.2022.04.008

[13] Wang YN, Koike T. Plantation and management of eucalyptus and broad-leaved trees in southwest China: Report of the IUFRO field study 2018. Forest Technology. 2019;**932**:30-33 (in Japanese)

[14] Makoto K, Nemilostiv YP, Zyryanova OA, Kajimoto T, Matsuura Y, Yoshida T, et al. Regeneration after forest fires in mixed conifer broad-leaved forests of the Amur region of Far Eastern Russia: The relationship between species specific traits against fire and recent fire regimes. Eurasian Journal of Forest Research. 2007;**10**:51-58. Available from: http://hdl.handle.net/2115/24486

[15] Qu LY, Makoto K, Choi DS, Quoreshi AM, Koike T. The role of ectomycorrhiza in boreal forest ecosystem. In: Osawa A, Zyryanova OA,

Matsuura Y, Kajimoto T, Wein RW, editors. Permafrost Ecosystem: Siberian Larch Forests. Springer Ecological Studies. 209. 2010. pp. 413-426

[16] Makoto K, Koike T. Charcoal ecology: Its function as a hub for plant succession and soil nutrient cycling in boreal forests. Ecological Research. 2021;**36**:4-12. DOI: 10.1111/1440-1703.12179

[17] Mao QZ, He BH, Ma M, Wang L, Koike T, Agathokleous E. Effects of biochar on Karst lime soil nutrients, soil microbial communities, and physiology of Sichuan pepper plants. Annals of Applied Biology. 2022;**181**:357-366. DOI: 10.1111/aab.12787

[18] Smith SE, Read D. Mycorrhizal Symbiosis. 3rd ed. Cambridge, MA: Academic Press; 2008. DOI: 10.1016/B978-0-12-370526-6.X5001-6

[19] Taniguchi T. Interactions between plants and mycorrhizal fungi associate with plant diversity in forests. Journal of Japanese Ecological Society. 2011;**61**:311-318 (in Japanese with English summary)

[20] Kitaoka S, Matsunami S, Wang YN, Fujita S, Makoto K, Hirata T, et al. Regulation of the growth of sprouting roots of black locust seedlings using root barrier panels. Journal of Forestry Research. 2022. DOI: 10.1007/s11676-022-01527-9

[21] Tobita H, Uemura A, Kitao M, Kitaoka S, Utsugi H. Interactive effects of elevated CO_2, phosphorus deficiency and soil drought on nodulation and nitrogenase activity in *Alnus hirsuta* and *Alnus maximowiczii*. Symbiosis. 2010;**50**:59-69. DOI: 10.1007/s13199-009-0037-7

[22] Wang XN, Agathokleous E, Qu LY, Watanabe M, Koike T. Effects of CO_2 and/or O_3 on the interaction between root of woody plants and ectomycorrhizae. Journal of Agriculture Meteorology. 2016;**72**:95-105. DOI: 10.2480/agrmet.D-14-00045

[23] Wang XN, Agathokleous E, Qu LY, Fujita S, Watanabe M, Tamai Y, et al. Effects of simulated nitrogen deposition on ectomycorrhizae community structure in hybrid larch and its parents grown in volcanic ash soil: The role of phosphorous. Science of the Total Environment. 2018;**618**:905-915. DOI: 10.1016/j.scitotenv.2017.08.283

[24] Brundrett MC, Tedersoo L. Evolutionary history of mycorrhizal symbioses and global host plant diversity. New Phytologist. 2018;**220**:1108-1115. DOI: 10.1111/nph.14976

[25] Cairney JWG. Evolution of mycorrhiza systems. Naturwissenschaften. 2000;**87**:467-475. DOI: 10.1007/s001140050762

[26] van der Heijden MGA, Martin FM, Selosse MA, Sanders IR. Mycorrhizal ecology and evolution: The past, the present, and the future. New Phytologist. 2015;**205**:1406-1423. DOI: 10.1111/nph.13288

[27] Floudas D, Binder M, (68 member), Hibbett DS. The Paleozoic origin of enzymatic mechanisms for decay of lignin reconstructed using 31 fungal genomes. Science. 2012;**336**:1715-1719. DOI: 10.1126/science.1221748

[28] Saito M, Marumoto T. Inoculation with arbuscular mycorrhizal fungi: The status quo in Japan and the future prospects. Plant and Soil. 2002;**244**:273-279. DOI: 10.1023/A:1020287900415

[29] Bücking H, Liepold E, Ambilwade P. The role of the mycorrhizal symbiosis in nutrient uptake of plants and the regulatory mechanisms underlying these

transport processes. IntecOpen. 2012. DOI: 10.5772/52570

[30] Matsuda Y, Kita K, Kitagami Y, Tanikawa T. Colonization status and community structure of arbuscular mycorrhizal fungi in the coniferous tree, *Cryptomeria japonica*, with special reference to root orders. Plant and Soil. 2021;**468**:423-438. DOI: 10.1007/s11104-021-05147-w

[31] Graham LE, Graham JM, Wilcox LW. Plant Biology. Upper Saddle River: Pearson Education, Inc.; 2003. p. 497

[32] Kayama M, Qu LY, Koike T. Elements and ectomycorrhizal symbiosis affecting the growth of Japanese larch seedlings regenerated on slopes of an active volcano in northern Japan. Trees-Structure and function. 2015;**29**:1567-1579. DOI: 10.1007/s00468-015-1238-8

[33] Courty P-E, Buée M, Diedhiou AG, Frey-Klett P, Tacon FL, Rineau F, et al. The role of ectomycorrhizal communities in forest ecosystem processes: New perspectives and emerging concept. Soil Biology & Biochemistry. 2010;**42**:679e692010. DOI: 10.1016/j.s

[34] Schulze E-D, Beck E, Buchmann N, Clemens S, Müller-Hohenstein K, Scherer-Lorenzen M. Plant Ecology. 2nd ed. Heidelberg: Springer Verlag; 2019

[35] Aber JD, Nadelhoffer KJ, Steudler P, Melillo JM. Nitrogen Saturation in Northern Forest Ecosystems: Excess nitrogen from fossil fuel combustion may stress the biosphere. Bioscience. 1989;**39**:378-386. DOI: 10.2307/1311067

[36] Ishii H, editor. Forest Ecology. Tokyo: Asakura-shuppan Publisher; 2019 (in Japanese)

[37] Nakaji T, Izuta T. Effects of Nitrogen Load on Asian Trees. In: Izuta T, editor. Air Pollution Impacts on Plants in East Asia. Tokyo: Springer; 2017. pp. 271-280. DOI: 10.1007/978-4-431-56438-6_18

[38] Ito K. Alternative method of forestry management- management of forest land owner under low price in timber. Sanrin (Forestry). 1660;**2022**:2-10 (in Japanese)

[39] Ohta T. Forest Saturation—Consideration on Land Transformation in Japan. Tokyo: NHK Books; 2012 (in Japanese)

[40] Takahashi E, Yone Y, Mori Y, Munemura H, Ide J, Sato T, et al. Analysis of present thinning conditions on the Sugi (*Cryptomeria japonica*) plantation forests in Okinoshima, Shimane Prefecture, Japan. Applied Forest Science. 2012;**21**:9-16 (in Japanese with English summary)

[41] Phillips RP, Brzostek E, Midgley MG. The mycorrhizal-associated nutrient economy: A new framework for predicting carbon–nutrient couplings in temperate forests. New Phytologist. 2022;**199**:41-51. DOI: 10.1111/nph.12221

[42] Kawahara A, An G-H, Miyakawa S, Sonoda J, Ezawa T. Nestedness in arbuscular mycorrhizal fungal communities along soil pH gradients in early primary succession: Acid-tolerant fungi are pH generalists. PLoS ONE. 2016;**11**:e0165035. DOI: 10.1371%2Fjournal.pone.0165035

[43] Mametani R, Fujimaki R, Tamashita T. Relationships between soil environmental factors and methane uptake by soil in a Japanese cedar plantation. Bulletin of Faculty of Life and Environmental Science of Shimane University. 2012;**17**:29-34 (in Japanese with English summary)

[44] Koga W, Suzuki A, Masaka K, Seiwa K. Conspecific distance-dependent seedling performance, and

replacement of conspecific seedlings by heterospecifics in five hardwood, temperate forest species. Oecologia. 2020;**193**:937-947. DOI: 10.1007/s00442-020-04725-6

[45] Seiwa K, Negishi Y, Eto Y, Hishit M, Masak K, Fukasawa Y, et al. Successful seedling establishment of arbuscular mycorrhizal-compared to ectomycorrhizal-associated hardwoods in arbuscular cedar plantations. Forest Ecology and Management. 2020;**468**:118155. DOI: 10.1016/j.foreco.2020.118155

[46] Wulantuya MK, Bayandala FY, Matsukura K, Seiwa K. Gap creation alters the mode of conspecific distance-dependent seedling establishment via changes in the relative influence of pathogens and mycorrhizae. Oecologia. 2020;**192**:449-462. DOI: 10.1007/s00442-020-04596-x

[47] Morikawa Y, Hayashi S, Negishi Y, Masuda C, Watanabe M, Watanabe K, et al. Relationship between the vertical distribution of fine roots and residual soil nitrogen along a gradient of hardwood mixture in a conifer plantation. New Phytologist. 2022. DOI: 10.1111/nph.18263

[48] Sakai T, Konno M, Hasegawa Y, Imaji A, Terabaru M, Nakamura R, et al. Role of mycorrhizal associations in tree spatial distribution patterns based on size class in an old-growth forest. Oecologia. 2019;**189**:971-980. DOI: 10.1007/s00442-019-04376-2

[49] Kim YS, Watanabe M, Imori M, Sasa K, Takagi K, Hatano R, et al. Reduced atmospheric CH_4 consumption by two forest soils under elevated CO_2 concentration in a FACE system in northern Japan. Japanese Journal of Atmospheric Environment. 2011;**46**:30-36. DOI: 10.11298/taiki.46.30

[50] Oikawa T. Simulation of forest carbon dynamics based on a dry–matter production model III.Effects of increasing CO_2 upon a tropical rain forest ecosystem. Botanical Magazine Tokyo. 1986;**99**:419-430

[51] Koike T, Watanabe M, Watanabe Y, Agathokleous E, Mao QZ, Eguchi N, et al. Ecophysiology of deciduous trees native to Northeast Asia grown under FACE (Free Air CO_2 Enrichment). Journal of Agricultural Meteorology. 2015;**71**:174-184. DOI: 10.2480 / agrmet.D-14-00020

[52] Norby RJ, Zak DR. Ecological Lessons from Free-Air CO_2 Enrichment (FACE) Experiments. The Annual Review of Ecology, Evolution, and Systematics. 2011;**42**:181-203. DOI: 10.1146/annurev-ecols

[53] Eguchi E, Funada R, Ueda T, Takagi K, Hiura T, Sasa K, et al. Soil moisture condition and growth of deciduous tree seedlings native to northern Japan raised under elevated CO_2 with a FACE system. Phyton -International Journal of Experimental Botany. 2005;**45**:133-138

[54] Ito M, Ohte N, Koba K. Methane flux characteristics in forest soils under an East Asian monsoon climate. Soil Biology & Biochemistry. 2009;**41**:388-395

[55] Kim YS, Makoto K, Takakai F, Shibata H, Satomura T, Takagi K, et al. Greenhouse gas emissions after a prescribed fore in white birch-dwarf bamboo stands in northern Japan, focusing on the role of charcoal. European Journal of Forest Research. 2011;**130**:1031-1044

[56] Makoto K, Minamiya Y, Kaneko N. Differences in soil type drive the intraspecific variation in the responses of an earthworm species and,

consequently, tree growth to warming. Plant and Soil. 2016;**404**:209-218. DOI: 10.1007/s11104-016-2827-z

[57] Kochian LV, Piñeros MA, Liu J, Magalhaes JV. Plant adaptation to acid soils: The molecular basis for crop aluminum resistance. Annual Review of Plant Biology. 2015;**66**:571-598. DOI: 10.1146/annurev-arplant-043014-114822

[58] Kayama M. Growth and nutrient acclimation of evergreen oak seedlings infected with *Boletus reticulatus* in infertile colluvial soil in warm temperate monsoon Asia: Evaluation of early growth. Forests. 2020;**11**:870-870. DOI: 10.3390/f11080870

[59] Kayama M, Yamanaka T. Growth characteristics of ectomycorrhizal seedlings of *Quercus glauca, Quercus salicina*, and *Castanopsis cuspidata* planted on acidic soil. Trees-Structure and Function. 2014;**28**:569-583. DOI: 10.1007/s00468-013-0973-y

[60] Labbé JL, Weston DJ, Dunkirk N, Pelletier DA, Tuskan GA. Newly identified helper bacteria stimulate ectomycorrhizal formation in *Populus*. Frontiers in Plant Science. 2014;**5**:579. DOI: 10.3389/fpls.2014.00579

[61] Maki T, Nomachi M, Yoshida S, Ezawa T. Plant symbiotic microorganisms in acid sulfate soil: Significance in the growth of pioneer plants. Plant and Soil. 2008;**310**:55-65. DOI: 10.1007/s11104-008-9628-y

[62] Watanabe T, Osaki M. Mechanisms of adaptation to high aluminum condition in native plant species growing in acid soils: A review. Communications in Soil Science and Plant Analysis. 2002;**33**:1247-1260. DOI: 10.1081/CSS-120003885

[63] Choi DS. Ecophysiological study of the growth of conifers in Korea in acidified soil with elevated CO_2: The role of ectomycorrhizal infection. Eurasian Journal of Forest Research. 2008;**11**:1-39. http://hdl.handle.net/2115/33063

[64] Zyryanova OA, Yaborov VT, Abaimov AP, Koike T, Sasa K, Terazawa M. Problems in the maintenance and sustainable use of forest resources in Priamurye in the Russian Far East. Eurasian Journal of Forest Research. 2005;**8**:54-63. Available from: http://hdl.handle.net/2115/22190

[65] Zyryanova OA, Yabarov VT, Tchikhacheva TL, Koike T, Makoto K, Matsuura Y, et al. The structure and biodiversity after fire disturbance in *Larix gmelinii* (Rupr.) Rupr. forests, northeastern Asia. Eurasian Journal of Forest Research. 2007;**10**:19-29. Available from: http://hdl.handle.net/211

[66] Makoto K, Tamai Y, Kim YS, Koike T. Buried charcoal layer and ectomycorrhizae cooperatively promote the growth of *Larix gmelinii* seedlings. Plant and Soil. 2010;**327**:143-152

[67] Makoto K, Choi DS, Hashidoko Y, Koike T. The growth of *Larix gmelinii* seedlings as affected by charcoal produced at two different temperatures. Biology and Fertility of Soils. 2011;**47**:467-472

[68] Kayama M, Nimpila S, Hongthong S, Yoneda R, Himmapan W, Noda I. Effect of biochar on the early growth characteristics of teak seedlings planted in sandy soil in northeast Thailand. Bulletin FFPRI. 2022;**21**:73-81. Available from: http:// www.jstage.jst.go.jp/article/ffpri/21/1/21_73/_pdf

[69] Makoto K, Hirobe M, DeLuca TH, Bryanin SV, Procopchuk VF, Koike T. Effects of fire-derived charcoal on soil properties and seedling regeneration in a recently burned *Larix gmelinii*/*Pinus*

sylvestris forest. Journal of Soils and Sediments. 2011;**11**:1317-1322

[70] Zhu JJ, Xu H, et al. Review on the ecological relationships between forest trees and ectomycorrhizal fungi. Chinese Journal of Ecology. 2003;**6**:70-76

[71] Qu LY, Shinano T, Quoreshi AM, Tamai Y, Osaki M, Koike T. Allocation of ^{14}C-Carbon in infected with ectomycorrhizae of two species of larch seedlings. Tree Physiology. 2004;**24**:1369-1376

[72] Chen YL, Shuie Liu S, Dell B. Mycorrhizal status of Eucalyptus plantations in south China and implications for management. Mycorrhiza. 2000;**17**:527-535. DOI: 10.1007/s00572

[73] Founoune H, Duponnois R, Bâ AM, Sall S, Branget I, Lorquin J, et al. Mycorrhiza helper bacteria stimulate ectomycorrhizal symbiosis of *Acacia holosericea* with *Pisolithus alba*. New Phytologist. 2002;**153**:81-89. DOI: 10.1046/j.0028-646X.2001.00284.x

[74] Frey-Klett P, Garbaye J, Tarkka M. The mycorrhiza helper bacteria revisited. New Phytologist. 2007;**176**:22-36. DOI: 10.1111/j.1469-8137.2

[75] Sun MQ, Wu XQ, Ye JR. Effects of Ectomycorrhizal Fungi on Germination and Growth of Pines. Journal of Nanjing Forestry University (Natural Science Edition). 2007;**50**(05):39-43. DOI: 10.3969/j.jssn.1000-2006.2007.05.009

[76] Agathokleous E, Kitao M, Komatsu M, Tamai Y, Saito H, Harayama H, et al. Effects of soil nutrient availability and ozone on container-grown Japanese larch seedlings and role of soil microbes. Journal of Forestry Research. 2020;**31**:2295-2311. DOI: 10.1007/s11676-019-01056-y

[77] Agathokleous E, Kitao M, Komatsu M, Tamai Y, Harayama H, Koike T. Single and combined effects of fertilization, ectomycorrhizal inoculation, and drought on container-grown Japanese larch seedlings. Journal of Forestry Research. 2022. DOI: 10.1007/s11676-022-01565-3

[78] Huang SY, Wu TG, et al. Effects of phosphorus addition and inoculation of mycorrhizal fungi on the growth and phosphorus utilization of Masson Pine container seedlings from different familie. Forest Research. 2021;**34**(5):142-151. DOI: 10.13275/j.cnki.lykxyj.2021.005.017

[79] Qu LY, Quoreshi AM, Koike T. Root growth characteristics, biomass and nutrient dynamics of seedlings of two larch species raised under different fertilization regimes. Plant and Soil. 2003;**255**:293-302

[80] Qu LY. Ecophysiological study on the natural regeneration of the two larch species with special references to soil environment in larch forests. Eurasian Journal of Forest Research. 2016;**19**:1-51. http://hdl.handle.net/211

[81] Wang Y, Qiu Q, Yang Z, et al. Arbuscular mycorrhizal fungi in two mangroves in South China. Plant and Soil. 2010;**331**(1):181-191. DOI: 10.1007/s11104-009-0244-2

[82] Yuan LH. Effect of arbuscular mycorrhizal fungi on the growth of *Elaeagnus moillis* seedling at different fertility levels. Chinese Agricultural Science Bulletin. 2010;**26**:173-177

[83] Song F, Li J, Zhang X. Characterization of expressed genes in the establishment of arbuscular mycorrhiza between *Amorpha fruticosa* and *Glomus mosseae*. Journal of Forestry Research. 2014;**25**:541-548. DOI: 10.1007/s11676-014-0493-7

[84] Song F, Qi D, Liu X, Kong XS, Gao Y, Zhou Z, et al. Proteomic analysis of symbiotic proteins of *Glomus mosseae* and *Amorpha fruticosa*. Scientific Report. 2016;**5**:18031. DOI: 10.1038/srep18031

[85] Jia Y, Song FQ. The Chemotaxis of Arbuscular Mycorrhizal to Rhizobia. Microbiology China. 2008;**35**(5):743-747. DOI: 10.3969/j.issn.0253-2654.2008.05.017

[86] André S, Neyra M, Duponnois R. Arbuscular mycorrhizal symbiosis changes the colonization pattern of *Acacia tortilis* spp. raddiana rhizosphere by two strains of rhizobia. Microbial Ecology. 2003;**45**:137-144

[87] Qu LY, Quoreshi AM, Iwase K, Tamai Y, Funada R, Koike T. *In vitro* ectomycorrhiza formation on larch seedlings with six different fungal species. Eurasian Journal of Forest Research. 2003;**6**:65-73

[88] Matyssek R, Lüttge U, Castell W. Evolution of Holobiont-Like Systems: From Individual to Composed Ecological and Global Units. Progress in Botany. 2022;**83**:203-248. DOI: 10.1007/124_2022_57

[89] Kayama M, Quoreshi AM, Kitaoka S, Kitahashi Y, Sakamoto Y, Maruyama Y, et al. Effects of deicing salt on the vitality and health of two spruce species, *Picea abies* Karst., and *P. glehnii* Masters planted along roadsides in northern Japan. Environmental Pollution. 2003;**124**:127-137. DOI: 10.1016/s0269-7491(02)00415-3

[90] Qu LY, Kitaoka S, Koike T. Factors controlling soil microbial respiration during the growing season in a mature larch plantation in Northern Japan. Journal of Soils and Sediments. 2018;**18**:661-668. DOI: 10.1007/s11368-017-1799-9

[91] Brunner I, Sperisen C. Aluminum exclusion and aluminum tolerance in woody plants. Frontiers in Plant Science. 2013;**4**:172. DOI: 10.3389/fpls.2013.00172

[92] Liu M, Chen G, et al. Research on EMC Seedlings of *Pinus massoniana* and Al Resistance in Southeast of Sichuan. Northern Horticulture. 2007;**11**:205-208. DOI: 10.3969/j.issn.1001-0009.2007.11.090

[93] Zheng WS, Fei YH, Huang Y. Soluble protein and acid phosphatase exuded by ectomycorrhizal fungi and seedlings in response to excessive Cu and Cd. Journal of Environmental Sciences. 2009;**21**(12):1667-1672. DOI: 10.1016/S1001-0742(08)62471-7

[94] Zhang CL, Wang W, Huang Y, Zhang HJ, Li MS. Growth of ectomycorrhizal fungi and mycorrhizal *Pinus bungeana* seedlings in response to Coal Cangue Stress. Scientia Silvae Sinicae. 2008;**44**(12):68-71. DOI: 10.11707/j.1001-7488.20081212

[95] Wu XQ, Sun MQ, Gao Y, Sheng JM, Ye JR. Effects of some ectomycorrhizas on Pine seedlings to disease resistance. Scientia Silvae Sinicae. 2007;**43**(6):88-93. DOI: 10.11707/j.1001-7488.20070616

[96] Meng FR, Shao JW, et al. Control effect of Ectomycorrhizal Fungi on damping-off disease of pine seedlings. Forest Science and Technology. 1991;**5**:27-31

[97] Tamai Y. Growth and ectomycorrhiza formation in container grown hybrid larch F_1 seedlings. Hoppo Ringyo (Northern Forestry). 2019;**70**:104-107 (in Japanese)

[98] Yang Q, Huang Y. Analyses of the Mycorrhizal Research State and Development in China Based on

SCI Database. Microbiology China, 2009;**36**(3):439-445

[99] Li T, Cui L, Song X, Cui X, Wei Y, Tang L, et al. Wood decay fungi: An analysis of worldwide research. Journal of Soils and Sediments. 2022;**22**:1688-1702. DOI: 10.1007/s11368-022-03225-9

[100] Collado E, Bonet JA, Alday JG, de Martínez Aragón J, de Miguel S. Impact of forest thinning on aboveground macrofungal community composition and diversity in Mediterranean pine stands. Ecological Indicators. 2021;**133**:108340. DOI: 10.1016/j.ecolind.2021.108340

[101] de Miguel S, Bonet JA, Pukkala T, de Martínez Aragón J. Impact of forest management intensity on landscape-level mushroom productivity: A regional model-based scenario analysis. Forest Ecology and Management. 2014;**30**:218-227. DOI: 10.1016/j.fore

[102] Kurata M. On the outline of the escape or microbe theory. Journal of Japanese Forestry Society. 1949;**31**:32-34. DOI: 10.11519/jjfs1934.31.1-2_32

[103] Sakamoto Y, Miyamoto T. Racodium snow blight in Japan. Forest Pathology. 2005;**35**:1-7. DOI: 10.1111/j.1439-0329.2004.00383.x

[104] Bače R, Svoboda M, Pouska V, Janda P, Červenka J. Natural regeneration in Central-European subalpine spruce forests: Which logs are suitable for seedling recruitment? Forest Ecology and Management. 2012;**266**:254-262. DOI: 10.1016/j.foreco.2011.11.025

[105] Miyamoto T. Regeneration and disease fungi. In: Koike T et al., editors. Physiological Ecology of Woody Plants. Tokyo: Kyoritsu-shupan Publisher. 2020. pp. 153-156 (in Japanese)

[106] Fukasawa Y, Kitabatake H. Which is the best substrate to regenerate? A comparative pot experiment for tree seedling growth on decayed wood and in soil. Forests. 2022;**13**:1036. DOI: 10.3390/f13071036 t

[107] Dong Y, Wang Z, Sun H, Yang W, Xu H. The response patterns of arbuscular mycorrhizal and ectomycorrhizal symbionts under elevated CO_2: A meta-analysis. Frontier of Microbiology. 2018;**11**:1248. DOI: 10.3389/fmicb.2018.01248

[108] Körner C, Asshoff R, Bignucolo O, Hattenschwile S, Keel SG, Pelaez-Riedl S, et al. Carbon flux and growth in mature deciduous forest trees exposed to elevated CO_2. Science. 2005;**309**:1360-1362. DOI: 10.1126/science.1113977

[109] Kivlin SN, Emery SM, Rudgers JA. Fungal symbionts alter plant responses to global change. American Journal of Botany. 2013;**100**:1445-1457. DOI: 10.3732/ajb.1200558

[110] Choi DS, Kayama M, Jin HO, Lee CH, Izuta T, Koike T. Growth and photosynthetic responses of two pine species planted under a polluted industrial region in Korea. Environmental Pollution. 2006;**139**:421-432. DOI: 10.1016/j.envpol.2005.06.006

[111] Choi DS, Quoreshi AM, Maruyama Y, Jin HO, Koike T. Effect of ectomycorrhizal infection on growth and photosynthetic characteristics of *Pinus densiflora* seedlings grown under elevated CO_2 concentrations. Photosynthetica. 2005;**43**:223-229. DOI: 10.1007/s11099-005-0037-7

[112] Wang XN, Qu LY, Mao Q, Watanabe M, Hoshika Y, Koyama A, et al. Ectomycorrhizal colonization and growth of the hybrid larch F_1 under elevated CO_2 and O_3. Environmental Pollution. 2015;**197**:116-126

[113] Hassani A, Azapagic A, Shokri N. Global predictions of primary soil salinization under changing climate in the 21st century. Nature Communications. 2021;**12**:6663. DOI: 10.1038/s41467-021-26907-3

[114] Sicard P, Agathokleous E, De Marco A, Paoletti E. Ozone-reducing urban plants: Choose carefully. Science. 2022;**377**:585. DOI: 10.1126/science.add9734

[115] Wang YN. Ecophysiological Study on the Effect of Elevated O_3 and Salinity Stress on the Growth and Photosynthesis of Three Larch Species. Sapporo, Japan: Graduate School of Agriculture of Hokkaido University; 2019. p. 26

[116] Sugai T, Wang YN, Watanabe T, Satoh F, Qu LY, Koike T. Salt stresses reduced the growth of two species larch seedlings under elevated ozone. Frontiers in Forests and Global Change. 2019;**18**. DOI: 10.3389/ffgc2019

Chapter 10

Analytical Trophodynamics Applied to Modeling Forest Dynamics with Carbon Cycling

Solange da Fonseca Rutz and Marcelo Santos Carielo

Abstract

Models based on analytical trophodynamics (AT) method have provided an analytical framework for modeling in ecology, including the dynamical flux of nutrients present in the soil for a fixed region. Dynamics occurring concurrently in different time scales are modeled. Through a mathematical treatment of the elements of both biotic and abiotic factors, is established Â stability and conservation laws for growing trajectories, whose solutions of the second-order differential systems equations known as Volterra–Hamilton systems. All solutions trajectories obtained to follow the biological principles of energy conservation. The tensors of AT were computed with the computational algebraic package FINSLER. Moreover, in this chapter, we present an overview of the last results and actual status of research in this area.

Keywords: forest modeling, mycorrhizal networks, finsler geometry, second-order differential equations, nonlinear connections

1. Introduction

Forest dynamics are closely related to the effects of climate change, biodiversity, air temperature regulation, precipitation patterns, carbon cycling, and fixation of essential nutrients for organic life such as nitrogen, phosphorus, and potassium [1]. Understanding the cycling of nutrients and carbon is essential to comprehend this process [2]. Such cycles present in forest ecosystems are influenced by biotic and abiotic effects, including ecological interactions and adaptations to environmental external changes [3, 4]. In [5], experimental results for species in Tropical Forests, show the role of the understory in the dynamics of interaction between different species of underground plants and microorganisms. In particular, it has already been shown that ectomycorrhiza fungi are responsible for the hyphae pathways that mediate transports in forest soil, increasing the rate of nutrients uptake, weathering of soil minerals, and also influencing the bacterial community in the rhizosphere [6]. Recent research [4], investigating the transfer of carbon between trees in a fixed area, under the effect of a specific microclimate, highlights the role of the rhizosphere in nutrient dynamics. Mycorrhizal networks play the role of mediators of interactions between trees and plants surrounding the understory [7–9].

The transfer of carbon and other nutrients between plants in the same place has been investigated, taking into account the mycorrhizal networks, which provide a means for the sending and receiving of nutrients to occur. Understanding quantitative aspects of the direction and magnitude, among other factors, of the transfer process directly affects the understanding of mechanisms related to the interaction and function of these plants [4, 6, 10].

Recent results [11] show that mycorrhizal fungi influence the coexistence and diversity of plants, being mediators of the symbiosis process of intra- and interspecific competition of plant communities. The different types of mycorrhizal fungi affect the level of nutrient sharing between neighboring plants and, consequently, their availability, acting as a regulator and stabilizer of the interaction mechanisms that lead to the coexistence of species.

To evaluate the concentration of nutrients present in the subsoil of regions where the different species are found, a long-term analysis under the effects of external agents, such as natural disasters and anthropic interference, is essential [12]. The long-term facilitation-type interactions have been shown to be more influential in the dynamics than the competition. This is characterized by the positive effect in which at least one of the local species benefits from interactions. A direct implication of this result is that plants with higher biomass may not prevail in the long term [13, 14].

Models based on the analytical trophodynamics (AT) method [15, 16] provide a new theoretical approach for a mathematical analysis of the carbon cycle dynamics and of other nutrients present in the soil, taking into account short-term and long-term as competition and facilitation, respectively. In addition, abiotic disruptive factors can also be integrated. AT has made it possible to model complex phenomena such as forest dynamics [5, 9, 17, 18]. Recent AT results describe the interaction between trees and the understory, including litter [19], succession forest [20], and the environmental impact of the transgenic crop [21].

Models based on TA can include external environmental effects in the local dynamics by analogy with continuum mechanics. In this approach, geometric quantities described through Finsler geometry and systems of second-order differential equations provide mathematical objects that allow the analysis of these dynamics. The algebraic manipulation of the geometric elements modeled is usually performed with the FINSLER package [22].

The effect of external forces on the local dynamics of plants affects their forms of interaction and functionality, causing stability scenarios to be altered. The eco-strain model [20] makes it possible to investigate the external effects of the surrounding space on the forest dynamics. Thus, transformations like that are important to deform the Huxley allometric space, in which the known classical ecological interactions (symbiosis, competition, and parasitism) occur so that in the new space there are optimal trajectories that conserve Medawar growth energy [16], being described by analytical expressions that modeling this mechanism.

The present work provides a review of the article [23], detailing biological and mathematical aspects of the constructed model, with the TA method. The mathematical model built using this approach took into account the concept of the potential of the mycorrhizal network [24] so that the solutions of the proposed model allowed us to describe the dynamics of carbon cycling for two species of trees in the British forest, corroborating experimental results. The trajectories of the proposed system of differential equations describe the dynamics of species in a production space disturbed by external agents. The results achieved in this model can be statistically tested at later times since it is expected that the structural behavior of the dynamics will continue to be valid for particular cases, in which the coefficients assume numerical values measured through field experiments.

In Section 1, we present an updated overview of the importance of the type of modeling. Sections 2 and 3 introduce the theoretical framework of the proposed approach. Examples of modeling applications with real data from biological experiments are shown in Section 4. We end with a description of the next steps to continue this research.

2. Volterra: Hamilton systems and Finsler and Wagner spaces

Given a coordinate chart (x^i, y^i) in the base space M of a slit tangent bundle TM^* consider the Volterra–Hamilton [25] system defined by[1]

$$\begin{aligned} \frac{dx^i}{dt} &= k_{(i)}y^i, \\ \frac{dy^i}{dt} &= G^i_{jk}y^jy^k + r^i_jy^j + e^i, \quad i,j,k = 1, \dots, n \end{aligned} \tag{1}$$

where $k_{(i)}$[2] is rate growth *per capita*, $G^i_{jk}(x,y)$ are positively homogeneous of degree zero in y^i representing the interaction between populations y^j and y^k, r^i_j[3] and e^i are, respectively, how the populations grow and react to environmental influence. The variable x is the variable of production of Volterra [26]. It measures the quantity produced by a population $N^i = \frac{dx^i}{dt}$ along an interval of time.

There are three possibilities for the interaction coefficients: (i) are all constants; (ii) depends explicitly of x^i; or (iii) depends on the ratio $\frac{y^i}{y^j}$ [16]. When (iii) occurs, the effects of *social interaction* [27] are included in modeling. Furthermore, a condition assumed in this work is that the population are proportional, that is, the ratios $\frac{N^i}{N^j}$ are constants. This is known as how *uniformity condition*.

Given an initial condition, solutions of the second order system (SODE) (1) provide trajectories that have the property of conserve growth energy [28] along the solutions curves in the correspondent production process space. In [16] is studied a two-dimensional case of this SODE model with classical ecological interactions between two populations. In this case, the G^i_{jk} will be constant along the time.

For the case that the coefficients G^i_{jk} are constant these equations define the coefficients of the Berwald connection [15, 29–31]. In this case, three of the eight SODEs, are elastic transformation and are denoted by A^i_{jk}. They are the three cases are the Minkowski space (M, F), where M is as before, and F is the Finsler norm such that with $ds = F(y)$, and the metric tensor[4]

$$g_{ij}(y) = \frac{1}{2}\dot{\partial}_i\dot{\partial}_j A^2_\alpha, \tag{2}$$

[1] In the system (1) was used the Einstein notation for repeated indices.

[2] The sub-index (i) in $k_{(i)}$ denote that this quantities are not summables over repeated indices.

[3] Without loss of generality for this work, we can assume that the organisms contribute for the dynamics similarly, that is $k_{(i)} \equiv 1$, and growth with same rate $r^i_j \equiv \lambda$.

[4] Symbols ∂_i and $\dot{\partial}_i$ indicates the partial derivative with respect x^i and y^i, respectively.

for

$$A_\alpha = \exp[\phi(x)]F(y), \tag{3}$$

where there are three possibilities for F, given by $F = F_\alpha(y)$, $\phi = \phi_\alpha(x)$, $\alpha \in \{1,2,3\}$, as stated in [20].

The ideal case of Huxley allometric space is one of these, and your geodesics are given by

$$\frac{d^2x^i}{ds^2} = 0, \tag{4}$$

for a parameter s is known as a *natural* parameter [16]. They are the perfect case of Huxley allometric space. Consequently, the solutions are straight lines $y^i = a^i s + b^i$, for arbitrary real constants of integration a^i and b^i.

As in forest ecosystems, it is common to have processes occurring on different time scales [7, 14], in the modeling developed with TA different scales are combined. While the parameter t indicates the real-time scale, the parameters s and p – introduced ahead – denote other time scales, in which forest dynamics phenomena studied here have patterns that preserve quantities of interest, obeying the principle of conservation of the energy in the allometric space. The Finslerian geodesics given by the metric tensor (2) preserve energy growth along trajectories in production process space.

Here transformations of the perfect case are *reversible* in the sense that one can be obtained from the other by a non-singular change of coordinates x^i into $\overline{x}^i$,

$$d\overline{x}^i = J^i_j(x)dx^j, \tag{5}$$

where the Jacobian $J^i_j = \left(\frac{\partial \overline{x}^i}{\partial x^j}\right)$ depends only on the Volterra production variable x^i.

The other five, of all eight SODEs mentioned above, are *plastic* deformation and denoted by $A^i_{jk} + W^i_{jk}$ where

$$W^i_{jk} = \delta^i_j \partial_k[\sigma(x)], \tag{6}$$

for a scalar function $\sigma(x)$ defined over M, and $\partial_k[\sigma(x)] = \sigma_k(x)$ is your gradient.

According to the TA approach, after a heterochronic transformation $\{\sigma(x), c^i(x,y)\}$, we obtain the following perturbed SODE from (1):

$$\begin{aligned} &\frac{d^2x^i}{dp^2} + \left[A^i_{jk} + \delta^i_j \sigma_k\right] \frac{dx^j}{dp}\frac{dx^k}{dp} = c^i, \\ &\frac{dp}{ds} = Q \exp\left[-\int_\gamma \sigma(r)\,dr\right], \\ &ds = \exp(\lambda t)\,dt, \end{aligned} \tag{7}$$

where Q is an arbitrary constant and γ is a solution curve. When $c^i \neq 0$, then the environmental effects act, such that the populations dynamics react through a time-sequence of growth along their trajectories [32].

Now the solutions of this new SODE (7), the autoparallels of a constant Wagner connection, are not geodesics [33]. In [34], it was demonstrated that the energy along these paths is conserved, and they provide a condition for optimality accordly division of labour principle [16, 28, 35].

Although model (7) includes the effect of environment in dynamics, it does not explain how soil carbon is periodically cycled and how it occurs in the case of interactions of paper birch and Douglas-fir trees [6]. The energy flux and mycorrhizal networks was introduced in the model by consider a proper form of the anholonomic frames $B^i_j(x, dx)$, depend on x^i and dx^i, that describe the plastic deformation. Through this element, defined below, the new version of SODE (7) will provide solutions that model the nutrient cycle as expected.

External events can have a high impact on forest dynamics, directly affecting soil organic matter, and leading most microorganisms to suffer drastic reductions in their population due to high temperatures [36]. In addition to the direct effects of short-term wildfires affecting the soil microbial communities, there are indirect abiotic effects. In the long term, there are changes in both microbial and plant communities, chemical changes at different soil levels, alterations in the microclime, reduction of overstory and understory plant diversity, soil microflora, and others [37]. In turn, such effects imply patterns of forest succession such that the vegetation cover present as well as the dynamics of this ecosystem are altered [38]. Although there are approaches based on Finsler geometries, different from TA, for modeling this aspect [39, 40], understanding the long-term effects of wildfires on forest dynamics is still a research area with many challenges [41–43].

3. Randers space

For investigating the effect of surround local environment, not considered by the initial Eco-strain model [20], in [23], it was assumed that there was an increase in the total intake of nutrients per unit of time, according to the biological event [44]. Mathematically, this corresponds to assume the following Finsler metric functional

$$\mathcal{F}_\alpha = \exp[\sigma(x)]A_\alpha + b_i(x)y^i, \tag{8}$$

and its norm $dS_\alpha = \mathcal{F}_\alpha(x, dx)$. The metric tensor in this case is denoted by $h_{\alpha\beta}(x, y)$. Using parameter p of Wagner SODE, we can obtain [23, 45] your geodesics,

$$\begin{aligned} &\frac{dy^i}{dp} + \left[A^i_{jk} + \delta^i_j\sigma_k(x) + \delta^i_k\sigma_j(x) - g_{jk}\sigma^i\right]y^jy^k = g^{im}(y)\left(\delta_m b_l - \delta_l b_m\right)y^l, \\ &y^i = \frac{dx^i}{dp}, \qquad \delta_m = \partial_m - A^r_m\dot{\partial}_r. \end{aligned} \tag{9}$$

The δ_m is known as Berwald delta derivative[5] and g^{im} denotes the inverse of metric tensor.

Another geometric element for the model proposed in [23] is the introduction of the *infectivity potential* represented by the real positive function $V(x)$. Considering how a function splits off the vector field (b_i) into $b_i(x) = \partial_i V(x) + \tilde{b}_i(x)$, the potential $V(x)$ is defined so that increase the total intake per unit of time because

$$dS_\alpha = \exp[\sigma(x)]A_\alpha + dV, \tag{10}$$

where $dV > 0$.

[5] Note that δ_m and δ^i_j denote, respectively, the Berwald delta derivative and the Kronecker delta.

In [23], the effect of a strained environment pressuring the forest dynamics was modeled by the Eco-strain model [20] where the Berwald frame $B_i^\alpha(x, dx)$ has been used, and additionally by the Holland's frame $U_i^\alpha(x, dx)$.

From the original allometric space described in terms of the metric tensor (2), the *relaxed state* $\delta_{\alpha\beta}$ and *strained state* g_{ij} are obtained so that

$$\begin{aligned} \delta_{\alpha\beta}(x,y) &= g_{ij}(x,y) B_\alpha^i B_\beta^j, \\ g_{ij}(x,y) &= \delta_{\alpha\beta} B_i^\alpha B_j^\beta. \end{aligned} \tag{11}$$

When the frame is a Jacobian matrix it is called *holonomic* (or *elastic*) and indicates a coordinate change that is reversible. Already when this does not occur, the frame is termed *anholonomic* (or *plastic*), and it is non-reversible.

Additionally, an already deformed space can be deformed again through a composition of frames. From the Holland's frame for Randers space, denoted by U_i^α and defined by

$$\begin{aligned} g_{ij}(x,y) &= U_i^\alpha U_j^\beta h_{\alpha\beta}, \\ h_{\alpha\beta}(x,y) &= g_{ij} U_\alpha^i U_\beta^j. \end{aligned} \tag{12}$$

obtain one of the plastic deformation of the Euclidean allometric space with a non-vanishing curl.

4. Applying the modeling in a concrete case

In [23], a model for carbon and nutrient cycling including understory and underground mycorrhizal networks was presented. This model extends the Eco-strain model [16]. Now, the results proposed allow to describe the periodicity of carbon cycling. The effect of mycorrhizal networks in the model proposed here shows that the nutrient dynamics described by the SODE trajectories (9), obtained by the TA method, now provide trajectories that model these dynamics.

As in the first SODE (1) where classical ecological interactions are modeled and the system provides trajectories that model the different dynamics under such interactions, SODE (9) described here provides trajectories that will describe the effect of carbon cycling, providing different interaction schemes according to the coordinates and assumed values for their coefficients. Although for different choices of these elements the dynamics effects can change, the geometric concepts of the TA [27] derived from this system allow the study of the stability, among other aspects, of the system according to this approach.

Using analytical trophodynamics [15, 16], a model was presented for two tree species (dominant and codominant) inhabiting a fixed region with a specific microclimate, defined by means of solar incidence, wind speed, air temperature, substrate, and pH. In this model, it was considered species interacting with the plant community in the local understory and with microorganisms from the rhizosphere, and of the subsoil of the soil-root interface [46, 47]. The results obtained extended other existing ones [20], in which the interaction coefficients had zero curvature. The new model constructed [23] took into account the potential for arbuscular infectivity defined by the mycorrhizal network, which contributes to the understanding of the interactions between trees and plants in the same region. Commonly, these interactions are complex, not being limited to the most usual in ecology: predation, parasitism, and symbiosis. Mycorrhizal networks, present in practically all subsoil, connect at least two local plants through the mycelia, which serves as a physical medium for the transmission of nutrients [1, 8]. This feature was also contemplated by the proposed mathematical model. It is also noteworthy that the results obtained corroborate experimental data [18, 44] that describe two species of trees – namely, the Douglas-fir (*Pseudotsuga menziesii*) and the paper birch (*Betula papyrifera*) – present in the same forest region of Canada's British Columbia. Thus, it was possible to mathematically investigate the nutrient dynamics in addition to classifying possible stability scenarios [15]. Additionally, through the concept of Eco-strain [20], the forest dynamics were studied [23] under the addition of external environmental effects that influence the flux of nutrients. Conceptually, this aspect is formulated by analogy with the mechanics of the continuous medium. Including the infectivity potential in the proposed model also allowed us to understand the plastic and elastic deformations in the considered allometric production space.

We emphasize that the model proposed in [23] also allow investigating interactions in scenarios that include rhizosphere microorganisms, abiotic soil variables, grafts, and mycorrhizal network. Through algebraic computation [22, 30], experiments were carried out that contributed to the validation of the results. In the Appendix, there are codes of computational experiments related to the following application examples.

Application 1. Assume that the two species of trees in a forest are both close as Douglas-fir and paper-birch in forests of British Columbia [44]. As mentioned above, the dynamic of ecological interactions among these involve the understories and mycorrhizal components. Using the indices #1 to denote Douglas-fir and #2 for paper-birch consider the following system [23], derived from the Huxley space defined by F_α and $\alpha = 1$, namely $F_1 = \sqrt{(y^1)^2 + (y^2)^2}$, that correspond to the Randers geodesics,

$$\begin{aligned}\frac{d^2X^1}{ds^2} &= \Omega_{12}Y^2, \\ \frac{d^2X^2}{ds^2} &= -\Omega_{12}Y^1,\end{aligned} \tag{13}$$

for $Y^i = \frac{dX^i}{ds}$, $b_1 = -c_2\left(x^2 - x^2_*\right)$, $b_2 = c_1\left(x^1 - x^1_*\right)$, $\Omega_{12} = c_1 + c_2$, and c_1, c_2 are constants. Therefore,

$$\left(Y^1\right)^2 + \left(Y^2\right)^2 = R^2, \tag{14}$$

where R is a constant. This solution shows that the net carbon flux varies periodically corroborating the biological experiments with these trees.

Application 2. Now, consider the Finsler functional defined by $\alpha = 2$, $F_2 = \sqrt{(y^1)^2 + \frac{2}{\lambda} + (y^2)^{\frac{2}{\lambda}}}$, so that, using normal coordinates $\overline{x}^i$, with rates $\frac{d\overline{x}^i}{dp} = N^i$ and the uniform condition to insure that $\frac{N^i}{N^j}$ are constants over a time interval or close to climax equilibrium, the Randers geodesics are given by

$$\begin{aligned}\frac{dN^1}{dp} &= \Omega_{12}\left(-g^{12}N^1 + g^{11}N^2\right), \\ \frac{dN^2}{dp} &= \Omega_{12}\left(-g^{22}N^1 + g^{12}N^2\right).\end{aligned} \tag{15}$$

In [23], it is shown that this system is also periodic both in p parameter and in real-time t. Moreover, the N-constant ratio occurs in original variables, $\frac{dx^1}{dx^2}$, but with different constants.

These two applications of the model presented here illustrate how the AT method allows a mathematical description of complex dynamics such as those involving plants under different effects of the external environment. The complex mycorrhiza's role in the interactions of plants and its effects on neighboring species, investigated under an analytical approach, can be tested from a statistical perspective in the next step.

Acknowledgements

The second author thanks Prof. Peter Antonelli and Prof. Solange Rutz for introducing him to such a rich and exciting area of research.

Conflict of interest

The authors declare no conflict of interest.

Appendix A

Computational experiments with FINSLER package [22] **Figures 1** and **2**.

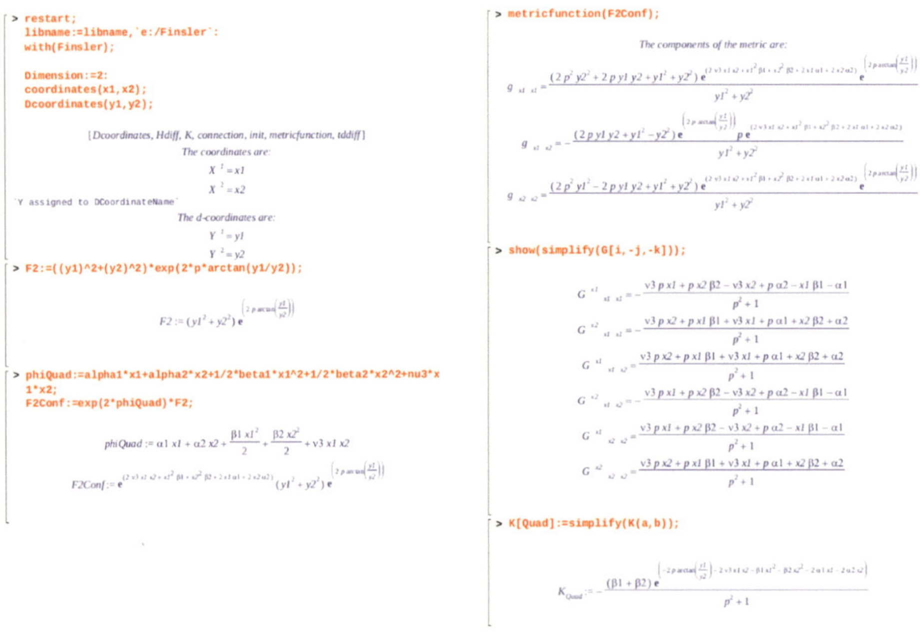

Figure 1.
Application 1 experiment codes.

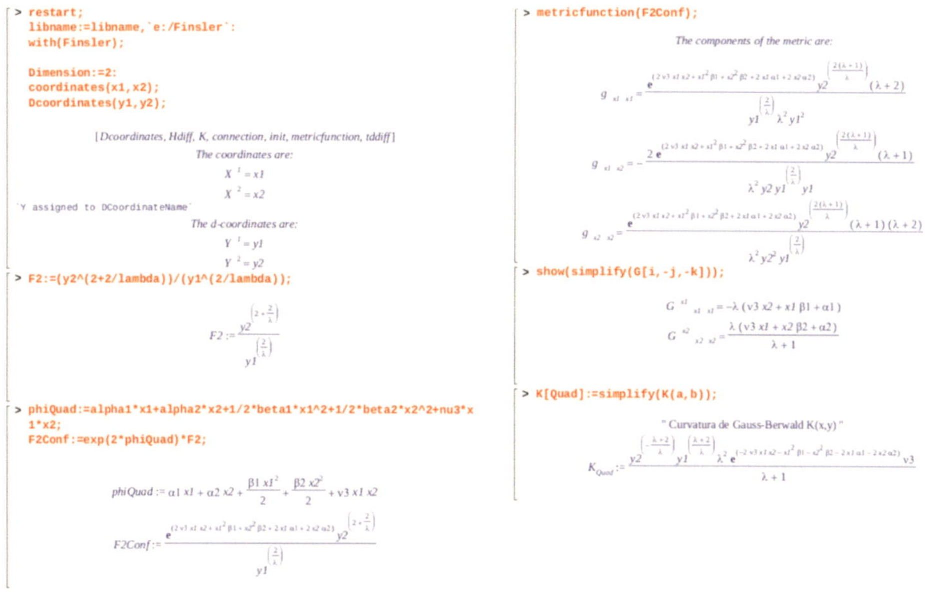

Figure 2.
Application 2 experiment codes.

Author details

Solange da Fonseca Rutz[1*†] and Marcelo Santos Carielo[2†]

1 Mathematics Department, Federal University of Pernambuco—UFPE, Recife, PE, Brazil

2 State University of Campinas—UNICAMP, IMECC, Campinas, SP, Brazil

*Address all correspondence to: solange.rutz@dmat.ufpe.br

† These authors contributed equally.

References

[1] Simard SW, Austin ME, editors. Climate Change and Variability. Rijeka, Croatia: Sciyo Publ; 2010

[2] Fitter AH, Graves JD, Watkins NK, Robinson D, Scrimgeour C. Carbon transfer between plants and its control in networks of arbuscular mycorrhizas. Functional Ecology. 1998;**12**(3):406-412

[3] Filotas E, Parrott L, Burton PJ, Chazdon RL, Coates KD, Coll L, et al. Viewing forests through the lens of complex systems science. Ecosphere. 2014;**5**(1):1-23

[4] Smith SE, Read D. Mycorrhizal Symbiosis. Third ed. San Diego: Academic Press; 2008

[5] Souza FM, Gandolf S, Rodrigues RR. Species-specific associations between overstory and understory tree species in a semideciduous tropical forest. Acta Botanica Brasilica. 2015;**29**:73-81

[6] Simard SW, Jones MD, Durall DM. Carbon and nutrient fluxes within and between mycorrhizal plants. In: Mycorrhizal Ecology. Berlin, Heidelberg: Springer; 2003. pp. 33-74

[7] Clemmensen KE, Bahr A, Ovaskainen O, Dahlberg A, Ekblad A, Wallander H, et al. Roots and associated fungi drive long-term carbon sequestration in boreal forest. Science. 2013;**339**(6127):1615-1618

[8] Newman EI. Mycorrhizal links between plants: Their functioning and ecological significance. In: Advances in Ecological Research. Vol. 18. London: Academic Press; 1988. pp. 243-270

[9] Selosse MA, Richard F, He X, Simard SW. Mycorrhizal networks: Des liaisons dangereuses?. Trends in Ecology & Evolution. 2006;**21**(11):621-628

[10] Eskelinen A, Harrison S, Tuomi M. Plant traits mediate consumer and nutrient control on plant community productivity and diversity. Ecology. 2012;**93**(12):2705-2718

[11] Tedersoo L, Bahram M, Zobel M. How mycorrhizal associations drive plant population and community biology. Science. 2020;**367**(6480)

[12] Sayer EJ. Using experimental manipulation to assess the roles of leaf litter in the functioning of forest ecosystems. Biological Reviews. 2005;**80**: 1-31

[13] Ariza C, Tielborger K. Biomass explains the intensity of facilitative-not competitive-interactions: three intraspecific tests with annuals. Web Ecology. 2012;**12**:49-55

[14] Baldrian P. Forest microbiome: Diversity, complexity and dynamics. FEMS Microbiology Reviews. 2017;**41** (2):109-130

[15] Antonelli PL, Ingarden RS, Matsumoto M. The Theory of Sprays and Finsler Spaces with Applications in Physics and Biology. Vol. 58. Dordrecht, The Netherlands: Springer Science & Business Media; 1993

[16] Antonelli PL, Bradbury RH. Volterra-Hamilton models in ecology and evolution of colonial organisms. In: Mathematical Biology and Medicine. Vol. 2. Singapore: World Scientific; 1996

[17] Pretzsch H. Forest Dynamics, Growth, and Yield. Berlin, Heidelberg: Springer-Verlag; 2009

[18] Simard SW, Beiler KJ, Bingham MA, Deslippe JR. Mycorrhizal networks: Mechanisms, ecology and modeling.

Fungal Biological Reviews. 2012;**26**: 39-60

[19] Antonelli PL, Rutz SF, Cantalice JBR. Carbon cycles in tree stands from KCC theory: Discounted production due litter decomposition. Nonlinear Studies International Journal. 2016;**23**:111-115

[20] Antonelli PL, Rutz SF. Eco-strain in model forests. Nonlinear Analysis: Real World Applications. 2009;**10**:576-588

[21] Antonelli PL, Rutz SF, Junior RV. Environmental analysis of impact of transgenic crops. International Journal of Applied Mathematics. 2013;**26**(4):515-524

[22] Rutz SF, Portugal R. FINSLER: A computer algebra package for Finsler geometries. Nonlinear Analysis. 2001;**47**: 6121-6134

[23] Rutz SF, Carielo MS, Antonelli PL. Mathematical models of forests: Understories and mycorrhizal networks. Nonlinear Analysis: Real World Applications. 2019;**47**:168-177

[24] Requena N, Jeffries P, Barea JM. Assessment of natural mycorrhizal potential in desertified semiarid ecosystems. Applied and Environmental Microbiology. 1996;**62**(3):842-847

[25] Antonelli PL, Bucataru I. Volterra–Hamilton production models with discounting: general theory and worked examples. Nonlinear Analysis: Real World Applications. 2001;**2**(3):337-356

[26] Volterra V. Principes de biologie mathématique, 1936. In: Mathematical Essays on Growth and the Emergence of Form. Winnipeg: Univ. Alberta Press; 1982. pp. 269-309

[27] Antonelli PL, Bradbury RH, Lin X. On Hutchinson's competition equations and their homogenization: A higher-order principle of competitive exclusion. Ecological Modeling. 1992a;**60**:309-320

[28] Beklemeshev W. Principles of comparative anatomy of invertebrates. In: Trans. from Russian. 3rd ed. Vol. 1. University of Chicago Press, Oliver and Boyd; 1969. p. 2

[29] Antonelli PL, Han B, Modayil J. New results in 2-dimensional constant sprays with an applications to heterochrony. Nonlinear Analysis. 1999;**37**:545-566

[30] Matsumoto M. Important Finsler Spaces [Part 8 - Chapter 3]. In: Antonelli PL, editor. Handbook of Finsler Geometry. Vol. 2. Dordrecht: Kluwer Academic Publishers; 2003

[31] Antonelli P. The Wagner theory of 2-dimensional constant sprays and its applications in evolutionary biology. Open Systems & Information Dynamics. 2003;**10**(1):65-87

[32] Antonelli PL. A mathematical theory of evolution by heterochrony in colonial invertebrates. Open Systems & Information Dynamics. 1992b;**1**(1):57-74

[33] Hashiguchi M. On Wagner's generalized Berwald space. Journal of the Korean Mathematical Society. 1975;**12** (1):51-61

[34] Antonelli PL, Han B, Modayil J. The Wagner theory of 2-dimensional constant sprays and its applications in evolutionary biology. Open Systems & Information Dynamics. 1999;**10**(1):65-87

[35] Wilson EO. Sociobiology. Cambridge, Mass: Harvard Univ. Press; 1975

[36] Dooley SR, Treseder KK. The effect of fire on microbial biomass: A

meta-analysis of field studies. Biogeochemistry. 2012;**109**(1):49-61

[37] Hart SC, DeLuca TH, Newman GS, MacKenzie MD, Boyle SI. Post-fire vegetative dynamics as drivers of microbial community structure and function in forest soils. Forest Ecology and Management. 2005;**220**(1–3): 166-184

[38] Treseder KK, Mack MC, Cross A. Relationships among fires, fungi, and soil dynamics in Alaskan boreal forests. Ecological Applications. 2004;**14**(6): 1826-1838

[39] Dehkordi HR. Applications of randers geodesics for wildfire spread modeling. Applied Mathematical Modeling. 2022;**106**:45-59

[40] Markvorsen S. A Finsler geodesic spray paradigm for wildfire spread modeling. Nonlinear Analysis: Real World Applications. 2016;**28**:208-228

[41] Finney MA. The challenge of quantitative risk analysis for wildland fire. Forest Ecology and Management. 2005;**211**(1–2):97-108

[42] O'Connor CD, Thompson MP, Rodríguez y Silva, F. Getting ahead of the wildfire problem: Quantifying and mapping management challenges and opportunities. Geosciences. 2016;**6**(3):35

[43] Tedim F, Leone V, Amraoui M, Bouillon C, Coughlan MR, Delogu GM, et al. Defining extreme wildfire events: Difficulties, challenges, and impacts. Fire. 2018;**1**(1):9. Available from: https://www.mdpi.com/2571-6255/1/1/9

[44] Philip L, Simard S, Jones M. Pathways for below-ground carbon transfer between paper birch and Douglas-fir seedlings. Plant Ecology & Diversity. 2010;**3**(3):221-233

[45] Antonelli PL, Bucataru I. On Holland's frame for Randers space and its applications in physics. In: Steps in Differential Geometry, Proceedings of the Colloquium on Differential Geometry, Debrecen. 2000

[46] Van der Hiejden MGA. Underground networking. Science. 2016; **352**(6283):290-291

[47] Klein T, Siegwolf R, Korner C. Belowground carbon trade among tall trees in a temperature forest. Science. 2016;**352**(6283):342-344

Chapter 11

Microbes as Artists of Life

Gayatri Sharma

Abstract

Scientists have been knocking the wood to ascertain the symbiotic relationships of tiny living creatures, that is, microorganisms with other beings such as plants, animals, insects, and humans. The concept of "symbiosis" got its existence in 1879, which means "living together." Microorganisms show a great deal of diverse interactions such as commensalism (moochers), mutualism (both benefitted), and parasitism (one benefitted and other unharmed) with other living beings and mutualism being the most common of all, thus forming a range of antagonistic to cooperative symbiotic relationships. These tiny creatures interact with plants by forming lichens (fungi and algae), mycorrhizae (plants and roots of higher plants), root noodles (*Rhizobium*) and acting as keyworkers in plant's rhizosphere promoting growth and development. Microbial community also extends itself to kingdom Animalia establishing relationships with phylum Mammalia including humans, animals, and the most abundant species of phylum Arthropoda, that is, insects such as termites, which have colonization of bacteria in gut to digest wood cellulose. Scientists have discovered that most studied organisms—mussels found in deep-sea hydrothermal vents too live in a mutualistic association whereby bacteria get protection and mussels get nutrition as bacteria use chemicals from hydrothermal fluid producing organic compounds.

Keywords: symbiosis, lichens, mycorrhizae, root nodulation, leghemoglobin, soil structure, PGPRs, siderophores, antibiosis, rhizosphere, insect vectors, mussels, deep-sea hydrothermal vents, nematodes, LAC bacteria, gut microbiota, bobtail squid, host-microbial interaction

1. Introduction

Since the birth of this universe, milky way, or be it an organism, nature had been always playing its role as a balancer to ensure the easy survival of every being. The law of nature "symbiosis" has got its existence since times gone and will continue as an eternal process in life. The greatest example in this context is the phenomenon of photosynthesis and respiration. Have you ever wondered how the living beings learnt the art to live in harmony? The answer lies in the fact that nature is the driving force behind this. Starting from microscopic bacteria to humans, all living beings are interlinked through a chain of interaction—symbiosis, that is, the art of living of different organisms with cooperation. The term "symbiosis" was coined by **Anton de Bary,** who was a German botanist, microbiologist, founding father of plant pathology and modern mycology in 1879. It has its origin from ancient Greek words "syn" meaning together and "bios" meaning life. It has been used to describe mutualistic beneficial relationship between different organisms and now can be broadly applied to different multitudes of relationships, that is, beneficial, harmful, or neutral.

Microbes are the most studied creatures; without them life would be nearly meaningless. The role they play in the ecosystem cannot be replaced by any other living organism. Fungi, bacteria, protozoa, and algae such as cyanobacteria have distinct structural organization, which displays a great deal of behavioral patterns serving various purposes. Nowadays, microbes are studied for their interactions in the environment and for utilizing it in research studies and also for the benefit of mankind.

"Life would not long remain possible in the absence of microbes"-Louis Pasteur [1].

This quote by Louis Pasteur (**Figure 1**) throws light on the microbial importance in our life. Thinking of an antimicrobial life is really a difficult task. From oceans to seas to land, microbes are ubiquitous. Prokaryotes account for nearly one-half of all global biomass. These cellular life forms such as *Prochlorococcus* and *Synechoccocus*, residing in oceans, generate oxygen from photosynthesis. If such sources disappear from earth's surface, aerobic life would be in danger [1].

1.1 Let us take an example of N-cycle

Nitrobacter, Nitrosomonas are involved in nitrogen fixation; without them whole N-cycle would be disturbed. Rhizobium species such as *Rhizobium meliloti, Rhizobium trifolii, R. japonicum, Rhizobium leguminosarum* nodulate alfalfa, red clover, soyabean,

Figure 1.
Louis Pasteur in his lab, painting by a. Edelfeldt in 1885 (modified as per reference number 1, taken from Wikipedia).

peas respectively. *Pseudomonas, and Bacillus* are involved in breakdown of proteins in the process called aminization. Apart from these, the roles played by denitrifers such as *Pseudomonas*, *Bacillus denitrificans* (facultative anaerobes), and *Thiobacillus denitrificans* (autotrophs) are responsible for reduction of nitrates and nitrites into oxides of N and nitrogen gas. For a moment, if we imagine the absence of any one of these microbes, the **N-cycle** would not be able to complete and the natural equilibrium will be disrupted. Similarly, phosphorus-solubilizing bacteria such as *Bacillus megatherium* and *Pseudomonas striatus* increase P availability under ideal conditions. What would be the fate of other cycles if microbes involved in them are removed? Simply, the answer would be imbalance in nature that would threaten the existence of all life forms. Hence, microbes are an important part of ecosystem, their respective works can only be done by them, and there is no substitute for their roles in this ecosystem. Being talented with different arts, they add colors to the life of other living beings with their healthy and positive interactions but at the same time can even act the other way. One must remember that these tiny creatures with their artistic behavior could be promising at one end and ominous on the other.

2. Microbes and Symbiosis

2.1 History

Although seventeenth century marked the discovery of microorganisms, their existence had been since the beginning of life on the earth. They have been the part of various types of interactions that are important for survival. These interactions, which are beneficial in the ecosystem, are referred to as what we call "**symbiosis**." Symbiosis has been law of nature as previously said. The discoveries and the research studies in its context have their origin since nineteenth century. The term "**symbiosis**" was used by **Albert Bernhard Frank** in **1877** to describe the mutualistic relationship in lichens. An integrative approach with a classification of "co-actions" was proposed by **Edward Haskell** in **1949** [2].

2.2 Microbial Symbiosis

Microorganisms have been always an important part of ecosystem. These microscopic entities perform the main role in soil and plant biology enhancing their properties and quality. Microbial symbiosis has been a topic of discussion among scientists with some putting their view forward that it should refer only to mutualism while others have a view point that it refers to all the biological interactions—mutualism, commensalism, and parasitism, but excluding predation (**Figure 2**). Mutualism is a type of interaction in which both organisms are mutually benefitted from each other and the survival of one would be nearly impossible without the other one. Commensalism is an interaction in which one organism is benefitted from the other and the other one is neither harmed nor benefitted. These can also be called as moochers. Parasitism involves interaction of a microbe as a parasite with the host. Hence, the microbes form important and useful symbiotic associations, and the phenomenon is called as **microbial symbiosis**.

Microbes have also been a major topic of phylogenetic studies, so evolutionarily, microbial symbiosis is an important survival mechanism, foe example, some free-living bacterium that was descended by mitochondria found a symbiotic relationship

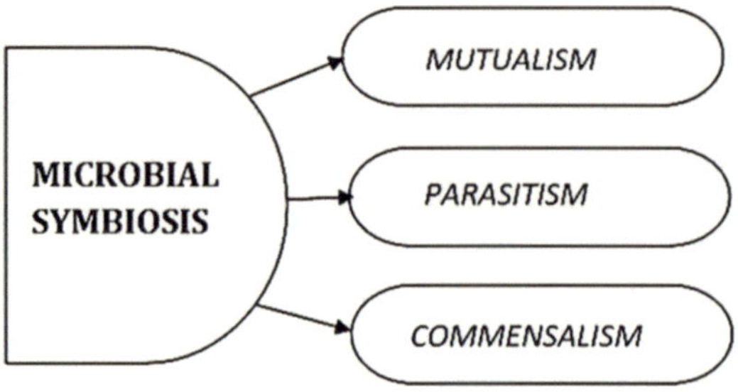

Figure 2.
Types of microbial symbiosis.

with a eukaryotic cell, and it proceeded toward permanency as this was mutually beneficial as a means of coexistence [3]. Microbial communities exist in various ways in nature; it can be complex, dynamic, or as simple as that of lichen. The most common example of this is of bacterial colonies present in the intestine of humans as well as animals participating in the process of digestion by breaking down complex food substances, synthesizing vital nutrients for the body such as riboflavin, vitamin B12, or other essential nutrients that are not synthesized by body itself.

These friendly microbial communities maintain natural balance and equilibrium in the body living inside the host. So, in this association, the host is benefitted by synthesis of important nutrients by these microflorae, and in the turn, bacteria are benefitted too since they get hospitable growth habitat to survive and food material to act on.

Definitely, one would wonder how beautifully these wonderful tiny creatures live their life artistically! These fascinating creatures have great associations with plants, animals, and insects and have coevolved by creating specific niches closely associated with them by adapting the way of living in the course of symbiotic interaction.

3. Microbial symbiosis with plants and in soil

Microbes are known to form symbiotic associations with plants interacting with its rhizosphere and phyllosphere, and their interactions in the soil are advantageous in maintaining soil properties. These interactions of microbes with organisms of kingdom **"Plantae"** are depicted below (**Figure 3**):

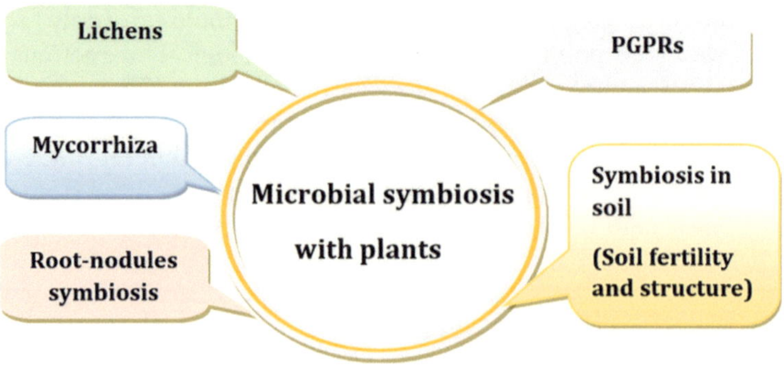

Figure 3.
Microbial symbiosis with plants.

3.1 Lichens

The simplest example of symbiosis is lichens. It is the mutualistic association of algae or cyanobacteria with fungus (ascomycetes/basidiomycetes) representing self-supporting symbiosis. In this, chlorophyll-containing partner. That is, algae provide simple carbohydrates and vitamins, which are absorbed by fungal component, and in turn fungi contribute by absorbing water from air and protecting algae as the upper thallus of fungi provides shade for underlying algae, some of which are photo-sensitive. Although appearing as a single plant-like organism, these lichens are made up of various cells of algae interwoven into matrix of fungal component. They cover 6–8% of earth surface, and approximately 20,000 different kinds of lichens have been reported till date having different structure and appearance—leaf-like, crust-like, powdery, filamentous, etc., growing on walls, trunks, in soil, etc. These organisms have the tendency to survive in a variety of conditions (extreme to mild) as they produce thallus of different forms, that is, **crustose, foliose, or fruticose** (**Figure 4**) that favors their growth in different climates; however, lichens have been economically important only as they produce various pigments that can be used as dyes. They are the best indicators of pollution due to their sensitive nature also they are widely used as food,

Uses of lichens:

- Most of the lichens such as *Bryoria fremontii* (horsehair lichen), *Umbilicaria spp.* and *Lasalia spp., and Umbilicaria esculenta,* are used as staple food and delicacies [4].
- Another use of lichens is in lichenometry, which is a technique to determine the age of exposed rock on the basis of size of lichen thalli, and lichens used for this purpose are *Xanthoria, Rhizocarpon, etc.* [4].
- Lichens produce metabolites such as usnic acid, which are similar to antibiotics and act as bactericidal agent against bacteria such as *E. coli* [4].

3.2 Mycorrhizae

Mycorrhizal interactions are associations of fungus (belonging to alomeromycota) with plant's rhizosphere, its root system playing an important role in enhancing plant nutrition and growth. This interaction can be seen as an important innovation of terrestrial plants in which plants being photosynthetic provide food to fungi and fungi provides water and nutrition to plants and also offer protection from various types of biotic and abiotic stresses (Smith and Read, 2008) [5]. Fungus can colonize vascular

Figure 4.
Types of lichens medicine, and in perfumes, for example, Cetraria islandica, used as appetite stimulating source, Evernia prumastri mostly used in perfume industry. The structures are undergoing fast evolution to serve different purposes.

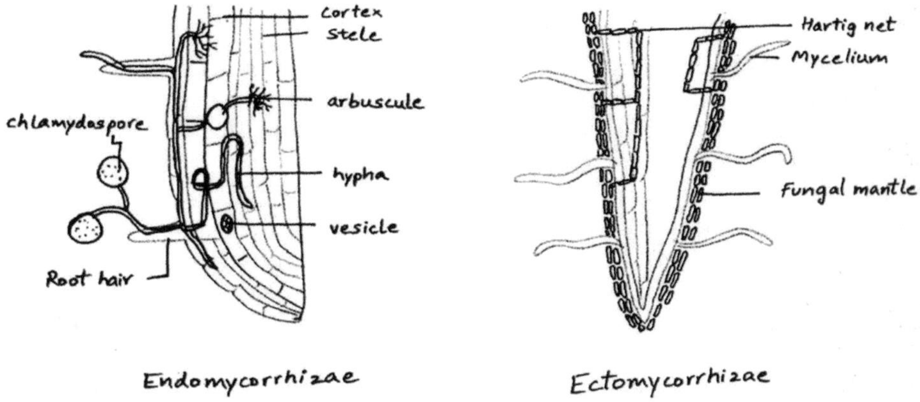

Figure 5.
Types of mycorrhizae – Endomycorrhizae and ectomycorrhizae.

plant's root system either intercellularly, for example, arbuscular mycorrhizal fungi or extracellularly. Hence, mycorrhizae are mainly divided into two groups (as shown in **Figure 5**)—endomycorrhizae (fungal hyphae penetrate cell wall and cell membrane) and ectomycorrhizae (fungal hyphae do not penetrate individual cells).

3.2.1 Endomycorrhizae

Endomycorrhizae are classified as arbuscular, ericoid, orchid, and monotropoid. **Arbuscular mycorrhizae** are the mycorrhizae in which hyphae penetrate plant cells producing vesicles or arbuscules as nutrient exchange medium as they increase the surface area for easy transfer of nutrients between them [5]. Fungi produce glomalin (glycoprotein) acting as a carbon store in soil. **Ericoid mycorrhizae** form dense coils of hyphae in the outermost epidermal layer of root cells and extend into the soil. They have saprotrophic nature enabling plants to receive nutrients. **Orchid mycorrhizae** are the mycorrhizal associations, which are myco-heterotrophic and form pelotons (coils) for nutrient exchange but do not have energy reserved [5]. **Monotropoid mycorrhizae** are mixotrophic and derive carbon from mycobiont component. This is not a mutualistic but a parasitic association [5].

3.2.2 Ectomycorrhizae

It refers to a mycorrhizal association in which fungal component forms sheath over the root surface or tip sending hyphae into intercellular spaces of cortex forming hartig-net. It includes **arbutoid** mycorrhizae. Arbutoids usually have fungal component belonging to basidiomycetes. They look similar to ericoid and monotropoid mycorrhizae and can be revealed only by anatomy tests. They proliferate mycelium into intercellular spaces forming dense hyphal complexes, hence are important type of ectomycorrhizal fungi. These often form extensive network in soil and leaf litter [5].

Another type of mycorrhizal association is ectendomycorrhizae in which hartig net is formed outside root tips but penetrates the cortical cells occasionally. These associations enhance nutrient and water uptake, increase crop quality and yield, increase seedling establishment, reduce fertilizer use, thereby increasing efficiency of the plant.

3.3 Root-nodules symbiosis

This is the very useful phenomenon in the leguminous plants. Legumes and N-fixing bacteria, that is, *Rhizobium* establishes root nodule symbiosis specifically in which bacterial species take part in interactions with root nodules of leguminous plants in rhizosphere. N-fixing bacteria (*Rhizobium* sp.) are stimulated by the flavonoids released by legumes in **low N condition,** and the bacteria respond by releasing nodulation factor, which initiates nodule formation, and the bacteria enter through deformed root hairs into the host plant root system forming infection thread and penetrate the cell wall. Infection initiates cell division in root cells and forms nodules (**Figure 6**). Functional nodules appear pinkish due to the presence of leghemoglobin. This type of symbiotic association helps in nitrogen fixation through reactions catalyzed by nitrogenase enzyme, thereby enhancing **N status** of soil as the bacteria convert atmospheric nitrogen to ammonium used up by the plant. For example, bacteria *Rhizobium leguminosarum, Bradyrhizobium, and Sinorhizobium* fix nitrogen in peas and soyabean, respectively. Except this, *Lotus japonicus* has also been used as a key model for leguminosae family for studying the mechanism of symbiosis with *Rhizobium sp.* taking part in nodule formation [6]. In this association, *Rhizobium* is benefitted by getting shelter and organic carbon for its growth.

Reaction of N-fixation is as follows (Eq. (1)):

$$N_2 + 8\,H^+ + 8\,e^- + 16\,ATP + 16\,H_2O\ {}_2\,NH_3 + H_2 + 16\,ADP + 16\,P_i \qquad (1)$$

3.4 Symbiosis in soil: soil fertility and structure

Soil microbes enhance fertility and structure of soil by interacting in various ways. Microbial diversity in soil performs various functions in the rhizosphere of the plant such as organic matter decomposition, seed germination, recycling nutrients, N-fixation, manages soil stability by performing various biochemical processes, and

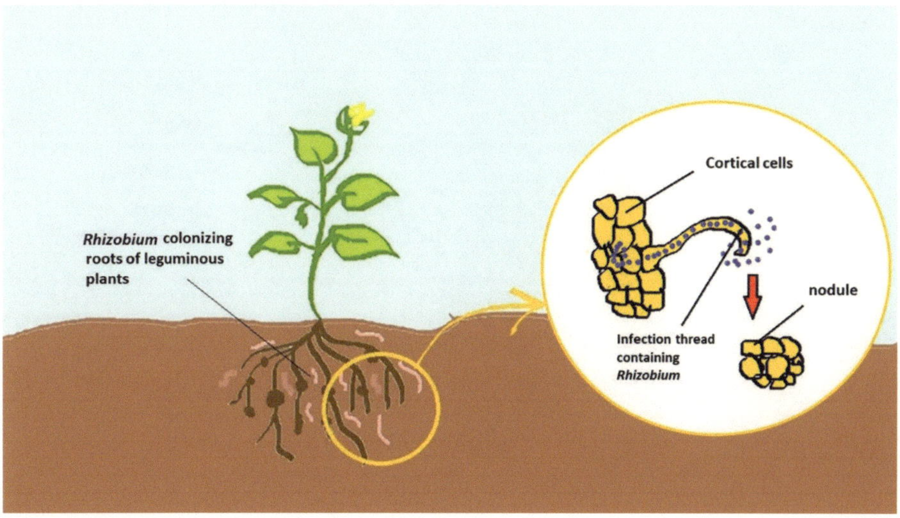

Figure 6.
Nitrogen fixation in leguminous plants by rhizobium.

themselves are benefitted with favorable growth conditions or shelter in soil. *Frankia* is associated with non-leguminous plants for N-fixation. *Azospirillum* is associated with plants as phosphate solubilizer.

3.5 Plant-growth-promoting rhizobacteria (PGPRs) and plant microbiome

PGPRs refer to plant-growth-promoting rhizobacteria, which are emerging as a leading topic of study since it is useful for sustainability of agriculture due to suppression of plant pathogens. These bacteria colonize roots of plants and thereby aim at improving the immune system of the plant against attack of disease-causing microbes. Basically, they offer biological control for the disease spread and are found in rhizosphere and rhizoplane (**Figure** 7).

Gray and Smith (2005) put forward the view that PGPRs range in their proximity to the root and association. They must colonize root, survive, and multiply in the rhizosphere or rhizoplane competing with other microflora so as to express plant growth promotion and protection activities. They act as rhizome mediators (degrade organic pollutants) phytostimulators, biofertilizers, and biopesticides [7]. For example, siderophores (low-molecular-weight iron chelators) produced by PGPRs to overcome iron deficiency allow solubilization of iron, *Pseudomonas fluorescens* produce siderophores pyoverdine that suppresses root rot disease, *Botrytis cinerea*, *Fusarium oxysporium*.

Various types of organic compounds such as ketones, pyrazine, and kanosamine are secreted by *Bacillus* sp. act as antagonists against pathogenic fungi. Actinomycetes, which are gram +ve spore-forming bacteria, too promote plant growth, act a biocontrol agent by inhibiting the proliferation of pathogens by degrading organic contaminants, breaking down dead plants. Beneficial effects of various rhizosphere microbes such as Firmicutes, *Pseudomonas*, and fungi belonging to *Trichoderma*, *Gliocladium*, etc., are recognized [8].

Siderophores are also involved in improvement of plant vigor as PGPRs (e.g., *Bacillus, Azotobacter, Azadirachta, and Burkholderia*) increase chlorophyll levels in

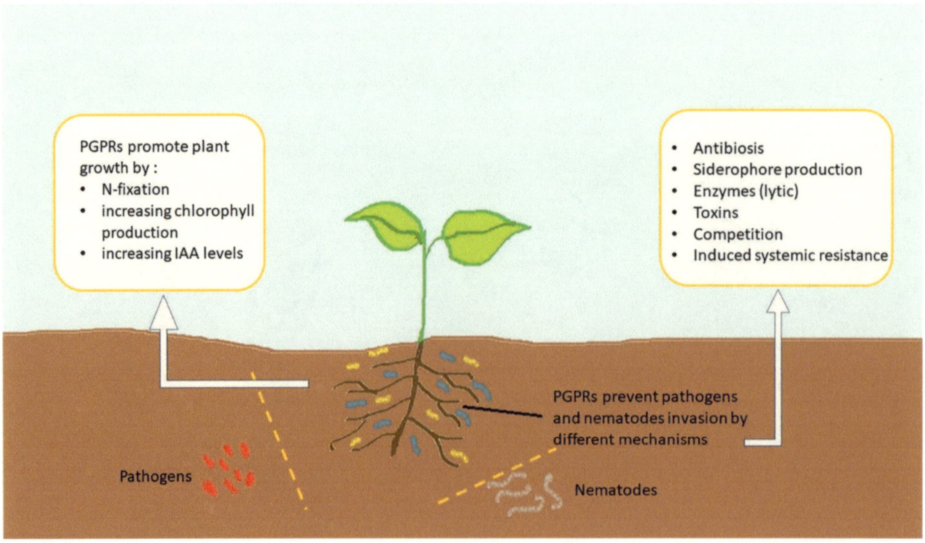

Figure 7.
Role of PGPRs in promoting plant growth.

plants. PGPRs such as *Pseudomanas, Bradyrhizobium, Enterobacter, and Agrobacterium* are responsible for production of IAA and promote root growth in plants. PGPRs also regulate salicylic acid pathway and jasmonate host plant stimulating resistance or defense response against various disease-causing microbes [8]. PGPRs have the potential to maintain growth of plant during stress conditions, for example, drought-tolerant type microbes in maize maintain water potential during stress.

Except this, PGPRs increase production of amino acids such as proline, production of sugars in plants, reduce antioxidant activity, and increase relative water content in maize. Soybean under drought stress has low content of chlorophyll, thereby photosynthesis decreases, but induction with *Pseudomonas putida H-2-3* overcomes the stress by balancing chlorophyll content and development of biomass [8]. PGPRs also ameliorate affected soils by secreting phytohormone and thereby improving plant health. The rhizosphere microbiome surely intercommunicates in an artistic way and serves as the major light for research purposes or new discoveries.

4. Microbial symbiosis with animals

Microbes form different symbiotic relationships with animals too. Having diverse nature of their interactions, they interact with different organisms of kingdom **"Animalea"** such as insects and mammals (animals and humans) (**Figure 8**).

4.1 Symbiosis of microbes with insects

"If insect is a plane, then microbes are its passengers." Insects and microbial symbiosis have proved a great synergy in the nature. With insect species accounting for about 80% of animal life on this wonderful planet, the earth, it is estimated that they have diverse population and represent the largest record of species with coleopterans as the highest of all known insect species. These species have undergone a great evolutionary history and adaptive success due to their potential for establishing symbiotic relationships with microbes—**endosymbionts**. Most common supportive example is the presence of microbiota in the insect gut (hindgut), which are acquired during their life and not inherited ones. Facultative and obligate anaerobic

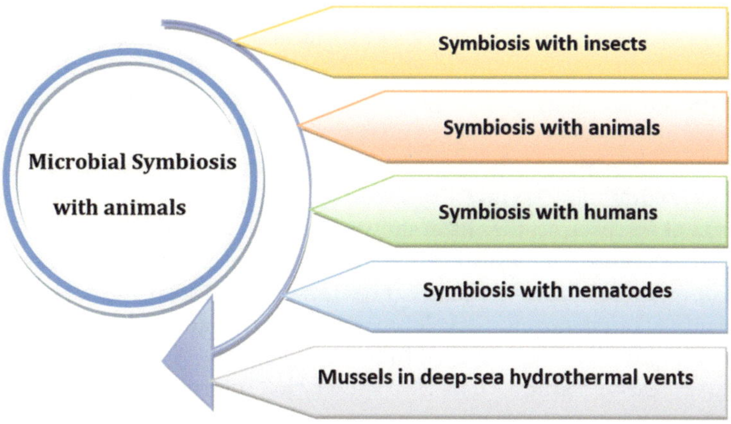

Figure 8.
Microbial symbiosis with animals.

prokaryotes such as **methanogens** and **protozoa** (in termites that help in digestion of wood cellulose) live in insects. Besides this, insects act as vectors for pathogenic microbes such as:

- *Trypanosoma* causing sleeping sickness transmitted by *Tsetse* flies
- *Plasmodium* causing malaria transmitted by female *Anopheles* mosquito
- *Leishmania* causing kala azar transmitted by Sand fly

Previous research studies conducted have shown the molecular basis of association between insect and microbes such as reactive oxygen species (immune effector) produced in the gut of *Drosophila melanogaster* and mosquito by dual-oxidases (DUOX) enzymes control microbiota in *Aedes* mosquito, but the population of microbes increases many folds if DUOX activity reduces. DUOX produces ROS, which is responsible for maintaining gut homeostasis in insect [9]. Even honey bees harbors gut bacteria to keep their hives healthy.

Pea aphids *(Acyrthosiphon pisum)* have an obligate symbiont *Buchnera aphidicola* that provides dietary nutrients lacking in insect host [10]. Gut microflora present in insect provides immunity, compensates nutrient deficiencies, and aids digestion. *Wolbachia* reflects a mutualistic relationship with many insects since it imparts resistance to its host against viruses. Molds of **Aspergillus, Cordyceps, Isaria,** etc., fungal species have also been associated with insect pests. Yeasts or yeast-like symbionts residing in the digestive tract of insects play a major role in metabolism and metamorphosis. Most popular is the interaction of insects and viruses, which also provides protective functions sometimes, for example, Pea aphid harboring *H.defensa* with *A. pisum* phage has great defensive power against parasitoid wasp attack as the phage infects bacteria, which further defends aphid against natural enemies [11]. **Hence, insects are key players in mutualistic symbiosis.**

4.2 Microbial symbiosis with animals

The most of the microbiota resides in the gut or gastrointestinal trait of animals, which are categorized as "luminal"—associated with lumen and others "mucosal," that is, associated with mucus lining the intestinal epithelium. Residing in the gut of animals, microbes function by metabolizing dietary food components ingested by the host organism producing short chain fatty acids and vitamins that fulfill the energy needs. Apart from this, they act as shield from attack of pathogens performing their roles in physiology and act as filters by detoxifying various compounds. Many microbes, for example, Bacteroides genera, alter expression of enzymes such as carbohydrate-degrading enzymes.

Many recent research studies have shown microbes as the key players in the physiological functions including even neural functions in animals. In case of bobtail squid, a light organ contains luminescent bacteria *(Photobacterium mandapamensis)*. This bacterium is important for development of light organ, and in cardinal fish, it is responsible for neural and anatomical functions [12].

Some latest research studies have also shown that microbes provide anti-predator defense by helping in camouflage and sponge-microbe symbiosis is emerging as model for putting light on animal-associated microbiota. Hence, associations of microbes with animals have unfolded and continue to unfold as a part of evolution.

4.3 Microbial symbiosis with humans

Human microbiome has always shown a remarkable range of art of living of microbes beginning from symbiosis to pathogenesis with human body being hub for multitude of various types of microbes—bacteria, fungi, viruses, etc. One of the hypotheses suggested that LAC bacteria and other bacteria colonizing in vagina are responsible for facilitating protective response and maintaining homeostasis. These symbiotic microbes such as *Lactobacillus salivarius* UCC118 release a toxin, that is, bacteriocin-ABP118, which is effective against food-borne pathogens including *Bacillus, Enterococcus, and Staphylococcus* species [13].

Another interesting interaction is represented by symbiotic microbes residing in skin that play a major role in priming T cells to respond to pathogens. *Staphylococcus epidermidis and Staphylococcus aureus,* which are coagulase negative staphylococci, belonging to Actinobaria such as *Corynebacterium,* are the primary colonizers of human skin [14]. The best example of microbial symbiosis with humans is the presence of gut and gastrointestinal microbiome influencing host metabolism and health (**Figure 9**). For example, *Faecalibacterium pracesnitzii* modulates eight urinary metabolites.

Humans are being considered as **"Superorganisms"** due to their diverse nature of interaction with microbial community [15]. So, gut bacteria have various metabolic proteins or agents that help in development of immunity and producing energy compounds from food left undigested by body, hence benefiting human health, and in turn, human gut provides conditions conducive to these microbes for their growth and development. Hence, trillions of microbes living in the human gut are known to play an important role in regulating metabolic functions, and this coexistence is considered to be the law of nature by various scientists.

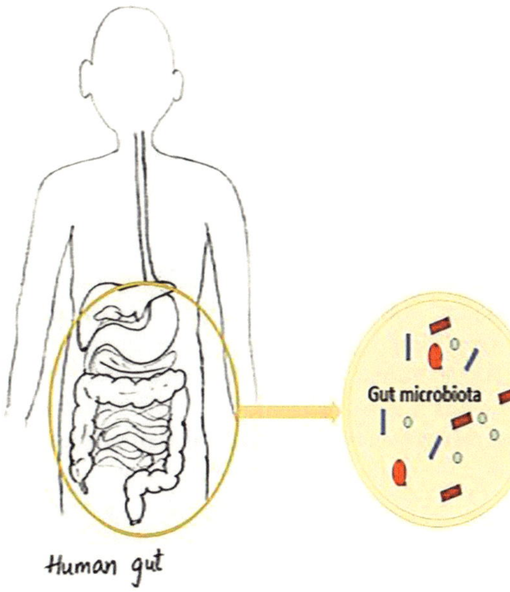

Figure 9.
Microbial symbiosis with humans (gut microbiota).

4.4 Microbial symbiosis with nematodes

Obligatory relationship is present between a bacterium (*Xenorhabdus nematophilus*) and a nematode (*Steinernema carpocapse*). Bacteria live inside the nematode producing toxins that nematode uses to kill the insects infected by it. Both the organisms are dependent on each other for efficient survival [3].

4.5 Mussels in deep-sea hydrothermal vents

Mussels are the most widely studied organisms as they are found in hydrothermal vents and are capable of forming a mutualistic association with bacteria. Bacteria act as a source of nutrition because bacteria use chemicals from hydrothermal fluid producing organic compounds that serve nutritional purposes, and in turn mussels provide bacteria with optimum conditions for growth by providing protection and other useful materials [16]. They too are perfect example of give–and-take relationship in this ecosystem.

5. Conclusion

Microbes have always been an integral part of ecosystem. Life would not be possible without these creatures, which perform important roles in even unnoticed reactions or phenomenon starting from different cycles to deep-sea hydrothermal vents. Blessed with distinct behaviors, they live in diverse habitats and climatic conditions performing functions that could not be taken over by any other organism. Since the present era is focusing on the role of microbes, everyday a new discovery of their role in biosphere leaves everyone awestruck and astonished at same time. Scientists and researchers have regarded microbes as helping hands that can maintain the balance of nature. Though they have small size, the role they play is commendable. The scope of microbial study is ever-increasing with increase in involvement and use of microbes, healthy and positive interactions among themselves surely lead to growth and development of microbiology sector. Studying the positive art of living among microbes, that is, symbiosis, one can exploit it for applications in plant and soil biome enhancing their efficiency and also maintain equilibrium in nature that is very essential for proper functioning and creation of healthy environment. Hence, this area of research has wide scope, which is applicable not only in department of microbiology but also in soils, botany, plant pathology, biochemistry, biology, environmental studies making it a subject of huge importance. They function as diverse beings by showing various types of interactions such as lichens, mycorrhizae, root nodulation, and plant-growth-promoting rhizobacteria with plants and various others with insects, animals, humans (gut microbiota), etc., playing different roles and functions. They serve as hub for research studies due to their amazing yet mysterial art of living. **Hence, symbiosis sustains life and microbes acting as beacon of artistic life in nature.**

Acknowledgements

This chapter acknowledges the work of researchers and professors whose published articles and journals have been a source of constant support, and it would remiss in not paying gratitude to all who motivated, assisted, and provided with content clarification, valuable suggestions, or advice in the project.

Acronyms and abbreviations

LAC	lactic acid bacteria
PGPR	plant-growth-promoting rhizobacteria
VAM	vesicular arbuscular mycorrhiza
+ve	positive
ATP	adenosine triphosphate
ADP	adenosine diphosphate
DUOX	dual-oxidase enzyme
ROS	reactive oxygen species

Appendices and nomenclature

N	nitrogen element
N_2	nitrogen gas
H	hydrogen
e^-	electrons
H_2O	water
NH_3	ammonia
H_2	hydrogen
P	phosphorus
P_i	inorganic phosphate

Author details

Gayatri Sharma
Punjab Agricultural University Ludhiana, Mohali, India

*Address all correspondence to: ujwalgayatri1150@gmail

References

[1] Gilbert JA, Neufeld JD. Life in a world without microbes. PLoS Biology. 2014;**12**(12):e1002020. DOI: 10.1371/journal.pbio.1002020

[2] Wikipedia contributors. Symbiosis. In Wikipedia, The Free Encyclopedia. 2022. Available from: https://en.wikipedia.org/w/index.php?title=Symbiosis&oldid=1126030010 [Accessed: December 10, 2022]

[3] Davey Reginald. What is a Symbiotic Microbial Community? News-Medical. 2022. Available from: https://www.news-medical.net/life-sciences/What-is-a-Symbiotic-Microbial-Community.aspx [Accessed: December 10, 2022]

[4] Wikipedia contributors. Lichen. In Wikipedia, The Free Encyclopedia. 2022. Available from: https://en.wikipedia.org/w/index.php?title=Lichen&oldid=1125607784 [Accessed: December 10, 2022]

[5] Wikipedia contributors. Mycorrhiza. In Wikipedia, The Free Encyclopedia. 2022. Available from: https://en.wikipedia.org/w/index.php?title=Mycorrhiza&oldid=1123670763 [Accessed: December 10, 2022]

[6] Stougaard J. Lotus Japonicus: A Model Plant for the Legume Family. Aalborg, Denmark: Elsevier A/S; 2017. DOI: 10.1016/B978-0-12-809633-8.06661-9

[7] Jeyanthi V, Kanimozhi S. Plant growth promoting Rhizobacteria (PGPR) – Prospective and mechanisms: A review. Journal of Pure and Applied Microiology. 2018;**12**:165-190. DOI: 10.22207/JPAM.12.2.34

[8] Carvalhais LC, Dennis PG. The Plant Microbiome. Brisbane, QLD, Australia: The University of Queensland, Springer; 2021. DOI: 10.1007/978-1-0716-1040-4

[9] Jang S, Mergaert P, Ohbayashi T, Ishigami K, Shigenobu S, Itoh H, et al. Dual oxidase enables insect gut symbiosis by mediating respiratory network formation. Proceedings of the National Academy of Sciences of the United States of America. 2021;**118**(10):e2020922118. DOI: 10.1073/pnas.2020922118

[10] Wilkinson TL, Koga R, Fukatsu T. Role of host nutrition in symbiont regulation: Impact of dietary nitrogen on proliferation of obligate and facultative bacterial endosymbionts of the pea aphid Acyrthosiphon pisum. Applied and Environmental Microbiology. 2007;**73**(4):1362-1366. DOI: 10.1128/AEM.01211-06

[11] Weldon SR, Strand MR, Oliver KM. Phage loss and the breakdown of a defensive symbiosis in aphids. Proceedings of the Biological Sciences. 2013;**280**(1751):20122103. DOI: 10.1098/rspb.2012.2103

[12] Eisthen HL, Theis KR. Animal-microbe interactions and the evolution of nervous systems. Philosophical Transactions of the Royal Society of London. Series B, Biological Sciences. 2016;**371**(1685):20150052. DOI: 10.1098/rstb.2015.0052

[13] Riley MA. Bacteriocins. In: Biology, Ecology, and Evolution, Encyclopedia of Microbiology. Third ed. Amherst, USA: Department of Biology, University of Massachusetts Amherst, Academic Press; 2009. DOI: 10.1016/B978-012373944-5.00065-1

[14] Byrd AL, Deming C, Cassidy SKB, Harrison OJ, Ng WI, Conlan S, et al. Staphylococcus aureus and

Staphylococcus epidermidis strain diversity underlying pediatric atopic dermatitis. Science Translational Medicine. 2017;**9**(397):eaal4651. DOI: 10.1126/scitranslmed.aal4651

[15] Sleator RD. The human superorganism - of microbes and men. Medical Hypotheses. 2010;**74**(2):214-215. DOI: 10.1016/j.mehy.2009.08.047

[16] Duperron S, Gaudron S, Laming S. A Mussel's life around Deep-Sea hydrothermal vents. Frontiers for Young Minds. 2019;7:76. DOI: 10.3389/frym.2019.00076

Section 3

Bioremediation

Chapter 12

The Role of V-Strategist Endophytes in Stimulating the Formation of Mycorrhizal Interactions and Soil Regeneration

Viktorija Oliferchuk, Nataliia Kendzora, Igor Shukel, Mariia Samarska and Oksana Olejniuk-Puchniak

Abstract

Endophyte, new species of yeast fungus, which belongs to the genera *Debariomycetaceae Vitasergia svidasoma Oliferchuk* PRJNA807518 was isolated from the fruiting body of *Tuber melanosporum* VS1223 (IMB F-100106). The preparation Mykovital was created on its basis. The possibility of regulation of soil fertility was established through the influence on the "bacteria-fungus-plant" system by stimulation of mycorrhizal formation. By the introduction of seedlings and saplings of trees and shrubs of endophyte species in rhysosphere applying the preparation Mykovital at the different types of devastated soils. Environmental efficiency of biological recultivation of devastated lands is determined during the cultivation of forest crops, which is proved by their biological sustainability and morphological indicators. The research shows the possibility to systematize microorganisms according to the strategies of their survival in ecosystems in such a way, that besides K, r, and L strategists it is necessary to introduce another notion about another group of microorganisms in ecology and soil microbiology—endophytes which are proposed to be named as V-strategies. These are fungi endophytes that are capable to restore and stimulate mycorrhizal symbiosis in the "bacteria-fungus-plant" system and function as provision of "heterotrophic" nutrition of plants on the Earth. Based on the research, a conceptual model of recultivation of devastated lands was proposed and priority of soil ecosystem support services was established.

Keywords: symbiotic nutrition, mycorrhiza, endophyte, *Vitasergia svidasoma* PRJNA807518, V-strategists

1. Introduction

The issue of increasing soil fertility is an ancient problem of human activity. These issues are also relevant in modern technologies of biological reclamation of devastated lands. Modern methods of growing forest crops and agricultural plants on these lands require specific approaches to increase the fertility of these soils. One of the approaches to increasing soil fertility is the method of stimulating mycorrhizal formation by influencing the "bacterium–fungus–plant" system through the introduction of an endophyte into the rhizosphere of the plant. The endophyte is introduced by using the drug "Mikovital." The ecological expediency and economic efficiency of the use of the drug "Mikovital" have been proven.

2. Symbiotic existence of plants

Symbiosis is a phenomenon that is the basis of the existence of living organisms in nature. There is a scientific idea about autotrophic existence of vegetation. However, only in case of consideration of symbiotrophic existence of plants, that is, compulsory participation of microorganisms in essential processes of plant organisms, it is possible to solve unsolved issues of their vital living and understand a range of important mechanisms that sustain life on the Earth. Mycorrhizal fungi play a crucial role in regulating terrestrial carbon dioxide (CO_2) fluxes [1–4]. They are obligate symbionts that form a relationship with plant roots known as mycorrhiza [5]. Mycorrhizal fungi obtain carbon from their host plants [6, 7] in exchange for the transfer of nutrients to the roots, which promotes plant growth. Previous studies have demonstrated the role of mycorrhiza in regulating plant productivity [8], water absorption [9, 10], litter decomposition [11, 12], root respiration [13, 14], soil respiration [15–17], stabilization of soil aggregates [18], and changes in soil carbon stocks [19], all of which affect above- and below-ground carbon dynamics in the ecosystem.

2.1 The role of "bacteria-: fungus- plant" system in providing life on the earth

The interaction of plants with microorganisms consists of a complex of complex processes that take place in two systems of different levels of organization: On the one hand, this is the microbial and fungal coenosis of the root zone of plants, and on the other hand, higher vascular plants, which are also interconnected in phytocenoses. As a result of such interaction, stable systems are formed, which play a key role in the functioning of natural edaphotopes and man-made agroecosystems.

The positive role of microorganisms in the root zone of plants is manifested in the transformation of organic residues, the synthesis of humus, the improvement of mineral nutrition of plants with nitrogen, phosphorus, and other macro- and microelements, the biocontrol of pathogens and pests, the production of biologically active substances that stimulate the growth and development of plants, and the detoxification of anthropogenic pollution. This ideal picture of positive interactions can be violated if the balance in the microbial-plant system is disturbed under the influence of natural or anthropogenic factors. In this connection, the tasks of correcting and regulating the functional structure and activity of microbial-plant systems arise.

One of the main ways of CO_2 input in the atmosphere is its emission from soils, or soil respiration, which is formed by root respiration and microbial decomposition of soil organic matter, dead plant residues, and organic substances produced by vegetative roots [20]. The sum of soil and above-ground respiration characterizes the gross respiration of the ecosystem, which, together with photosynthesis, forms the balance of CO_2 in the ecosystem, or net ecosystem exchange (Net Ecosystem Exchange, NEE). Therefore, the balance in the "plant-fungus-bacterium" system is of key importance for the deposition of carbon in the soil ecosystem and for maintaining the CO_2 balance in the atmosphere. The heterotrophic nutrition of plants has been studied quite widely for the last 150 years, and recent studies show that 98% of plants on Earth are mycotrophic, that is, those that live due to symbiosis with fungi and bacteria [5]. The ability to grow plants in sterile conditions, as well as in the conditions of the use of chemical means of protection and plant care, which have a detrimental effect on the soil microflora, for many years formed the opinion that the participation of microorganisms in the vital activity of plants is not mandatory, and only in recent decades scientists from many countries of the world thoroughly study mycorrhizal symbiosis and the functioning of the "bacterium-fungus-plant" system. The fallacy of such a judgment is caused by the fact that "sterile plants," as well as plants treated with chemical agents, always have metabolic products of endophytes, epiphytes, and rhizosphere bacteria inside their seeds, and also contain endophyte cells themselves.

Thus, the role of mycorrhizal symbiosis in the functioning of Earth's ecosystems can be represented by a diagram (**Figure 1**).

Based on our vision, both autotrophic nutrition and heterotrophic nutrition of plants are provided by mycorrhiza, which is formed mainly on the roots of plants but can also be formed with other parts of it. Mycorrhizal symbiosis affects the functioning of the soil biota and is responsible for sustaining life on the Earth, the interaction of the geological with biological cycle of substances, regulation of the composition of the atmosphere and hydrosphere, regulation of the intensity of biosphere processes, the formation of humus, and provides a protection in relation to the lithosphere.

2.2 Factors contributing to the stimulation of heterotrophic nutrition of plants stimulate the formation of plant mycorrhizae

Bacteria contribute to the formation of mycorrhizal symbiosis. They are called helleras and stimulate the growth of fungi at the stage prior to symbiosis, increasing the probability of root contact with fungi. The study of bacteria and filamentous

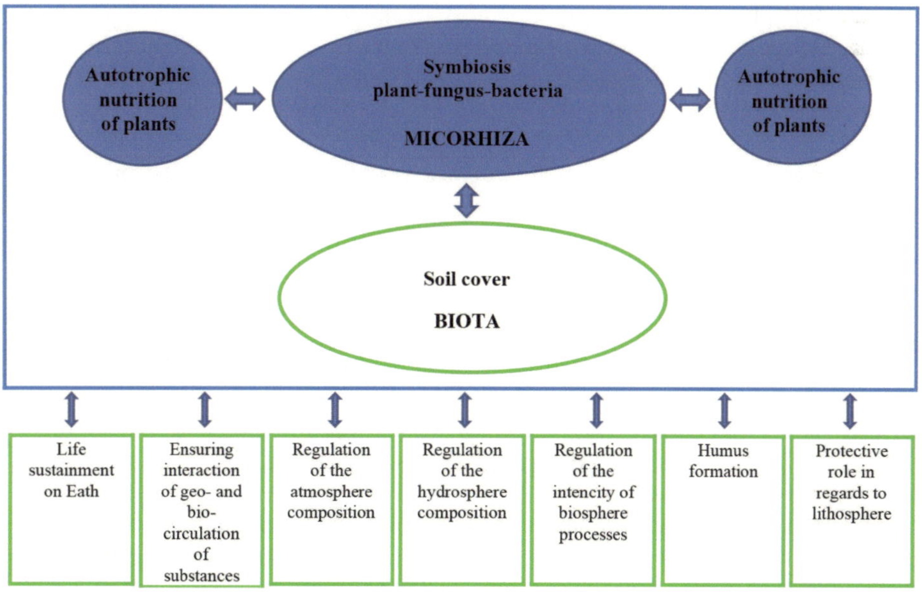

Figure 1.
The role of mycorrhizal symbiosis in functioning of the earth ecosystem.

fungi living in plants has been widely studied and continues to be studied. The biology of another agent-stimulating mycorrhizal formation, namely endophytic yeasts, remains less clear [21]. Recent studies indicate the potential of endophytic yeasts for their use in industry and agriculture. Endophytic yeasts have significant advantages over bacterial and filamentous endophytes because they can be easily cultivated, stored for long periods of time, and introduced into crops [21].

The use of endophytic yeasts is an effective way to reduce fertilizer and water use in agriculture and potentially increase yields. Their application is particularly promising in the field of heavy metal pollution remediation, and as biocontrol and bioregulatory agents to protect plants from pathogens.

A number of authors describe the great variability of endophytic yeast communities based on the plant host, as well as the influence of various biotic and abiotic factors that influence the composition of these communities. The authors express the opinion that among endophytic yeasts, ascomycetes mainly dominate [22].

Our studies agree with the studies of a number of authors and testify to the positive influence of a new species of endophyte belonging to the Debariomycetaceae family, the ascomycete *Vitasergia svidasoma* PRJNA807518, on the composition of soil mycobiota and plant productivity [23–25].

Most of the knowledge about the plant growth-promoting properties of endophytes is derived from the study of bacterial endophytes, as they have been studied in common ecological niches with yeast fungi in the phytosphere. In addition, there is an evidence of the formation of a mixed biofilm containing both bacteria and yeast [26, 27].

Several growth-promoting properties of endophytic yeasts include phytohormone production, stress relief, pathogen protection, and increased plant nutrient uptake. This action is caused by the production of IAA, the production of siderophores, and the activity of ACC deaminase.

The production of plant hormones provides a direct way to stimulate plant growth by endophytes. Auxins and gibberellins have many plant growth-promoting properties, including promoting root growth and stem elongation, as well as cell proliferation and elongation. In particular, a number of authors indicate the production of indole-3-acetic acid (IAA) by endophytic yeast [28–30].

One of the characteristics of endophytic yeast is the promotion of plant tolerance to stress [31]. In addition, endophytic yeasts produce polyamines, compounds that play an important role in plant growth and are involved in many cellular processes, including the synthesis of macromolecules, as well as cell growth, survival, and stress resistance. Yeast can also generally promote plant growth [32].

2.3 Industrial and ecological use of endophytic yeasts

Given the plant growth-stimulating properties of endophytic yeasts, they can be widely used in agriculture. The production of indole-acetic acid, a phytohormone that stimulates plant growth, by three yeast strains is described. To the best of our knowledge, the yeast strains presented in this study were the first endophytic yeast strains isolated from *Populus* species [33]. In addition, another strain of *Rhodotorula* was shown to increase germination of cucumber seedlings both by inoculation and by treatment with filtered supernatant [34]. The authors emphasize that the use of these yeast endophytes in agriculture can reduce water and fertilizer consumption, and it is important to study them in the future, given the growing population and increased demand for food.

There are also data that endophytic yeasts can be used for bioaugmented phytoremediation of heavy metals, as using only plants for this process affects their growth and often harms the plants. The use of endophytic yeast showed that seedlings of *Brassica* sp. treated with the endophyte *Rhodotorula* sp. showed enhanced growth and increased transformation of Cd, Pb, Zn, and Cu [35].

Another promising direction of the use of endophytic yeasts in agriculture is their use as biocontrol. The described strain of *Rhodotorula rubra*, isolated from rice, showed strong suppression of various types of pathogens and toxin-producing species of *Fusarium* [34]. Antifungal activity of poplar endophyte *Rhodotorula graminis* against common fungal pathogen *Rhizoctonia solani* is also known [36].

Endophytic yeasts can be used for the metabolism of various chemicals that are difficult to produce because they possess a significant range of metabolic processes and are characterized by a wide range of tolerance to environmental conditions [28].

An important characteristic of endophytic yeast is the ability to colonize the host plant. The endophytic yeast *R. graminis* is able to colonize the entire plant, including leaf and xylem tissue, when the plant is inoculated from the roots [36]. The yeast can colonize the plant passively by simply growing inside the root until the population gains access to the xylem tissue and from there colonize the rest of the plant with the flow of water that passes through the plant. Like fungi and bacteria, endophytic yeasts have the enzymatic ability to destroy plant cell walls [37–39].

2.4 The role of endophytic yeast V-strategists *Vitasergia svidasoma* PRJNA807518 in stimulating the productivity of forest crops

In 2002, we isolated an endophyte from the fruiting body of the Perigord truffle, which was identified as a new yeast species *Vitasergia svidasoma* PRJNA807518 [23]. Based on the isolated endophyte, the preparation "Mykovital" was created, which received state registration, the Organic Standard certificate, and all the necessary ecological and technical documentation, which made it possible to use it in the studied areas of forests, fields, and pomological gardens of Ukraine [24, 25].

2.5 Features of tuber melanosporum ascomycetes and yeasts symbiotic with them

Truffles are one of the most famous ectomycorrhizal fungi in the world. There is little information on the ecological mechanisms of ectomycorrhizal synthesis of truffles *in vitro*. Truffles are filamentous fungi belonging to the genus *Tuber*, found naturally in various climate zones of the Northern Hemisphere. Some truffle species, such as the Perigord black truffle (*T. melanosporum*) and the white truffle (*T. magnatum*), are found in Europe, the Chinese black truffle (*T. indicum*)—in Asia, and the Peking truffle (*T. lyonii*)—in North America, especially known for the organoleptic qualities of their fruiting bodies and are among the most expensive culinary delicacies in the world.

The *Tuber* melanosporum genome contains only 7500 protein-coding genes, with very rare polygenic families. The structural organization of the genome of *T. melanosporum* is strikingly different from other ascomycetes. It lacks large sets of carbohydrate-degrading enzymes, but some of them are involved in the degradation of plant cell walls during colonization.

Black truffle is characterized by extremely low allergenicity, as well as the absence of mycotoxin synthesis enzymes, and the overexpression of various flavor-related enzymes in the fruiting body. Among the latter, processes of sulfur assimilation and enzymes of interconversion of S-amino acids are active. These include by-products of methyl sulfide volatiles found in truffles and enzymes involved in the breakdown of amino acids *via* the Ehrlich Pathway, which produces the well-known truffle volatiles and flavors, such as 2-methyl-1-butanal.

Another feature of truffle-fruiting bodies is their symbiosis with yeasts. A total of 29 yeast strains isolated from ascocarps of black and white truffles (*Tuber melanosporum* Vitt. and *Tuber magnatum* Pico, respectively) are described. The strains were identified as *Candida saitoana*, *Debaryomyces hansenii*, *Cryptococcus* sp., *Rhodotorula mucilaginosa*, and *Trichosporon moniliiforme* [40].

Scientists attribute the complex aroma of truffles to the additional role of yeast symbionts, which share similar characteristics as the ecosystem of the fruiting body.

Volatile organic compounds (VOCs) released by fruiting bodies belong to several chemical classes (aldehydes, alcohols, esters, lactones, terpenes, and sulfur compounds). Truffle ascocarps are interesting ecological niches because they form a symbiosis with the root systems of various tree species.

Some yeast strains, such as *Cryptococcus humicolus*, which are constantly present on the surface of mature truffles, have shown extracellular cellulolytic, chitinolytic, and proteolytic activities that may promote ascocarp growth and nutrition during the saprophytic stage, or promote fungal ascospore dispersal [41].

2.6 The effect of the preparation "Mykovital" on the growth and development of seedlings of trees and shrubs

The new yeast species *Vitasergia svidasoma* PRJNA807518, which is the active agent of the Mykovital preparation, is widely used to stimulate the mycorrhizal formation and plant productivity. An experiment on growing planting material was conducted in a semi-controlled environment. Seedlings of trees and shrubs common in phytoremediation were used: *Quercus robur* L., *Pinus sylvestris* L., *Betula pendula* Roth., *Sorbus aucuparia* L., *Prunus divari-cata* Ledeb., *Robinia pseudoacacia* L., *Hippophea rhamnoides* L., and *Rosa canina* L. During the experiment, treatment of plant roots with Mykovital was used. Seedlings planted without treatment served as a control.

The ameliorative role of the inoculation of tree and shrub seedlings on sulfur-containing man-made soils of the Yavoriv sulfur quarry was investigated during two growing seasons in semi-productive conditions by planting seedlings in boxes with soil selected from embryonic soils. Seedlings care and measurement of their morphometric indicators of growth and development were carried out throughout the growing season. The growth and development of the planted seedlings were evaluated by the viability of the plants and their growth in height during the first and second years of cultivation.

The results of a two-year experiment on the cultivation of experimental plants in two versions, treated and untreated with the preparation Mykovital, are summarized in **Table 1**.

The survival rate of seedlings of experimental breeds is on average 31.25 ± 2.00% in 2019 and 30.25 ± 1.80% in 2020. Treatment of seedlings with Mykovital contributes to a significant increase in their survival rate. Thus, in 2019, the average rate of survival of experimental breeds was 85.37 ± 4.56% and 91.12 ± 3.54% in 2020. That is, the increase in survival of seedlings of experimental breeds in 2019 was an average of 2.7 times, and in 2020 3.0 times more compared to the control (not treated with Mykovital). At the same time, it should be pointed out that the survival rate varied both within a single breed and in a separate growing season (**Table 2**). In both, the

Name of species	Untreated root		Treated root	
	Years of observation		Years of observation	
	2019	**2020**	**2019**	**2020**
Common oak	21	21	85	88
Scots pine	35	37	89	86
White acacia	39	27	94	96
Betula pendula	36	38	96	96
Prunus cerasifera (cherry plum)	34	32	88	98
Sorbus aucuparia	34	35	79	87
Sea-buckthorn	26	27	74	89
Dog rose	25	25	88	89

Table 1.
Survival of planting material with treatment of plant roots with the preparation Mykovital, and without treatment, %.

Name of species of seedlings	Ratio of treated/untreated	
	2019 year	**2020 year**
Common oak	4.04	4.1
Scots pine	2.5	2.3
White acacia	2.4	3.5
Betula pendula	2.6	2.5
Prunus cerasifera (cherry plum)	2.5	3.0
Sorbus aucuparia	2.3	2.4
Sea-buckthorn	2.8	3.2
Dog rose	3.5	3.6

Table 2.
The survival rate of fitomeliorating cultures treated and untreated with the preparation Mykovital, times.

first and second growing seasons, the highest survival rate was observed for common oak seedlings—4.04 and 4.1 times more, respectively. The highest rate of survival of pine was 2.5 times in 2019 and 2.3 times in 2020. Dog rose, acacia, and other plants actively survived.

The obtained results confirm the effectiveness of the method of improving the survival and growth of seedlings when treated with Mykovital, which significantly activates the processes of survival and growth of tree and shrub seedlings on technologically polluted and degraded soils of the sulfur quarry. Patents have been issued for this method.

In addition, to determine the effect of treatment with the preparation Mykovital on the survival of forest crops, the study of the effect of treatment with the preparation of endophyte—a stimulator of mycorrhizal formation on the growth of seedlings in height was given. The results of measuring observations are summarized in **Table 3**.

According to the results of observations, a positive effect of the treatment of seedlings with the preparation Mykovital was established (at a high level of accuracy

Name of species of seedlings	Growth in height, cm				Growth of treated/growth of untreated, times
	Untreated		Treated		
	Zh ± m	p., 95%	Zh ± m	p., 95%	
Common oak	4.19 ± 0.18	0.37	10.64 ± 0.23	0.48	2.5
Scots pine	5.02 ± 0.16	0.32	10.33 ± 0.19	0.38	2.1
White acacia	10.71 ± 0.15	0.31	31.81 ± 0.42	0.87	3.0
Betula pendula	13.38 ± 0.50	1.04	34.40 ± 0.65	1.35	2.6
Prunus cerasifera (cherry plum)	6.76 ± 0.23	0.48	22.41 ± 0.21	0.44	3.3
Sorbus aucuparia	7.87 ± 0.22	0.46	23.29 ± 0.25	0.52	3.0
Sea-buckthorn	5.99 ± 0.21	0.44	14.62 ± 0.20	0.42	2.4
Dog rose	4.69 ± 0.21	0.44	14.54 ± 0.25	0.52	3.1

Table 3.
The preparation Mykovital treatment effect on the growth of seedlings.

(p 95% < 1.35%). The increase in the height of the studied species reached an average of 2.7 times compared to the control. In terms of species, as well as survival, and the difference is also observed in the growth of seedlings in height. The main reason may be the genetic condition of a specific species regarding the rate of formation of mycorrhizal symbiosis, which depends on the work of signaling mechanisms in the "plant-fungus-bacterium" system.

Thus, the growth of pine seedlings increased by 2.1 times, sea buckthorn by 2.5 times, oak by 2.5 times, birch by 2.6 times, rowan by 3.0 times, Robinia pseudoacacia by 3, 0 times, dog rose –3.1 times, and plum seedlings –3.3 times. The increase in height of seedlings treated with mycorrhiza is two to three times higher than the control. At the same time, it was established that deciduous species respond more actively to the application of the spore preparation of mycorrhizae compared to Scots pine. Shrub species and the introducer, Robinia pseudoacacia react, most intensively to the treatment of the root system of *Vitasergia svidasoma* PRJNA807518 seedlings.

Similar data were obtained in the characteristics of seedling growth in height in the first growing year after transplanting (**Table 4**).

According to **Table 4**, it can be seen that the average height of the seedlings of all studied breeds increased by 55.77% compared to the control. And within the boundaries of specific species, the height of seedlings reaches the following values: seedlings of Scots pine by 35.70%, sea buckthorn by 43.11%, oak by 48.55%, birch by 61.76%, rowan by 64%, 43%, pseudoacacia Robinia by 68.78%, dog rose by 68.97%, and plum seedlings by 69.12%.

The given data testify to the positive effect of mycorrhizal symbiosis on the growth and productivity of plants in the conditions of man-made soils polluted with sulfur. According to the research results, the promising use of the preparation for stimulating mycorrhizal processes Mykovital with the active agent—*Vitasergia svidasoma* PRJNA807518 in the amount of cell titer 7x109 to increase the resistance of seedlings and saplings of trees and shrubs to the adverse factors of sulfur pollution of the soil has been established.

Name of species of seedlings	Height of the plant, cm			
	Untreated		Treated	
	H ± m	p., 95%	H ± m	p., 95%
Common oak	13.76 ± 0.20	0.37	20.44 ± 0.25	0.48
Scots pine	15.64 ± 0.16	0.33	21.22 ± 0.16	0.34
White acacia	31.07 ± 0.14	0.33	52.44 ± 0.39	0.79
Betula pendula	32.95 ± 0.47	0.98	54.18 ± 0.60	1.24
Prunus cerasifera (cherry plum)	22.41 ± 0.16	0.34	37.90 ± 0.15	0.32
Sorbus aucuparia	25.08 ± 0.16	0.32	40.58 ± 0.19	0.39
Sea-buckthorn	20.62 ± 0.16	0.33	29.52 ± 0.19	0.40
Dog rose	14.66 ± 0.15	0.31	24.76 ± 0.20	0.41

Table 4.
The height of planting material of meliorative species treated and untreated with the preparation Mykovital, cm.

2.7 The influence of the preparation Mykovital on the chemical composition of plants

Mycorrhiza causes the maximum influence on the vitality of the host plant, exactly where it reaches the greatest development. Since the statistical survival of seedlings was low on the lands disturbed by sulfur mining due to the lack of mycorrhizal formation and nutrients necessary for growth, we compared them with artificially mycorrhized seedlings. When examining the root systems of uninfected seedlings that took root in cultures on embryos, mycorrhiza was found. The rest of the uninfected showed signs of a lack of nutrients, characteristic of nitrogen and phosphorus starvation, and a lack of trace elements. The mycorrhized plants differed not only in a better appearance, greater height, and weight, but also had twice as many short roots.

Thus, the surface area of the root system of the plant increases, which gives it the possibility of effective nutrition (**Table 5**).

The results of the chemical analysis showed that the seedlings treated with Mykovital contain significantly more nitrogen, phosphorus, and potassium than in non-mycorrhizal plants, and the difference is well observed in absolute dry weight.

Indicator	Seedlings		Reliability of experiment
	Treated	Untreated	
Raw weight, mg	1242	598	P > 0.01
Absolute dry weight, mg	345	155	P > 0.01
Nitrogen, % to absolute dry weight	1.78	1.88	P > 0.01
Nitrogen in single seedling, mg	5.75	2.87	P > 0.01
Phosphorus, % to absolute dry weight	0.185	0.097	P > 0.01
Phosphorus in a single seedling, mg	0.60	0.15	P > 0.01
Potassium, % to absolute dry weight	0.66	0.62	P > 0.01
Potassium in a single seedling, mg	2.17	0.96	P >0.01

Table 5.
Chemical composition of scots pine seedlings treated and untreated with the preparation Mykovital.

Indicators	05.07.2019		20.08.2020	
	Treated	Untreated	Treated	Untreated
Short mycorrhiza roots, %	68.6	1.2	70.2	9.5
Absolute dry weight of plant, mg	207.9	160.2	337.2	180.6
Absolute dry weight of root, mg	72.3	73.0	127.4	74.6
Absolute dry weight of sampling, mg	135.6	87.2	209.8	106.0
Ratio of absolute dry weight of roots to samplings	0.53	0.84	0.61	0.71
Nitrogen, % to absolute dry weight	1.49	1.23	1.60	1.20
Nitrogen in single seedling, mg	3.1	2.0	5.4	2.2
Phosphorus, % to absolute dry weight	0.13	0.08	0.21	0.07
Phosphorus in a single seedling, mg	0.27	0.13	0.72	0.13
Potassium, % to absolute dry weight	0.45	0.55	0.63	0.45
Potassium in a single seedling, mg	1.03	0.88	2.12	0.81

Table 6.
Chemical composition of scots pine seedlings treated and untreated with the preparation Mykovital.

In addition, the content of nitrogen, phosphorus, and potassium in seedlings during the season was determined. The results are summarized in **Table 6**.

It was established that mycorrhized seedlings accumulate a much larger amount of mineral nutrients in terms of the absolute dry weight of plants (**Table 6**).

The main nutrients such as nitrogen, phosphorus, and potassium were in low concentrations in non-mycorrhizal seedlings, while the absolute and percentage content of these nutrients increased in seedlings treated with mycorrhizal preparation. Over time, plants that were not treated with a stimulator of mycorrhizal formation began to show symptoms of starvation and died, while mycorrhizal ones continued to grow.

The given information indicates that the use of the proposed methods of biological restoration and improvement of survival of plants during afforestation of devastated lands ensures higher survival and growth of forest crops contributes to effective biological restoration of devastated lands due to stimulation of mycorrhizal formation, formation of plant cover, and rhizosphere remediation of the territory.

2.8 Recovery of the "bacterium-fungus-plant" system with the participation of the stimulator of mycorrhizal formation—the fungus *Vitasergia svidasoma* PRJNA807518

The conceptual model of the use of the proposed biological method on various types of devastated lands includes the use of natural self-recovery of the soil and plant cover and optimization of the soil environment by improving the properties of the soil system, biodetoxification, and biodecontamination through the expansion of populations of soil microorganisms and the use of phytomeliorants with a simultaneous effect on the biological and inorganic components of the soil.

The disadvantages of the proposed methods are that the genetic material of microorganisms (bacteria and fungi) is introduced into the ecosystem without a preliminary study of the micro- and mycoflora of these soils. Usually, the introduction of a living biological preparation is aimed at solving a local problem, for example, to stop the development of pathogenic microorganisms or toxin-producing microorganisms;

besides, there are no data on the ability of microorganisms to adapt to different types of devastated soils and their contamination with toxic substances and their effectiveness. In the theory and practice of such influence on the soil ecosystem, there is the term "biocontrol," which fully corresponds to the content, since control over pathogens and toxin-producing agents occurs due to changes in chemical and biochemical interactions with the environment and due to natural antagonism between species and their metabolites. However, this does not consider how other groups of microorganisms react to such actions, such as K-strategists. This is a group of bacteria and fungi that are able to maintain a stable population level in climactic systems when soil conditions are relatively stable and stress reactions are absent. These microorganisms fully and economically consume external resources and provide themselves with a stably high population density. The impact of separately introduced strains on r-strategists is also important, which develops in the soil when a large amount of readily available substrates are received and have a high growth rate, due to which there are outbreaks of an increase in the number of populations that quickly assimilate the substrate. They develop with particular activity at the initial stages of successions during the settlement of some substrate, for example, during the assimilation of harvest residues or during the degradation of wood during the formation of forest plantations. Although the membership of fungi or bacteria to a certain strategy is not absolute, because each species has the ability to assimilate substrates and can withstand certain stressful situations, it is still necessary to carry out the distribution of microorganisms according to strategies in relation to each individual microbial group in a certain ecological situation, which was done by us while studying the structure of soil micromycetes of different degrees of degradation of forest soils and agrocenoses.

We did not focus our attention on "biocontrol" but studied the integral system "plant-fungus-bacterium" and methods of its bioregulation, which involves, in our case, the introduction of an endophyte into the soil ecosystem through the rhizosphere of the plant, which triggers self-recovery and self-regulation mechanisms in microecosystems and is, in a way, a "catalyst" of natural processes that restore the natural mycorrhizal triple symbiosis and the general mycorrhizal network in the ecosystem. We showed the efficiency and productivity of this process by using and testing a species new to science, belonging to the *Debariomycetaceae* family—the endophyte *Vitasergia svidasoma* PRJNA807518, isolated from the fruiting body of black truffle, which, in fact, stimulates the processes of mycorrhizal formation.

Back in 1966, the famous American ecologist Barry Commoner outlined modern environmental problems in the form of certain aphorisms as well-known laws. One of his laws is "Nature knows best." This law prompts us to the need for reasonable regulation by already existing principles in natural ecosystems. We only need a systematic process of deep knowledge and understanding of the functioning of these complex systems and the search for new mechanisms and discoveries in their reasonable management.

Our research indicates that mycorrhizal associations are key in the plant-fungus-bacterium triple symbiosis. In spring, when the vegetation process begins, the plant releases exudates into the rhizosphere, the basis of which are sugars. At the same time, the endophyte fungus with which the plant is inoculated, according to the technology, affects the species composition of microbial-fungal associations and stimulates the processes of their functioning. Effective growth and development of mycorrhizal associations begin. Mycorrhizal fungi secrete trehalose into the medium, a disaccharide that is synthesized in the cells of many fungi and enters the environment of the mycorrhizosphere and rhizosphere of plants as part of mycelial exudates. It is believed that trehalose is one of the factors that determine the composition of bacterial communities. All life forms on the Earth—the transfer of inorganic substances within and between different carbon reservoirs is the basis of all organic substances and the source of hydrocarbons for plants. An effective process of photosynthesis can occur only with a sufficient amount of carbon. For this, plants need 50% of carbon and only 5–7% of minerals, including macroelements (phosphorus, nitrogen, and potassium) and microelements (cobalt, zinc, magnesium, iodine, and iron).

The result of the tested approach ensures the adaptation of trees to the soils of devastated lands by influencing the formation of plant associations with an effective plant metagenome and mycorrhizal formation due to the regulatory mechanisms of *Vitasergia svidasoma* PRJNA807518 resistant to the environmental conditions that occur on lands in need of restoration.

The implementation of the method involves the performance of the biological stage of the work that is, on the prepared area, planting deciduous and coniferous trees, the roots of which are treated with the mycorrhizal preparation Mykovital.

2.9 Conceptual model of restoration of disturbed lands

Restoration of biological properties of devastated soils and optimization of its biochemical processes is related to restoration of chemical and physical parameters of the soil, which requires the use of an appropriate system of agrotechnical measures. The use of V-strategies to restore and create a mycorrhizal network in the ecosystem will contribute to the stimulation of signaling systems between the plant, fungus, and bacteria, especially at the initial stages of restoration of degraded soils, will ensure higher resistance of plants to natural and man-made stresses, and increase their productivity. Achieving compliance of the state characteristics of the restored area with regulatory requirements is recognized as a successful result of the measures (**Figure 2**).

The conducted research gives reason to conclude that, in forest biocenoses, the proposed biotechnology will give an opportunity to influence the formation of tree stands, and, in agroecosystems, it will be possible to create self-regulating phytocenoses. The viability of forest crops, their growth, and development contribute to the faster closing of forest meliorative crops (at least for 1 year) and the reduction of the cost of crops production and care. The proposed approach of biological restoration is ecologically feasible and economically justified and has remarkable social significance.

The developed and proposed preparation provides accelerated growth and development of plants due to activation of soil microflora, and improvement of nitrogen and phosphorus nutrition. The application of this method of restoration not only accelerates the process of soil formation, but also improves the ecological conditions in the area of application: It contributes to the creation of a green landscape, the improvement of the air environment, and the return of disturbed lands to land use.

The conducted experiments made it possible to verify the possibility of creating cultural phytocenoses with minimal costs for improving the properties of the substrate.

The restoration of various types of devastated soils, the use of natural self-healing of the soil and plant cover, and the optimization of the soil environment by improving the properties of the soil system, biodetoxification, and biodecontamination through the stimulation of symbionts—endophytes, and the creation of a general mycorrhizal network in the ecosystem are a promising dynamic direction in land use.

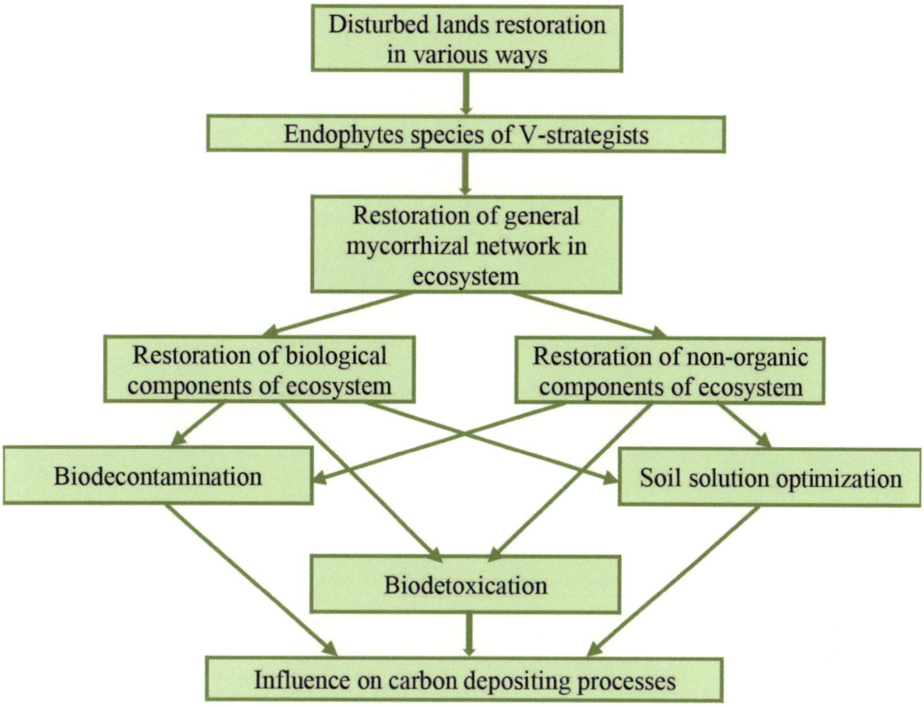

Figure 2.
Conceptual model of restoration of disturbed lands.

3. Conclusions

- The possibility of regulating soil fertility through the influence on the "bacterium-fungus-plant" system by stimulating mycorrhizal formation through the introduction of an endophyte species into the rhizosphere of the plant through the use of the preparation "Mykovital" on various types of devastated soils has been established;

- The ecological effectiveness of biological restoration of devastated lands during the cultivation of forest crops and agricultural plants with the use of the preparation "Mykovital" was confirmed by their biological stability and morphophysiological indicators, and an increase in the productivity of tree and shrub seedlings was established;

- A conceptual model for the restoration of devastated lands based on endophytes V-strategists is proposed;

- A shift in the paradigm of thinking in regard to the priority of autotrophic nutrition of plants is proposed and the equivalence of "heterotrophic" nutrition of plants and the role of mycorrhizal symbiosis in the functioning of the Earth's ecosystem is indicated.

Author details

Viktorija Oliferchuk[1*], Nataliia Kendzora[2], Igor Shukel[3], Mariia Samarska[1] and Oksana Olejniuk-Puchniak[3]

1 Department of Ecology, Ukrainian National Forestry University, Lviv, Ukraine

2 Department of Forestry and Botanical Research of the Botanical Garden, Ukrainian National Forestry University, Lviv, Ukraine

3 Department of Landscape Architecture, Garden and Park Management and Urban Ecology, Ukrainian National Forestry University, Lviv, Ukraine

*Address all correspondence to: oliferchuk@nltu.edu.ua

References

[1] Fitter AH, Heinemeyer A, Staddon PL. The impact of elevated CO2 and global climate change on arbuscular mycorrhizas: A mycocentric approach. New Phytologist. 2000;**147**:179-187

[2] Rygiewicz PT, Andersen CP. Mycorrhizae alter quality and quantity of carbon allocated below ground. Nature. 1994;**369**:58-60

[3] Treseder KK, Allen MF. Mycorrhizal fungi have a potential role in soil carbon storage under elevated CO2 and nitrogen deposition. New Phytologist. 2000;**147**:189-200

[4] Zhu YG, Miller RM. Carbon cycling by arbuscular mycorrhizal fungi in soil-plant systems. Trends in Plant Science. 2003;**8**:407-409

[5] Allen MF, Swenson W, Querejeta JI, Egerton-Warburton LM, Treseder KK. Ecology of mycorrhizae: A conceptual framework for complex interactions among plants and fungi. Annual Review of Phytopathology. 2003;**41**:271-303

[6] Hobbie EA. Carbon allocation to ectomycorrhizal fungi correlates with belowground allocation in culture studies. Ecology. 2006;**87**:563-569

[7] Johnson D, Leake JR, Ostle N, Ineson P, Read DJ. In situ 13 CO2 pulse-labelling of upland grassland demonstrates a rapid pathway of carbon flux from arbuscular mycorrhizal mycelia to the soil. New Phytologist. 2002;**53**:327-334

[8] van der Heijden MGA, Streitwolf-Engel R, Riedl R, Siegrist S, Neudecker A, Ineichen K, et al. The mycorrhizal contribution to plant productivity, plant nutrition and soil structure in experimental grassland. New Phytologist. 2006;**172**:739-752

[9] Auge RM. Water relations, drought and vesicular-arbuscular mycorrhizal symbiosis. Mycorrhiza. 2001;**1**:3-42

[10] Duddridge JA, Malibari A, Read DJ. Structure and function of mycorrhizal rhizomorphs with special reference to their role in water transport. Nature. 1980;**287**:834-836

[11] Cornelissen JHC, Aerts R, Cerabolini B, Werger MJA, van der Heijden MGA. Carbon cycling traits of plant species are linked with mycorrhizal strategy. Oecologia. 2001;**129**:611-619

[12] Langley JA, Hungate BA. Mycorrhizal controls on belowground litter quality. Ecology. 2003;**84**:2302-2312

[13] Burton AJ, Pregitzer KS, Ruess RW, Hendrik RL, Allen MF. Root respiration in north American forests: Effects of nitrogen concentration and temperature across biomes. Oecologia. 2002;**131**:559-568

[14] Hawkes CV, Hartley IP, Ineson P, Fitter AH. Soil temperature affects carbon allocation within arbuscular mycorrhizal networks and carbon transport from plant to fungus. Global Change Biology. 2008;**14**:1181-1190

[15] Heinemeyer A, Hartley IP, Evans SP, De la Fuente JAC, Ineson P. Forest soil CO2 flux: Uncovering the contribution and environmental responses of ectomycorrhizas. Global Change Biology. 2007;**13**:1786-1797

[16] Langley JA, Johnson NC, Koch GW. Mycorrhizal status influences the rate but not the temperature sensitivity

of soil respiration. Plant and Soil. 2005;**2005**(277):335-344

[17] Moyano FE, Kutsch WL, Rebmann C. Soil respiration fluxes in relation to photosynthetic activity in broad-leaf and needle-leaf forest stands. Agricultural and Forest Meteorology. 2008;**148**:135-143

[18] Rillig MC, Wright SF, Eviner VT. The role of arbuscular mycorrhizal fungi and glomalin in soil aggregation: Comparing effects of five plant species. Plant and Soil. 2002;**238**:325-333

[19] Treseder KK, Egerton-Warburton LM, Allen MF, Cheng YF, Oechel WC. Alteration of soil carbon pools and communities of mycorrhizal fungi in chaparral exposed to elevated carbon dioxide. Ecosystems. 2003;**6**:786-796

[20] Rodrigo Vargas DD, Baldocchi JI, Querejeta PS, Curtis NJ, Hasselquist IA, Janssens MF, et al. Ecosystem CO2 fluxes of arbuscular and ectomycorrhizal dominated vegetation types are differentially influenced by precipitation and temperature. New Phytologist. 2010;**185**:226-236. DOI: 10.1111/j.1469-8137.2009.03040

[21] Joubert P, Doty S. Endophytic Yeasts: Biology, Ecology and Applications. Springer International Publishing AG, part of Springer Nature; 2018. DOI: 10.1007/978-3-319-89833-9_1

[22] Prior R, Mittelbach M, Begerow D. Impact of three different fungicides on fungal epi- and endophytic communities of common bean (Phaseolus vulgaris) and broad bean (Vicia faba). Journal of Environmental Science and Health. Part. B. 2017;**52**(6):376-386. DOI: 10.1080/03601234.2017.1292093

[23] Oliferchuk V, Fedorovych D, Samarska M, Bunetsky V, Samborskyy M, Kachor A, et al. Changes in the structure of myco - and microbiocenosis of soil when using immobilized on biochar strains of fungi and bacteria as an example of ecosystem maintenance services. Ecological Engineering & Environmental Technology. 2022;**6**:442-452. DOI: 10.12912/27197050/152522

[24] Oliferchuk VP, Oliferchuk SP. A Complex Biologically Active Preparation for Regulating the Development and Growth of Plants Based on a Spore Suspension of Mycorrhizal Fungi "Mykovital". Patent for invention № 111174 (19) UA (51) IPC A01 N 63/04(2006. 01) C12N 1/14 (2006.01)

[25] Oliferchuk VP, Oliferchuk SP, Diner TV. Method for Restoring and Increasing Soil Fertility According to the Principle of Bioregulation in Microbial and Mycocenoses. Patent for invention №124179 (19) UA (51) IPC A01B 79/02 (2006.01) A01N 63/30 (2020.01) C05F 11/08 (2006.01); 2006

[26] Bandara WM, Seneviratne G, Kulasooriya SA. Interactions among endophytic bacteria and fungi: Effects and potentials. Journal of Biosciences. 2006;**31**(5):645-650

[27] Firrincieli A, Otillar R, Salamov A, Schmutz J, Khan Z, Redman RS, et al. Genome sequence of the plant growth promoting endophytic yeast Rhodotorula graminis WP1. Frontiers in Microbiology. 2015;**6**:978. DOI: 10.3389/fmicb.2015.00978

[28] Doty SL. Endophytic yeasts: Biology and applications. In: Aroca R, editor. Symbiotic Endophytes. Springer: Berlin; 2013. pp. 335-343

[29] Moller L, Lerm B, Botha A. Interactions of arboreal yeast endophytes: An unexplored discipline. Fungal Ecology. 2016;**22**:73-82. DOI: /10.1016/j.funeco.2016.03.003

[30] Nassar AH, El-Tarabily KA, Sivasithamparam K. Promotion of plant growth by an auxin-producing isolate of the yeast Williopsis saturnus endophytic in maize (Zea mays L.) roots. Biology and Fertility of Soils. 2005;**42**(2):97-108. DOI: 10.1007/s00374-005-0008-y

[31] Khalifa AY, Alsyeeh AM, Almalki MA, Saleh FA. Characterization of the plant growth promoting bacterium, Enterobacter cloacae MSR1, isolated from roots of non-nodulating Medicago sativa. Saudi Journal of Biological Sciences. 2016;**23**(1):79-86. DOI: 10.1016/j.sjbs.2015.06.008

[32] Takahashi T, Kakehi J. Polyamines: Ubiquitous polycations with unique roles in growth and stress responses. Annals of Botany. 2010;**105**(1):1-6. DOI: 10.1093/aob/mcp259

[33] Xin G, Glawe D, Doty SL. Characterization of three endophytic, indole-3-acetic acid-producing yeasts occurring in Populus trees. Mycological Research. 2009;**113**(9):973-980. DOI: 10.1016/j.mycres.2009.06.001

[34] Akhtyamova N, Sattarova RK. Endophytic yeast Rhodotorula rubra strain TG-1: Antagonistic and plant protection activities. Biochem Physiol: Open Access. 2013;**02**(1). DOI: 10.4172/2168-9652.1000104

[35] Wang W, Deng Z, Tan H, Cao L. Effects of Cd, Pb, Zn, Cu-resistant endophytic Enterobacter sr CBSB1 and Rhodotorula sp. CBSB79 on the growth and phytoextraction of brassica plants in multimetal contaminated soils. International Journal of Phytoremediation. 2013;**15**(5):488-497. DOI: 10.1080/15226514.2012.716101

[36] Kandel SL, Firrincieli A, Joubert PM, Okubara PA, Leston ND, McGeorge KM, et al. An in vitro study of bio-control and plant growth promotion potential of Salicaceae endophytes. Frontiers in Microbiology. 2017;**8**:386. DOI: 10.3389/fmicb.2017.00386

[37] Kumar A, Krishnankutty R, Droby S, Singh V, Singh S, White J. Entry, colonization, and distribution of endophytic microorganisms in plants. Microbial Endophytes. Functional Biology and Applications. Woodhead Publishing. 2020. pp. 1-33. DOI: 10.1016/B978-0-12-819654-0.00001-6

[38] Santoyo G, Moreno-Hagelsieb G, Orozco-Mosqueda Mdel C, Glick BR. Plant growth-promoting bacterial endophytes. Microbiological Research. 2016;**183**:92-99. DOI: 10.1016/j.micres.2015.11.008

[39] Uzma F, Konappa NM, Chowdappa S. Diversity and extracellular enzyme activities of fungal endophytes isolated from medicinal plants of Western Ghats, Karnataka. Egypt Journal of Basic and Applied Sciences. 2016;**3**(4):335-342. DOI: 10.1016/j.ejbas.2016.08.007

[40] Buzzini P, Gasparetti C, Turchetti B, Cramarossa MR, Vaughan-Martini A, Martini A, et al. Production of volatile organic compounds (VOCs) by yeasts isolatedfrom the ascocarps of black (Tuber melanosporum Vitt.) and white (Tuber magnatum Pico) truffles. Archives of Microbiology. 2005;**184**:187-193. DOI: 10.1007/s00203-005-0043-y

[41] Zacchi L, Vaughan-Martini A, Angelini P. Yeast distribution in a truffle–field ecosystem. Annals of Microbiology. 2003;**53**(3):275-282